Collins

The Shangh Maths Project

For the English National Curriculum

Teacher's Guide 3A

Teacher's Guide Series Editor: Amanda Simpson

Practice Books Series Editor: Professor Lianghuo Fan

Authors: Caroline Clissold, Sarah Eaton, Linda Glithro, Steph King and Cherri Moseley

HarperCollins PUBLISHERS
Since 1817

William Collins' dream of knowledge for all began with the publication of his first book in 1819.

A self-educated mill worker, he not only enriched millions of lives, but also founded a flourishing publishing house. Today, staying true to this spirit, Collins books are packed with inspiration, innovation and practical expertise. They place you at the centre of a world of possibility and give you exactly what you need to explore it.

Collins. Freedom to teach.

Published by Collins
An imprint of HarperCollins*Publishers*
The News Building
1 London Bridge Street
London
SE1 9GF

Browse the complete Collins catalogue at
www.collins.co.uk

978-0-00-819721-6

Teacher's Guide Series editor: Amanda Simpson

Practice Books Series editor: Professor Lianghuo Fan

Authors: Caroline Clissold, Sarah Eaton, Linda Glithro, Steph King and
          Cherri Moseley

British Library Cataloguing in Publication Data

A catalogue record for this publication is available from the British Library.

Publishing Manager: Fiona McGlade
In-house Editor: Nina Smith
In-house Editorial Assistant: August Stevens
Project Manager: Emily Hooton
Copy Editor: Tracy Thomas
Proofreader: Jo Kemp
Cover designer: Kevin Robbins and East China Normal University Press Ltd
Internal design: 2Hoots Publishing Services Ltd
Typesetting: 2Hoots Publishing Services Ltd
Illustrations: QBS
Production: Rachel Weaver

Printed and bound by CPI Group (UK) Ltd, Croydon, CR0 4YY

Photo acknowledgements
The publishers wish to thank the following for permission to reproduce photographs. Every effort has been made to trace copyright holders and to obtain their permission for the use of copyright materials. The publishers will gladly receive any information enabling them to rectify any error or omission at the first opportunity.

(t = top, c = centre, b = bottom, r = right, l = left)

p197 Quackdesigns/Shutterstock, p197 br Hudyma Natallia/Shutterstock, p197 tr Hudyma Natallia/Shutterstock, p203 bl BlueRingMedia/Shutterstock, p203 cr robuart/Shutterstock, p206 bl Studio_G/Shutterstock, p211 c BlueRingMedia/Shutterstock, p299 t Nemanja Cosovic/Shutterstock.

Contents

The Shanghai Maths Project: an overview

The Shanghai Maths Project is a collaboration between Collins and East China Normal University Press Ltd., adapting their bestselling maths programme, 'One Lesson, One Exercise', for England, using an expert team of authors and reviewers. This carefully crafted programme has been continually reviewed in China over the last 24 years, meaning that the materials have been tried and tested by teachers and children alike. Some new material has been written for The Shanghai Maths Project but the structure of the original resource has been preserved and as much original material as possible has been retained.

The Shanghai Maths Project is a programme from Shanghai for Years 1–11. Teaching for mastery is at the heart of the entire programme, which, through the guidance and support found in the Teacher's Guides and Practice Books, provides complete coverage of the curriculum objectives for England. Teachers are well supported to deliver a high-quality curriculum using the best teaching methods; pupils are enabled to learn mathematics with understanding and the ability to apply knowledge fluently and flexibly in order to solve problems.

The programme consists of five components: Teacher's Guides (two per year), Practice Books (two per year), Shanghai Learning Book, Homework Guide and Collins Connect digital package.

In this guide, information and support for all teachers of primary maths is set out, unit by unit, so they are able to teach The Shanghai Maths Project coherently and confidently, and with appropriate progression through the whole mathematics curriculum.

Practice Books

The Practice Books are designed to serve as both teaching and learning resources. With graded arithmetic exercises, plus varied practice of key concepts and summative assessments for each year, each Practice Book offers intelligent practice and consolidation to promote deep learning and develop higher order thinking.

There are two Practice Books for each year group: A and B. Pupils should have ownership of their copies of the Practice Books so they can engage with relevant exercises every day, integrated with preparatory whole-class and small group teaching, recording their answers in the books.

The Practice Books contain:

- chapters made up of units, containing small steps of progression, with practice at each stage
- a test at the end of each chapter
- an end-of-year test in Practice Book B.

Each unit in the Practice Books consists of two sections: 'Basic questions' and 'Challenge and extension questions'.

We suggest that the 'Basic questions' be used for all pupils. Many of them, directly or sometimes with a little modification, can be used as starting questions, for motivation or introduction or as examples for clear explanation. They can also be used as in-class exercise questions – most likely for reinforcement and formative assessment, but also for pupils' further exploration. Almost all questions can be given for individual or peer work, especially when used as in-class exercise questions. Some are also suitable for group work or whole-class discussion.

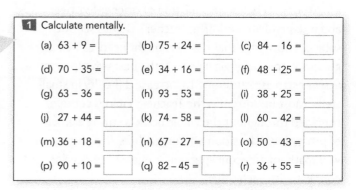

1 Calculate mentally.

(a) $63 + 9 =$ (b) $75 + 24 =$ (c) $84 - 16 =$

(d) $70 - 35 =$ (e) $34 + 16 =$ (f) $48 + 25 =$

(g) $63 - 36 =$ (h) $93 - 53 =$ (i) $38 + 25 =$

(j) $27 + 44 =$ (k) $74 - 58 =$ (l) $60 - 42 =$

(m) $36 + 18 =$ (n) $67 - 27 =$ (o) $50 - 43 =$

(p) $90 + 10 =$ (q) $82 - 45 =$ (r) $36 + 55 =$

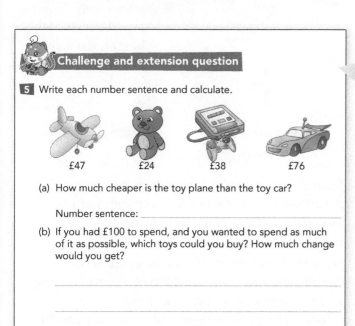

Challenge and extension question

5 Write each number sentence and calculate.

£47   £24   £38   £76

(a) How much cheaper is the toy plane than the toy car?

Number sentence: _____

(b) If you had £100 to spend, and you wanted to spend as much of it as possible, which toys could you buy? How much change would you get?

All pupils should be given the opportunity to solve some of the 'Challenge and extension questions', which are good for building confidence, but they should not always be required to solve all of them. A general suggestion is that most pupils try about 40–60 per cent of the 'Challenge and extension questions'.

Unit tests sometimes include questions that relate to content in the 'Challenge and extension questions'. This is clearly shown in the diagnostic assessment grids provided in the Teacher's Guides. Teachers should make their own judgements about how to use this information since not all pupils will have attempted the 'Challenge and extension question'.

Teacher's Guides

Theory underpinning the Teacher's Guides

The Teacher's Guides contain everything teachers need in order to provide the highest quality teaching in all areas of mathematics, in line with the English National Curriculum. Core mathematics topics are developed with deep understanding in every year group. Some areas are not visited every year, though curriculum coverage is in line with Key Stage statutory requirements, as set out in the National curriculum in England: mathematics programmes of study (updated 2014).

There are two Teacher's Guides for each year group: one for the first part of the year (Teacher's Guide 3A) and the other for the second (Teacher's Guide 3B).

The Shanghai Maths Project is different from other maths schemes that are available in that there is no book called a 'textbook'. Lessons are a mixture of teacher-led, peer and independent work. The Teacher's Guides set out subject knowledge that teachers might need, as well as guidance on pedagogical issues – the best ways to organise activities, to ask questions, to increase difficulty in small steps. Most importantly, the Teacher's Guides contain, threaded throughout the whole book, a strong element of professional development for teachers, focusing on the way mathematics concepts can be enabled to develop and connect with each other.

The Shanghai Maths Project Teacher's Guides are a complete reference for teachers working with the Practice Books. Each unit in the Practice Book for each year group is set out in the corresponding Teacher's Guide over a number of pages.

Most units will need to be taught over more than one lesson – some might need three lessons. In the Practice Books, units contain a great deal of learning, densely packed into a few questions. If pupils are to be able to tackle and succeed with the Practice Book questions, they need to have been guided to learn new mathematics and to connect it to their existing knowledge.

This can only be achieved when teachers are able to break down the conceptual learning that is needed and to provide relevant and high quality teaching. The Teacher's Guides show teachers how to build up pupils' knowledge and experience so they learn with understanding in small steps. This way, learning is secure, robust and not reliant on memorisation.

The small steps that are necessary must be in line with what international research tells us about conceptual growth and development. The Shanghai Maths Project embodies that knowledge about conceptual development and about teaching for mastery of mathematics concepts and skills. The way that difficulty is varied, and the same ideas are presented in different contexts, is based on the notion of 'teaching with variation'. 'Variation' in Chinese mathematics carries particular meaning as it has emerged from a great deal of research in the area of 'variation theory'. Variation theory is based on the view that, 'When a particular aspect varies whilst all other aspects of the phenomenon are kept invariant, the learner will experience variation in the varying aspect and will discern that aspect. For example, when a child is shown three balls of the same size, shape, and material, but each of a different color: red, green and yellow, then it is very likely that the child's attention will be drawn to the color of the balls because it is the only aspect that varies.' (Bowden and Marton 1998, cited in Pang & Ling 2012)

In summary, two types of variation are necessary, each with a different function; both are necessary for the development of conceptual understanding.

Variation

Conceptual

Function – this variation provides pupils with multiple experiences from different perspectives.

'multi-dimensional variation'

Procedural

Function – this variation helps learners:
- aquire knowledge step by step
- develop pupils' experience in problem solving progressively
- form well-structured knowledge.

'developmental variation'

Teachers who are aiming to provide conceptual variation should vary the way the problem is presented without varying the structure of the problem itself.

The problem itself doesn't change but the way it is presented (or represented) does. Incorporation of a Concrete–Pictorial–Abstract (CPA) approach to teaching activities provides conceptual variation since pupils experience the same mathematical situations in parallel concrete, pictorial and abstract ways.

CPA is integrated in the Teacher's Guides so teachers are providing questions and experiences that incorporate appropriate conceptual variation.

Procedural variation is the process of:
- forming concepts logically and/or chronologically (i.e. scaffolding, transforming)
- arriving at solutions to problems
- forming knowledge structures (generalising across contexts).

In the Practice Book there are numerous examples of procedural variation in which pupils gradually build up knowledge, step by step; often they are exposed to patterns that teachers should guide them to perceive and explore.

It is this embedded variation that means that when The Shanghai Maths Project is at the heart of mathematics teaching throughout the school, teachers can be confident that the curriculum is of the highest order and it will be delivered by teachers who are informed and confident about how to support pupils to develop strong, connected concepts.

Teaching for mastery

There is no single definition of mathematics mastery. The term 'mastery' is used in conjunction with various aspects of education – to describe goals, attainment levels or a type of teaching. In teaching in Shanghai, mastery of concepts is characterised as 'thorough understanding' and is one of the aims of maths teaching in Shanghai.

Thorough understanding is evident in what pupils do and say. A concept can be seen to have been mastered when a pupil:
- is able to interpret and construct multiple representations of aspects of that concept
- can communicate relevant ideas and reason clearly about that concept using appropriate mathematical language
- can solve problems using the knowledge learned in familiar and new situations, collaboratively and independently.

Within The Shanghai Maths Project, mastery is a goal, achievable through high-quality teaching and learning experiences that include opportunities to explore, articulate thinking, conjecture, practise, clarify, apply and integrate new understandings piece by piece. Learning is carefully structured throughout and across the programme, with Teacher's Guides and Practice Books interwoven – chapter by chapter, unit by unit, question by question.

Since so much conceptual learning is to be achieved with each of the questions in any Practice Book unit, teachers are provided with guidance for each question, breaking down the development that will occur and how they should facilitate this – suggestions for teachers' questions, problems for pupils, activities and resources are clearly set out in an appropriate sequence.

In this way, teaching and learning are unified and consolidated. Coherence within and across components of the programme is an important aspect of The Shanghai Maths Project, in which Practice Books and Teacher's Guides, when used together, form a strong, effective teaching programme.

Promoting pupil engagement

The digital package on Collins Connect contains a variety of resources for concept development, problem solving and practice, provided in different ways. Images can be projected and shared with the class from the Image Bank. Other resources, for pupils to work with directly, are provided as photocopiable resource sheets at the back of the Teacher's Guides, and on Collins Connect. These might be practical activities, games, puzzles or investigations, or are sometimes more straightforward practice exercises. Teachers are signposted to these as 'Resources' in the Unit guidance.

Coverage of the curriculum is comprehensive, coherent and consolidated. Ideas are developed meaningfully, through intelligent practice, incorporating skilful questioning that exposes mathematical structures and connections.

Shanghai Year 3 Learning Book

Shanghai Year 3 Learning Books are for pupils to use. They are concise, colourful references that set out all the key ideas taught in the year, using images and explanations pupils will be familiar with from their lessons. Ideally, the books will be available to pupils during their maths lessons and at other times during the school day so they can access them easily if they need support for thinking about maths. The books are set out to correspond with each chapter* as it is taught and provide all the key images and vocabulary pupils will need in order to think things through independently or with a partner, resolving issues for themselves as much as possible. The Year 3 Learning Book might sometimes be taken home and shared with parents: this enables pupils, parents and teachers to form positive relationships around maths teaching that is of great benefit to children's learning.

---

* Note that because Chapter 6 in Year 3 is a Consolidation and Enhancement Chapter, there is no Chapter 6 in the Year 3 Learning Book.

How to use the Teacher's Guides

Teaching

Units taught in the first half of Year 3:

Contents

Teacher's Guide 3A sets out, for each chapter and unit in Practice Book 3A, a number of things that teachers will need to know if their teaching is to be effective and their pupils are to achieve mastery of the mathematics contained in the Practice Book.

Each chapter begins with a chapter overview that summarises, in a table, how Practice Book questions and classroom activities suggested in the Teacher's Guide relate to National Curriculum statutory requirements.

Chapter overview

Area of mathematics	National Curriculum statutory requirements for Key Stage 2	Shanghai Maths Project reference
Number – Multiplication and division	Year 3 Programme of study: Pupils should be taught to: ■ recall and use multiplication and division facts for the 3, 4 and 8 multiplication tables	Year 3, Units 2.2, 2.10, 2.11. 2.12, 2.13
	■ write and calculate mathematical statements for multiplication and division using the multiplication tables that they know, including for two-digit numbers times one-digit numbers, using mental and progressing to formal written methods	Year 3, Units 2.1, 2.2, 2.3, 2.4, 2.5, 2.6, 2.7, 2.8, 2.10, 2.11, 2.12, 2.13
	■ solve problems, including missing number problems, involving multiplication and division, including positive integer scaling problems and correspondence problems in which n objects are connected to m objects.	Year 3, Units 2.1, 2.2, 2.3, 2.4, 2.5, 2.7, 2.8, 2.9 2.10, 2.11, 2.12, 2.13
Number – Multiplication and division	Year 4 Programme of study: Pupils should be taught to: ■ recall multiplication and division facts for multiplication tables up to 12 × 12	Year 3, Units 2.1, 2.2, 2.3, 2.4, 2.5, 2.6, 2.7, 2.8, 2.9, 2.10, 2.11, 2.12, 2.13
	■ use place value, known and derived facts to multiply and divide mentally, including: multiplying by 0 and 1; dividing by 1; multiplying together three numbers	Year 3, Units 2.10, 2.11, 2.12, 2.13
	■ recognise and use factor pairs and commutativity in mental calculations	Year 3, Unit 2.12

It is important to note that the National Curriculum requirements are statutory at the end of each Key Stage and that The Shanghai Maths Project does fulfil (at least) those end of Key Stage requirements. However, some aspects are not covered in the same year group as they are in the National Curriculum Programme of Study – for example, end of Key Stage 1 requirements for 'Money' are achieved in Year 2 and 'Money' is not taught again in Year 2.

All units will need to be taught over 1–3 lessons. Teachers must use their judgement as to when pupils are ready to move on to new learning within each unit – it is a principle of teaching for mastery that pupils are given opportunities to grasp the learning that is intended before moving to the next variation of the concept or to the next unit.

All units begin with a unit overview, which has four sections:

Conceptual context – a short section summarising the conceptual learning that will be brought about through Practice Book questions and related activities. Links with previous learning and future learning will be noted in this section.

Conceptual context

Prior to counting sets of objects, learning the numbers to 10 and comparing them, there are several skills that need to be developed. These include being able to process information presented visually and link it to known facts about objects, to understand how abstract tokens can be used to represent objects, and compare sets of different objects through subitising or counting.

At this stage, pupils need hands-on experiences – being able to handle objects as they count them adds an important physical and concrete dimension to counting. Number is an abstract concept and pupils need to experience it and see it represented in many different ways. This will help them to form meaningful number concepts that will form vital foundations for their mathematical thinking. Although the questions in the Practice Book provide necessarily pictorial representations, the experiences that you provide in your teaching should focus where possible on the 'concrete' aspect of the Concrete Pictorial Abstract approach. Pupils should be provided with opportunities to count, match and sort physical objects, both everyday objects and abstract tokens (for example, plastic counters).

It is important that, through these activities and questions, pupils have the opportunities to learn that:

a) objects do not always have to be counted – often it is possible to see how many there are without counting them individually (subitising)

b) each object that is counted is only given one number name (one-to-one correspondence)

c) numbers occur in a fixed order – the same order every time

d) it does not matter in which order objects are counted or how they are arranged (conservation of number)

e) the last number said is the total number of objects counted.

Learning pupils will have achieved at the end of the unit

- Pupils will have practised generating a subtraction number sentence from a story (Q1)
- Pupils will have revisited recording part–whole relationships in abstract representations such as bar models (Q1)
- Pupils will have used concrete and pictorial bar models to reinforce understanding of subtraction using addition bonds (Q1, Q2, Q3)
- Pupils will have consolidated using a variety of representations for recording subtraction, including pictures (Q1), bar models (Q1, Q2, Q3), mapping format (Q2), number sentences (Q1, Q2, Q3)
- Pupils will have linked the mapping format with the bar model (Q2)
- Pupils will have revisited and reinforced the correct language for subtraction (Q2, Q3)
- Pupils will have revisited recording addition and subtraction in a number sentence (Q3)

This list indicates how skills and concepts will have formed and developed during work on particular questions within this unit.

These are resources useful for the lesson, including photocopiable resources supplied in the Teacher's Guide. (Those listed are the ones needed for 'Basic questions' – not for 'Challenge and extension questions'.)

This is a list of vocabulary necessary for teachers and pupils to use in the lesson.

Resources

small world toys, e.g. domestic animals, wild animals, sea creatures, vehicles; sets of natural objects, e.g. shells, leaves, fruits, vegetables; beads; buttons; coloured counters; coins; sets of objects made from different materials, e.g. plastic/metal/wood; 2-D/3-D shapes, interlocking cubes, hoops or sorting rings
Resource 1.1.2 'Can you see...?'

Vocabulary

sort, sets, group, similar, same, different, criteria, characteristic, odd one out

The Shanghai Maths Project: an overview

After the unit overview, the Teacher's Guide goes on to describe how teachers might introduce and develop necessary, relevant ideas and how to integrate them with questions in the Practice Book unit. For each question in the Practice Book, teaching is set out under the following headings:

What learning will pupils have achieved at the conclusion of Question X?

This list responds to the following questions: Why is this question here? How does this question help pupils' existing concepts to grow? What is happening in this unit to help pupils prepare for a new concept about …? This list of bullet points will give teachers insight into the rationale for the activities and exercises and will help them to hone their pedagogy and questioning.

What learning will pupils have achieved at the conclusion of Question 1?

- Pupils will have practised processing information presented visually as pictures and connecting it with concepts that they hold about everyday objects.
- Pupils will have practised talking aloud in full sentences to describe characteristics, similarities, differences and connections related to everyday objects.

Activities for whole-class instruction

- Pupils have practised asking what is the same and what is different about groups of objects. The learning for this question requires them to find one object that does not fit a group.
- Prepare sets of five or six concrete objects (or pictures) that are similar and add one that is different. Choose scenarios that are appropriate for your pupils. Possible ideas are:
  - 4–5 farm animals and one tractor
  - 4–5 vegetables and one fruit
  - 4–5 boats and one shark
  - 4–5 sets of shoes and one pair of socks
  - 4–5 fish and one crab
  - 4–5 identical birds flying and one perched
- Encourage the pupils to work in pairs to identify the group and recognise the one that does not belong through discussion.
- Choose a pupil to express this verbally, for example: *This is a group of (farm) animals. The one that does not belong is the tractor because it is not a (farm) animal.*

  Repeat what the pupil has said and introduce the term 'odd one out'. Say: *This is a group of (farm) animals. The odd one out is the tractor because it is not a (farm) animal.*
- Challenge pairs of pupils to use small world resources to make their own group of objects that includes an 'odd one out'. Invite pairs to identify and explain the odd one out.
- Allow pupils to work through Question 3 in the Practice Book recognising the group and identifying the object that does not belong to the group. Listen to them explaining their reasoning.

Activities for whole-class instruction

This is the largest section within each unit. For each question in the Practice Book, suggestions are set out for questions and activities that support pupils to form and develop concepts and deepen understanding. Suggestions are described in some detail and activities are carefully sequenced to enable coherent progression. Procedural fluency and conceptual learning are both valued and developed in tandem and in line with the Practice Book questions. Teachers are prompted to draw pupils' attention to connections and to guide them to perceive links for themselves so mathematical relationships and richly connected concepts are understood and can be applied.

The Concrete–Pictorial–Abstract (CPA) approach underpins suggestions for activities, particularly those intended to provide conceptual variation (varying the way the problem is presented without varying the structure of the problem itself). This contributes to conceptual variation by giving pupils opportunities to experience concepts in multiple representations – the concrete, pictorial and the abstract. Pupils learn well when they are able to engage with ideas in a practical, concrete way and then go on to represent those ideas as pictures or diagrams, and ultimately as symbols. It is important, however, that a CPA approach is not understood as a one-way journey from concrete to abstract and that pupils do not need to work with concrete materials in practical ways if they can cope with abstract representations – this is a fallacy. Pupils of all ages do need to work with all kinds of representations since it is 'translating' between the concrete, pictorial and abstract that will deepen understanding, by rehearsing the links between them and strengthening conceptual connections. It is these connections that provide pupils with the capacity to solve problems, even in unfamiliar contexts.

In this section, the reasons underlying certain questions and activities are explained, so teachers learn the ways in which pupils' concepts need to develop and how to improve and refine their questioning and provision.

Usually, for each question, the focus will at first be on whole-class and partner work to introduce and develop ideas and understanding relevant to the question. Once the necessary learning has been achieved and practised, pupils will complete the Practice Book question, when it will be further reinforced and developed.

Same-day intervention

Pupils who have not been able to achieve the learning that was intended must be identified straight away so teachers can try to identify the barriers to their learning and help pupils to build their understanding in another way. (This is a principle of teaching for mastery.) In the Teacher's Guide, suggestions for teaching this group are included for each unit. Ideally, this intervention will take place on the same day as the original teaching. The intervention activity always provides a different experience from that of the main lesson – often the activity itself is different; sometimes the changes are to the approach and the explanations that enable pupils to access a similar activity.

Same-day intervention

- Provide worksheets to practise drawing numerals that use arrows to show the correct direction required for each stroke. Check which numerals individual pupils find difficult and provide focused support.
- Write numbers on white paper using a pink or blue highlighter pen. Give pupils a yellow highlighter and ask them to trace the numbers carefully. If they trace accurately, the colour will change from pink to orange, or from blue to green.
- Use PE small apparatus, for example, beanbags, quoits, or spot markers, for pupils to play counting games. Place the apparatus a sensible distance away. Give pairs of pupils a hoop in which to place items and then challenge them to run and collect a number of objects, for example, six beanbags. Choose a pupil to set a new challenge.

Same-day enrichment

- Split pupils into pairs. One pupil should make a group of everyday objects (all the same) to represent a number less than 10.
- Their partner should then use plastic counters to make a group to match that number.
- Encourage pupils to explain how they know that the groups are the same.

Same-day enrichment

For pupils who do manage to achieve all the planned learning, additional activities are described. These are intended to enrich and extend the learning of the unit. This activity is often carried out by most of the class while others are engaged with the intervention activity.

Lessons might also have some of the following elements:

Information point

Inserted at points where it feels important to point something out along the way.

(i) This question allows pupils to identify links between objects according to their characteristics. Although these characteristics are not number-based, perceiving characteristics – and therefore similarities and differences – is a vital skill for mathematics. Pupils who get into good habits and are practised at noticing similarities and differences will be able to see connections more readily and be able to solve problems in future. Seeing similarities generally is also a pre-requisite for being able to say 'I have matched these two groups together because they both show 4' – that is, to match quantities that are represented in different ways.

All say ...

Phrases and sentences to be spoken aloud by pupils in unison and repeated on multiple occasions whenever opportunities present themselves during, within and outside of the maths lesson.

 How many more are needed to make the sticks the same. All say the sentence together.

Look out for ...

Common errors that pupils make and misconceptions that are often evident in a particular aspect of maths. Do not try to prevent these but recognise them where they occur and take opportunities to raise them in discussion in sensitive ways so pupils can align their conceptual understanding in more appropriate ways.

(Look out for) ... left-handed pupils may need additional support in forming numerals, as they cannot easily see what they are writing. They may find it easier to write an 8 in reverse. Similarly, pupils with dyspraxia will find this challenging. Ensure that such pupils are allowed to demonstrate their true cognitive abilities and are not penalised for immature motor control. These pupils should be receiving support to help with handwriting and as their handwriting skills improve, so will their ability to write clear numerals.

Within the guidance there are many prompts for teachers to ask pupils to explain their thinking or their answers. The language that pupils use when responding to questions in class is an important aspect of teaching with The Shanghai Maths Project. Pupils should be expected to use full sentences, including correct mathematical terms and language, to clarify the reasoning underpinning their solutions. This articulation of pupils' thinking is a valuable step in developing concepts, and opportunities should be taken wherever possible to encourage pupils to use full sentences when talking about their maths.

Ideas for resources and activities are for guidance; teachers might have better ideas and resources available. The principle guiding elements for each question should be 'What learning will pupils have achieved at the conclusion of Question X?' and the 'Information points'. If teachers can substitute their own questions and tasks and still achieve these learning objectives they should not feel concerned about diverging from the suggestions here.

The Shanghai Maths Project: an overview

Planning

The Teacher's Guides and Practice Books for Year 3 are split into two volumes, 3A and 3B, one for each part of the year.

- Teacher's Guide 3A and Practice Book 3A cover Chapters 1–6.
- Teacher's Guide 3B and Practice Book 3B cover Chapters 7–10.

Each unit in the Practice Book will need 1–3 lessons for effective teaching and learning of the conceptual content in that unit. Teachers will judge precisely how to plan the teaching year but, as a general guide, they should aim to complete Chapters 1–4 in the autumn term, Chapters 4–8 in the spring term and Chapters 8–10 in the summer term.

The recommended teaching sequence is as set out in the Practice Books.

Statutory requirements of the National Curriculum in England 2013 (updated 2014) are fully met, and often exceeded, by the programme contained in The Shanghai Maths Project. It should be noted that some curriculum objectives are not covered in the same year group as they are in the National Curriculum Programme of Study – however, since it is end of Key Stage requirements that are statutory, schools following The Shanghai Maths Project are meeting legal curriculum requirements.

A chapter overview at the beginning of each chapter shows, in a table, how Practice Book questions and classroom activities suggested in the Teacher's Guide relate to National Curriculum statutory requirements.

Level of detail

Within each unit, a series of whole-class activities is listed, linked to each question. Within these are questions for pupils that will:

- structure and support pupils' learning, and
- aid teachers' assessments during the lesson.

Questions and questioning

Within the guidance for each question are sequences of questions that teachers should ask pupils. Embedded within these is the procedural variation that will help pupils to make connections across their knowledge and experience and support them to 'bridge' to the next level of complexity in the concept being learned.

In preparing for each lesson, teachers will find that, by reading the guidance thoroughly, they will learn for themselves how these sequences of questions very gradually expose more of the maths to be learned, how small those steps of progression need to be, and how carefully crafted

the sequence must be. With experience, teachers will find they need to refer to the pupils' questions in the guidance less, as they learn more about how maths concepts need to be nurtured and as they become skilled at 'designing' their own series of questions.

Is it necessary to do everything suggested in the Teacher's Guide?

Activities are described in some detail so teachers understand how to build up the level of challenge and how to vary the contexts and representations used appropriately. These two aspects of teaching mathematics are often called 'intelligent practice'. If pupils are to learn concepts so they are long-lasting and provide learners with the capacity to apply their learning fluently and flexibly in order to solve problems, it is these two aspects of maths teaching that must be achieved to a high standard. The guidance contained in this Teacher's Guide is sufficiently detailed to support teachers to do this.

Teachers who are already expert practitioners in teaching for mastery might use the Teacher's Guide in a different way from those who feel they need more support. The unit overview provides a summary of the concepts and skills learned when pupils work through the activities set out in the guidance and integrated with the Practice Book. Expert mastery teachers might, therefore, select from the activities described and supplement with others from their own resources, confident in their own 'intelligent practice'.

Assessing

Ongoing assessment, during lessons, will need to inform judgements about which pupils need further support. Of course, prompt marking will also inform these decisions, but this should not be the only basis for daily assessments – teachers will learn a lot about what pupils understand through skilful questioning and observation during lessons.

At the end of each chapter, a chapter test will revisit the content of the units within that chapter. Attainment in the text can be mapped to particular questions and units so teachers can diagnose particular needs for individuals and groups. Analysis of results from chapter tests will also reveal questions or units that caused difficulties for a large proportion of the class, indicating that more time is needed on that question/unit when it is next taught.

Shanghai Year 3 Learning Book

As referenced on page vii, The Shanghai Maths Project Year 3 Learning Book is a pupil textbook containing the Year 3 maths facts and full pictorial glossary to enable children to master the Year 3 maths programmes of study for England. It sits alongside the Practice Books to be used as a reference book in class or at home.

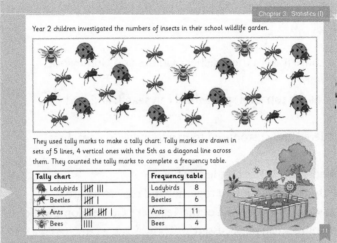

Maths facts correspond to the chapters in the Practice Books for ease of use.

Key models and images are provided for each mathematical concept.

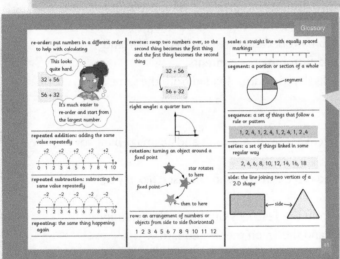

A visual glossary defines the key mathematical vocabulary children need to master.

Homework Guides

The Shanghai Maths Project Homework Guide 3 is a photocopiable master book for the teacher. There is one book per year, containing a homework sheet for every unit, directly related to the maths being covered in the Practice Book unit. There is a 'Learning Together' activity on each page that includes an idea for practical maths the parent or guardian can do with the child.

Homework is directly related to the maths being covered in class.

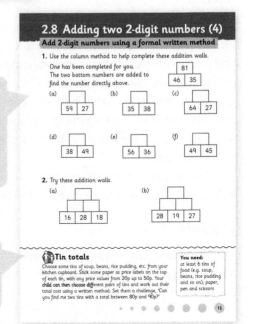

An idea for practical maths the parent or guardian can do with the child

The Shanghai Maths Project: an overview

Collins Connect

Collins Connect is the home for all the digital teaching resources provided by The Shanghai Maths Project.

The Collins Connect pack for The Shanghai Maths Project consists of four sections: Teach, Resources, Record, Support.

Teach

The Teach section contains all the content from the Teacher's Guides and Homework Guides, organised by chapter and unit.

- The entire book can be accessed at the top level so teachers can search and find objectives or key words easily.
- Chapters and units can be re-ordered and customised to match individual teachers' planning.
- Chapters and units can be marked as complete by the teacher.
- All the teaching resources for a chapter are grouped together and easy to locate.
- Each unit has its own page from which the contents of the Teacher's Guide, Homework Guide and any accompanying resources can be accessed.
- Teachers can record teacher judgements against National Curriculum attainment targets for individual pupils or the whole class with the record-keeping tool.
- Units from the Teacher's Guide and Homework Guide are provided in PDF and Word versions so teachers can edit and customise the contents.
- Any accompanying resources can be displayed or downloaded from the same page.

Resources

The Resources section contains 35 interactive whiteboard tools and an image bank for front-of-class display.

- The 35 maths tools cover all topics, and can be customised and used flexibly by teachers as part of their lessons.
- The image bank contains the images from the Teacher's Guide, which can support pupils' learning. They can be enlarged and shown on the whiteboard.

Record

The Record section is the home of the record-keeping tool for The Shanghai Maths Project. Each unit is linked to attainment targets in the National Curriculum for England, and teachers can easily make records and judgements for individual pupils, groups of pupils or whole classes using the tool from the 'Teach' section. Records and comments can also be added from the 'Record' section, and reports generated by class, by pupil, by domain or by National Curriculum attainment target.

- View and print reports in different formats for sharing with teachers, senior leaders and parents.
- Delve deeper into the records to check on the progress of individual pupils.
- Instantly check on the progress of the class in each domain.

Support

The Support section contains the Teacher's Guide introduction in PDF and Word formats, along with CPD advice and guidance.

Chapter 1
Revising and improving

Chapter overview

Area of mathematics	National Curriculum statutory requirements for Key Stage 2	Shanghai Maths Project reference
Number – number and place value	Year 3 Programme of study: Pupils should be taught to:	
	■ recognise the place value of each digit in a two-digit number (tens, ones)	Year 3, Units 1.1, 1.2, 1.3, 1.4, 1.5, 1.6, 1.7
	■ identify, represent and estimate numbers using different representations, including the number line	Year 3, Units 1.1, 1.2, 1.3, 1.4, 1.5, 1.6, 1.7
	■ read and write numbers to at least 100 in numerals and in words	Year 3, Units 1.1, 1.2, 1.3, 1.4, 1.5, 1.6, 1.7
	■ use place value and number facts to solve problems.	Year 3, Units 1.1, 1.2, 1.3, 1.4, 1.5, 1.6, 1.7
Number – addition and subtraction	Year 3 Programme of study: Pupils should be taught to:	
	■ solve problems with addition and subtraction: • using concrete objects and pictorial representations, including those involving numbers, quantities and measures	Year 3, Units 1.1, 1.2, 1.3, 1.4, 1.5
	• applying their increasing knowledge of mental and written methods	Year 3, Units 1.4
	■ add and subtract numbers using concrete objects, pictorial representations, and mentally, including: • a two-digit number and ones	Year 3, Units 1.1, 1.2, 1.3, 1.4, 1.5
	• a two-digit number and tens	Year 3, Units 1.1, 1.3, 1.4
	• two two-digit numbers	Year 3, Units 1.1, 1.2, 1.3, 1.4
	• adding three one-digit numbers	
	■ show that addition of two numbers can be done in any order (commutative) and subtraction of one number from another cannot	Year 3, Units 1.1, 1.2, 1.3, 1.4, 1.5
	■ recognise and use the inverse relationship between addition and subtraction and use this to check calculations and solve missing number problems.	Year 3, Units 1.1, 1.2, 1.3, 1.4, 1.5

Number – multiplication and division	Year 3 Programme of study: Pupils should be taught to:	
	■ recall and use multiplication and division facts for the 2, 5 and 10 multiplication tables, including recognising odd and even numbers	Year 3, Units 1.6, 1.7
	■ calculate mathematical statements for multiplication and division within the multiplication tables and write them using the multiplication (×), division (÷) and equals (=) signs	Year 3, Units 1.6, 1.7
	■ show that multiplication of two numbers can be done in any order (commutative) and division of one number by another cannot	Year 3, Units 1.6, 1.7
	■ solve problems involving multiplication and division, using materials, arrays, repeated addition, mental methods, and multiplication and division facts, including problems in contexts.	Year 3, Units 1.6, 1.7
Measurement	Year 3 Programme of study: Pupils should be taught to:	
	■ recognise and use symbols for pounds (£) and pence (p); combine amounts to make a particular value	Year 3, Units 1.1, 1.2, 1.3
	■ solve simple problems in a practical context involving addition and subtraction of money of the same unit, including giving change.	Year 3, Units 1.1, 1.2, 1.3
Area of mathematics	**National Curriculum statutory requirements for Key Stage 2**	**Shanghai Maths Project reference**
Number – addition and subtraction	Year 3 Programme of study: Pupils should be taught to:	
	■ add and subtract numbers mentally, including: ● a three-digit number and ones ● a three-digit number and tens ● a three-digit number and hundreds	
	■ add and subtract numbers with up to three digits, using formal written methods of columnar addition and subtraction	Year 3, Unit 1.2
	■ estimate the answer to a calculation and use inverse operations to check answers	Year 3, Units 1.1, 1.2
	■ solve problems, including missing number problems, using number facts, place value, and more complex addition and subtraction.	Year 3, Units 1.1, 1.2
Number – multiplication and division	Year 3 Programme of study: Pupils should be taught to:	
	■ recall and use multiplication and division facts for the 3, 4 and 8 multiplication tables	Year 3, Units 1.6, 1.7
	■ write and calculate mathematical statements for multiplication and division using the multiplication tables that they know, including for two-digit numbers times one-digit numbers, using mental and progressing to formal written methods	Year 3, Units 1.6, 1.7
	■ solve problems, including missing number problems, involving multiplication and division, including positive integer scaling problems and correspondence problems in which n objects are connected to m objects.	Year 3, Units 1.6, 1.7

Unit 1.1
Revision for addition and subtraction of 2-digit numbers

Conceptual context

This is the first of three units revising addition and subtraction of two-digit numbers. Pupils begin by revising addition and subtraction of two numbers. This is then extended to three numbers and mixed operations. The focus is on working mentally, so place value and partitioning, counting on, commutativity and the inverse relationship between addition and subtraction are all revisited to support calculating. Base 10 resources are used, but pupils are also encouraged to visualise these resources to develop mental tools to support their thinking.

Learning pupils will have achieved at the end of the unit

- Pupils will have applied their understanding of place value to support addition and subtraction (Q1, Q2)
- Pupils will have revisited and further explored counting on to add and subtract (Q1)
- Visualisation of apparatus such as a number line will have been practised in order to develop mental tools for thinking (Q1)
- Knowledge of the commutative law will have been applied to reorder addition calculations (Q2)
- Pupils will have extended their understanding of commutativity to use with mixed calculations and those involving adding or subtracting two numbers (Q3, Q4)
- Pupils will have revisited and further developed their understanding of the bar model to find missing numbers (Q5)
- Understanding of the inverse relationship between addition and subtraction will have been applied to find missing numbers (Q5)
- Pupils will have applied their understanding of addition and subtraction to solve problems in the context of money (Q6)

Resources

number lines; number cards (11–40); place value mats and counters; 100 square; base 10 apparatus; toy money; a simple menu or **Resource 3.1.1** Mia's Burger Bar

Vocabulary

addition, subtraction, difference, recombine, commutative, operation, part/whole model, bar model

Question 1

1 Calculate mentally.

(a) 63 + 9 = ☐ (b) 75 + 24 = ☐ (c) 84 − 16 = ☐

(d) 70 − 35 = ☐ (e) 34 + 16 = ☐ (f) 48 + 25 = ☐

(g) 63 − 36 = ☐ (h) 93 − 53 = ☐ (i) 38 + 25 = ☐

(j) 27 + 44 = ☐ (k) 74 − 58 = ☐ (l) 60 − 42 = ☐

(m) 36 + 18 = ☐ (n) 67 − 27 = ☐ (o) 50 − 43 = ☐

(p) 90 + 10 = ☐ (q) 82 − 45 = ☐ (r) 36 + 55 = ☐

What learning will pupils have achieved at the conclusion of Question 1?

- Pupils will have applied their understanding of place value to support addition and subtraction.
- Pupils will have revisited and explored using counting on to add and subtract.
- Visualisation of apparatus such as a number line will have been practised in order to develop mental tools for thinking.

Activities for whole-class instruction

- Show pupils the subtraction 73 − 43 = ☐ and ask them what they notice. Pupils should recognise that both numbers have the same digit in the ones place, so the difference must be a multiple of 10.
- Ask pupils how they could find the difference mentally. Share ideas. Pupils may suggest partitioning 73 into 70 and 3, subtracting 40 from 70 to give 30 and subtracting 3 from 3 to give 0. Recombining the two parts gives a difference of 30.
- Pupils may also suggest keeping 73 whole, subtracting 40 then 3 to find the difference of 30.
- Explain that for calculations such as this one, they could count up from 43 to 73, for example by counting on in tens, 53, 63, 73, the difference is 30. Draw a number line on the board, mark 43 and draw in the three jumps of 10 from 43 to 73.
- Record some similar calculations and ask pupils to count up to find the difference. Provide number lines for support if necessary.
- Go on to explain that the same approach can be used for other subtractions. Write 83 − 68 = ☐ on the board.
- This time ask pupils to picture a number line to support their thinking. Ask: *How many to get from 68 to 70?*

And from 70 to 80? And from 80 to 83? Agree that the difference is 2 + 10 + 3, 15. Draw the matching number line to support those who found it hard to visualise.

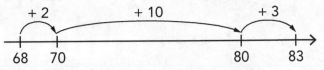

- Point out that you chose to jump to each successive multiple of 10, to help you keep track of what you needed to add to get from 68 to 83.
- Work through a few more examples together. Ensure that some are better solved using other strategies and model those, for example 70 − 10 = ☐, 45 + 15 = ☐, 63 + 19 = ☐
- Pupils are now ready to complete Question 1 in the Practice Book.

Same-day intervention

- Show pupils the subtraction 40 − 32 = ☐. Give pupils a 0 to 100 number line marked in ones or tens and ask them to mark 32.
- Check that pupils can also see 40. Remind them that 40 − 32 is asking them to find the difference between 40 and 32 or, in other words, how many more to get from 32 to 40.
- Draw in a jump from 32 to 40 and count how many spaces within the jump, by counting on from 32 to 40, 8. So 40 − 32 = 8.
- Count from 32 to 40 and from 40 to 32 to remind pupils that it does not matter whether you count forwards or back, the difference is still 8.
- Ask pupils if they noticed anything else about the calculation. If necessary, point out that since 2 + 8 = 10, they could use their number bonds knowledge to help them find the difference when the minuend is a multiple of 10.
- Show pupils the subtraction 50 − 35 = ☐. Ask them to use the number line to find the difference as they did before. Support pupils to draw a jump from 35 to 40 and another from 40 to 50. Pupils should label their jumps, then complete the subtraction 50 − 35 = 15. Explain that pupils could have used their number bonds for 10, this time 5 + 5 = 10, to help them find the difference.
- Work through some similar subtractions, encouraging pupils to use the number line initially but to move on to working mentally.

Same-day enrichment

- Ask pupils to look at the subtraction calculations in Question 1 in the Practice Book and sort them into groups according to the method used to find the difference, for example counting in tens, counting on, using number bonds to 10 or something else. Some calculations may have required a combination of approaches.

- Ask: *Why is this method most suitable for this calculation?* (For example, counting in tens when both the minuend and subtrahend have the same number of ones.) Repeat the question for other methods.

Question 2

2 Add three numbers.

(a) 21 + 23 + 50 =

(b) 36 + 44 + 13 =

(c) 13 + 30 + 37 =

(d) 23 + 20 + 40 =

(e) 67 + 25 + 12 =

(f) 42 + 26 + 24 =

(g) 60 + 18 + 11 =

(h) 21 + 44 + 12 =

(i) 35 + 16 + 40 =

(j) 34 + 35 + 14 =

What learning will pupils have achieved at the conclusion of Question 2?

- Pupils will have applied their understanding of place value to support adding three two-digit numbers.

- Knowledge of the commutative law will have been applied to reorder addition calculations.

Activities for whole-class instruction

- Shuffle a set of two-digit number cards (11 to 40) and turn over the top three cards. Ask pupils to add the three numbers together.

- Confirm the total, then ask pupils to help you write a guide to adding three numbers mentally. This might include:

  – Starting with the greatest number of tens, add all the tens together. Remember the number of tens.

  – Starting with the greatest number of ones, add all the ones together.

  – Add the ones to the tens to find the total.

- Turn over the next three cards and ask pupils to work through the guide to check that it works.

- Having agreed a list of hints, ask pupils if this would change if one of the numbers was a multiple of 10 with no ones. Replace one of the three numbers to test pupils' ideas.

- Agree that if one of the numbers was a multiple of 10 with no ones, the process is the same but adding the ones together should be easier because one of the ones values is zero.

- Pupils are now ready to complete Question 2 in the Practice Book.

Same-day intervention

- Shuffle a set of two-digit number cards (11 to 40) and turn over the top three cards. Ask pupils to work together to make each number using base 10 apparatus. Label each number with place value cards too.

- Add all the tens together by collecting the base 10 ten sticks and counting them in tens. Then collect all the place value tens cards, counting in tens to add them together. Confirm that the total is the same.

- Repeat with the ones. Add the ones and tens together to find the total.

- Record the calculation and total horizontally.

- Work along the addition, partitioning each number to add the tens then add the ones mentally. Confirm that the total is the same.

- Turn over the next three cards and arrange them in a horizontal addition calculation. Work along the addition together, partitioning to add. Repeat for a third set of numbers, asking pupils to work independently. Check and agree the total together.

Same-day enrichment

- Give pupils a total such as 63 and explain that this was the total after adding three different two-digit numbers together.

- Ask pupils to work systematically to identify what those numbers could have been, for example beginning with 10 as one of the numbers and exploring what the other two numbers could be.

- After 5 minutes, share possible additions. Ask pupils to identify some rules for the set of numbers, for example one or three of the numbers must be odd, or the largest number used must be 42 (63 – 10 – 11 = 42, so 42 + 10 + 11 = 63).

- Challenge pupils to work together to find all possible solutions. Remind them that the order the numbers are written in does not matter.

Question 3

> **3** Subtract three numbers.
>
> (a) $72 - 33 - 12 = \boxed{}$ (b) $54 - 27 - 18 = \boxed{}$
>
> (c) $67 - 21 - 22 = \boxed{}$ (d) $68 - 45 - 18 = \boxed{}$
>
> (e) $65 - 56 - 3 = \boxed{}$ (f) $86 - 35 - 24 = \boxed{}$
>
> (g) $88 - 12 - 51 = \boxed{}$ (h) $79 - 19 - 26 = \boxed{}$
>
> (i) $77 - 56 - 6 = \boxed{}$ (j) $89 - 42 - 13 = \boxed{}$

What learning will pupils have achieved at the conclusion of Question 3?

- Pupils will have reviewed the commutative law for subtraction.
- Pupils will have extended their understanding of commutativity to use with mixed calculations and those involving adding or subtracting two numbers.

Activities for whole-class instruction

- Show pupils a subtraction such as $75 - 27 - 16 = \boxed{}$ and ask them what they notice.
- Check that pupils recognise that they are being asked to subtract two amounts from 75. 75 is the minuend; 27 and 16 are subtrahends; which must both be subtracted from 75 to find the difference.
- Give pupils place value counters to find the difference. Pupils should make 75, then exchange 1 ten for 10 ones to enable them to subtract 27 to leave 48. They then subtract a further 16 to find the difference of 32. Alternatively, they could subtract 16 and then subtract 27. They key understanding is that they must subtract both subtrahends from the minuend.
- Draw a bar model to illustrate the same calculation: $75 - 27 - 16 = 32$. Check that pupils can identify the minuend and subtrahends.

75		
27	16	32

- Show pupils a similar bar model with three parts, one of them unknown:

68		
32	7	?

- Ask: *How could you find the value of the unknown part? Can pupils tell you that the two known parts must be subtracted from the whole? Do they describe them as subtrahends subtracted from the minuend?*

- Introduce an imaginary pupil 'Eric' and draw him on the board. Tell pupils that Eric is confident that he knows how to work out the answer to this problem.

> $32 - 7 = 25$
>
> Then
>
> $68 - 25 = 43$
>
> Answer is 43

- Ask: *Did Eric do the same as us? Did he get the same answer as you? What happened?*
- Pupils should discuss with partners and share ideas with the class. Can they explain that Eric has subtracted one subtrahend from another, instead of subtracting them both from the minuend to find the difference, the missing part? Refer back to the bar model to reinforce that 68 is the minuend, the quantity being subtracted from; 32 and 7 are subtrahends, quantities to be subtracted from the minuend to give the difference.
- Pupils are now ready to complete Question 3 in the Practice Book.

Same-day intervention

- Choose a calculation from Question 3, for example $88 - 12 - 51 = \boxed{}$. Check that pupils recognise that they need to subtract both 12 and 51 from 88. Name each number: 88 is the minuend, 12 and 51 are both subtrahends. The unknown value is the difference.
- Ask pupils to make 88 using base 10 blocks.
- Ask them to subtract 12, recording this as $88 - 12 = \boxed{}$. Confirm that the difference is 76.
- Now ask pupils to subtract 51 from what they have left, recording this as $88 - 12 - 51 = \boxed{}$. Confirm that the final difference is 25.
- Show pupils how they worked along the calculation from left to right, making 88 then subtracting 12, following that by subtracting 51 to find the solution, 25.

88		
12	51	25

- Draw and label the corresponding bar model together. Remind pupils that drawing a bar model can help them to see the calculation more clearly because it shows the whole and the parts. 88 is the minuend; 12 and 51 are subtrahends; which are subtracted from 88 to give a difference of 25.

- Choose a second example and work through it in the same way. Reinforce working from left to right.

- Choose a third example that does not involve any exchanging, for example 67 − 21 − 22. Work through this together mentally, first calculating 67 − 21 = 46 and recording the 46 just above the second subtraction sign. Check that pupils can see that the rest of the calculation is 46 − 22 and that the difference is 24, 67 − 21 − 22 = 24.

Same-day enrichment

- Remind pupils that subtraction is not commutative, that they cannot subtract in any order.

- Choose a calculation from Question 3, for example 68 − 45 − 18. Explain that the calculation is asking them to subtract 45 and 18 from 68. While they cannot do part of the calculation, 45 − 18, first because they are then not subtracting 18 from 68, they can combine subtract 45 and subtract 18 into subtract 63.

- Ask pupils to check that 68 − 45 − 18 = 68 − 63.

- Ask pupils to revisit Question 3 to write each calculation in this format.

Question 4

> 4 Complete these mixed addition and subtraction calculations.
>
> (a) 87 + 10 − 27 = ☐   (b) 57 + 31 − 27 = ☐
>
> (c) 54 − 52 + 17 = ☐   (d) 96 − 63 + 41 = ☐
>
> (e) 74 − 47 + 15 = ☐   (f) 88 − 46 + 55 = ☐
>
> (g) 54 − 33 + 21 = ☐   (h) 36 − 18 + 9 = ☐
>
> (i) 77 + 18 − 34 = ☐   (j) 64 + 23 − 19 = ☐

What learning will pupils have achieved at the conclusion of Question 4?

- Pupils will have applied what they know about commutativity to solve mixed addition and subtraction calculations.

Activities for whole-class instruction

- Remind pupils that a number sentence is just like a written sentence, we read along it from left to right.

- Explain that in Year 3 this is also the order in which we carry out the calculation. So if the first operation we come to is addition, that is what we do first. If the first operation is subtraction, that is what we do first.

(i) Although the order of the calculation in a mixed calculation can be changed without changing the total (70 − 30 + 40 = 80 and 70 + 40 − 30 = 80), this is not always the case. Reordering a mixed calculation could take pupils into negative numbers before they understand how to calculate with them. Pupils may also think they can select and solve part of the number sentence first if they can reorder mixed calculations. For example, 70 − 30 + 40 = ☐; 30 + 40 = 70, 70 − 70 = 0, when 70 − 30 + 40 = 80. To prevent these situations arising, it makes sense to follow the order of the number sentence until pupils have a greater understanding of the number system.

- Show pupils the number sentence 70 − 30 + 40 = ☐. Tell pupils that you have made each number a multiple of 10 to make the addition and subtraction straightforward. Point at a large-scale number line or 100 square, while pupils say the 'All say…' together.

 All say…   70 − 30 = 40, 40 + 40 = 80. 70 − 30 + 40 = 80.

- Pupils are now ready to complete Question 4 in the Practice Book.

Same-day intervention

- Show pupils the mixed calculation, 59 − 23 + 18

- Ask pupils to make 59 in base 10 blocks. Together, first subtract 23, leaving 36. Ask pupils to draw this on a number line. They should mark 59 and then draw a jump from 59 to 36, labelling the jump −23. If necessary, pupils may make two (or three) jumps. Subtract 20, marking a jump from 59 to 39, labelling it −20. Then draw a second jump to show −3. Draw the jump from 39 to 36 and label it −3. Use three jumps if pupils need to partition 23 into 10, 10 and 3.

- Now ask pupils to add 18 using base 10 blocks to find total of 54. Ask pupils to draw this on their number line. Since they have already drawn jumps above the number line, ask them to draw a jump (or jumps) to show +18 below the number line. Pupils could draw a single jump to 54 labelling it +18, or a jump of 10 and a jump of 8.

- Ask pupils to read the mixed calculation together, moving a finger along the number line to show how they carried out each part to reach the total of 54.

- Choose a calculation from Question 4 to work through together, working from left to right. Give pupils the choice of base 10 blocks or a number line. Check that everyone has the same answer and ask two pupils to demonstrate their method.

- Repeat with a further mixed calculation to help pupils develop confidence. Can they use a different method from the one they used in the previous calculation?

Same-day enrichment

- Show pupils the outline calculation
  $99 - \triangle + \pentagon = 54$ and ask them what they notice.

- Although the minuend and difference are given, how much to subtract or add is not given.

- Tell pupils they must decide what are some of the possible numbers that could go in $\triangle$ and $\pentagon$. Ask: *What is the first thing you need to work out?* Pupils should see that they must determine the difference between 99 and 54 as that will be the total value subtracted from 99, so the two numbers they select must give a total of that number. In this example, they must subtract 45 from 99 to equal 54, so $\triangle$ and $\pentagon$ must total 45. For example, $99 - 46 + 1 = 54$ and $99 - 47 + 2 = 54$.

- Pupils should explore finding pairs of numbers for $\triangle$ and $\pentagon$ that:
  - are both even
  - are both odd
  - are both the same or explain why not, if they cannot find appropriate pairs.

- In order to generate other start and end numbers, give pupils two-digit number cards. After shuffling, pupils should turn over the top two cards. These are the minuend and difference for their mixed calculation. Challenge them to find numbers for $\triangle$ and $\pentagon$.

Question 5

5 Fill in the boxes.

(a) ☐ + 41 = 56   (b) 82 − ☐ = 55

(c) ☐ − 36 = 39   (d) 36 + ☐ = 71

(e) 80 − ☐ = 29   (f) 34 = ☐ − 34

What learning will pupils have achieved at the conclusion of Question 5?

- Pupils will have revisited and further developed their understanding of the bar model to find missing numbers.

- Understanding of the inverse relationship between addition and subtraction will have been applied to find missing numbers.

Activities for whole-class instruction

- Revisit the part/whole model as illustrated by the bar model.

- Show pupils a generic bar model and ask them to work in pairs to explain to each other how they would find out an unknown whole or an unknown part.

whole	
part	part

- Share ideas. Agree that if the whole is unknown, pupils can add both parts together to find the sum. If a part is unknown, pupils could subtract or count on to find the unknown part.

- Show pupils some missing number addition sentences and ask them to draw the matching bar model to help them to decide how to find the unknown number:

$28 + 46 = \square$, solve using $28 + 46 = \square$

whole	
28	46

$\square + 37 = 68$, solve using $68 - 37 = \square$

68	
part	37

$37 + \square = 68$, solve using $68 - 37 = \square$

- Move on to some missing number subtraction sentences and again ask pupils to draw the matching bar model to help them to decide how to find the unknown number:

$74 - 18 = \square$, solve using $74 - 18 = \square$

74	
part	18

$74 - \square = 18$, solve using $74 - 18 = \square$

$\square - 53 = 29$, solve using $29 + 53 = \square$

whole	
29	53

- Pupils are now ready to complete Question 5 in the Practice Book.

Same-day intervention

- Draw a generic bar model with unequal parts

whole	
part	part

- Choose an addition calculation that pupils can easily solve, such as $12 + 6 = 18$, and label the bars.

18	
6	12

- Use a piece of card to cover up one of the values in the bar model and discuss how to find out the unknown number. Begin by covering up 18 and agreeing that they would add 12 and 6 to find the missing whole. Record the calculation as a missing number calculation, $12 + 6 = \square$.

- Cover 12 and ask pupils how they would find the missing part. Record their calculations as $\square + 6 = 18$, $18 - \square = 6$ and $18 - 6 = \square$. Talk about how the bar really helps them to see what it is they do not know and how they might find out.

- Repeat, covering up the 6.

Same-day enrichment

- Ask pupils to look at the last part of Question 5, $34 = 68 - 34$.

- Explain that a pupil in another class said that if two of the numbers in a calculation are the same, the third number is double that number. (Focus on addition and subtraction only.)

- Ask pupils to investigate if this is always, sometimes or never true.

- Check that pupils notice that $34 - 34 = 0$ and since 0 is not the double of 34, the third number is only sometimes the double of the two identical numbers.

Question 6

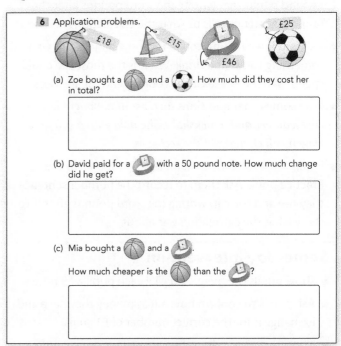

What learning will pupils have achieved at the conclusion of Question 6?

- Pupils will have applied their understanding of addition and subtraction to solving money problems.

Activities for whole-class instruction

- Show pupils a simple menu, for example **Resource 3.1.1** Mia's Burger Bar

- Explain that Sumi has £5. Ask: *Can she afford to buy a cheeseburger and a milkshake?* (£4 + £2 = £6, so Sumi does not have enough money to buy both.) Check that pupils recognise that they calculate in exactly the same way as they usually do, so the calculation is 4 + 2 = 6 (pounds). Remind pupils that writing the word pounds after the number sentence tells us what the calculation was about.

- Ask pupils what Sumi could buy if she had £5 and wanted something to eat and drink. What is the most she could spend? What is the least she could spend? Share ideas.

- Pose some other questions such as: *Sumi bought a cheeseburger and a milkshake. She paid with a £10 note. How much change did she get?*

- Pupils are now ready to complete Question 6 in the Practice Book. Ask them to record the number sentence they use and solve it, writing the word pounds after it to show what the calculation was about.

Same-day intervention

- Show pupils the menu and give each pupil a £5 note.

- Ask pupils to confirm how much money they have and exchange it for the correct number of £1 coins.

- Now ask each pupil to choose one item to buy on the menu. As pupils pay you for their purchase, ask the rest of the group to record the calculation that shows how much money the pupil has left, for example if buying a milkshake, £5 − £2 = £3, the pupil has £3 left.

- Explain that you would like to buy a cheeseburger and a milkshake. Ask pupils how much this would cost and if you could pay with a £5 note. Agree that the total is £6 so you do not have enough.

- Collect the £1 coins and give each pupil a £5 note. Ask pupils to choose two items to buy with their £5. Ask them to record an addition calculation for their purchase, to check that they have enough money. They can then subtract this total from £5 to find out how much change they would get.

- Let pupils take it in turns to make their purchase and receive their change to check that their calculations are correct.

Same-day enrichment

- Ask pupils to find the total cost of the four items in Question 6.

- Explain that Sam is saving up to buy these items. If he saves all his pocket money of £10 a week, how many weeks will it take him to buy all four items? Ask pupils to advise Sam about which items to buy when so that he gets his items as soon as possible.

Challenge and extension question

Question 7

7 Write + or − in each circle to make each equation true.
(a) 1 ◯ 2 ◯ 3 ◯ 5 ◯ 4 = 1
(b) 1 ◯ 2 ◯ 3 ◯ 4 ◯ 5 ◯ 6 = 1

This question asks pupils to insert a + or − symbol between each number so that the answer to each calculation is 1. Remind pupils that they must work along the number sentence from left to right, carrying out each operation in turn to reach a solution. Pupils may need to use a number line or counters for support. They may also need several attempts before they find the correct solution. Rather than telling pupils that they cannot subtract 5 from 3, you could take the opportunity to introduce some pupils to negative numbers, using a number line from −10 to +10, with 0 in the middle.

Unit 1.2
Addition and subtraction (1)

Conceptual context

This is the second of three units revising addition and subtraction of two-digit numbers. Pupils are introduced to a line format of the bar model for part/whole relationships. Some pupils will find this model quicker and easier to draw and visualise. Pupils will also revisit the column method in this unit, giving teachers the opportunity to check that pupils can use the method accurately and with understanding before extending to three-digit numbers in Chapter 7.

Learning pupils will have achieved at the end of the unit

- A range of methods for adding or subtracting a single-digit number mentally will have been revisited (Q1)
- Flexibility in choice of method will have been developed through considering how the numbers affect the chosen method of addition or subtraction (Q1)
- The column method for addition and subtraction will have been practised (Q2)
- The bar model representation will have been expanded to include a line model version (Q3)
- Pupils will have revised how to use the bar and line models to support problem solving (Q3)
- Pupils will have been introduced to recording the answer to a word problem in a word sentence after completing the relevant number sentence (Q4)

Resources

ten-frames; counters; digit cards; base 10 apparatus; number cards (11–40); timer

Vocabulary

addition, subtraction, method, partition, column method, exchange, part/whole model, bar model, line model, bracket or brace, solution, answer

Question 1

> **1** Calculate mentally.
>
> (a) 35 + 8 = ☐
>
> (b) 54 + 5 = ☐
>
> (c) 91 − 9 = ☐
>
> (d) 63 − 8 = ☐
>
> (e) 35 + 65 = ☐
>
> (f) 42 + 9 = ☐
>
> (g) 53 − 6 = ☐
>
> (h) 74 − 60 − 3 = ☐
>
> (i) 40 + 50 − 5 = ☐
>
> (j) ☐ − 35 = 20
>
> (k) ☐ + 81 = 81
>
> (l) ☐ − 7 + 5 = 44

What learning will pupils have achieved at the conclusion of Question 1?

- A range of methods for adding or subtracting a single-digit number mentally will have been revisited.
- Flexibility in choice of method will have been developed through considering how the size a number affects the method of addition or subtraction.

Activities for whole-class instruction

- Ask pupils to share different ways of adding a single-digit number to a two-digit number mentally.
- List methods, for example count on; use known number facts; compensation; partition the two-digit number into tens and ones, add the ones together then add to the tens; use number bonds for 10 to partition the single-digit number to make the two-digit number a multiple of 10, then add on the rest. Pupils may suggest other methods.
- Ask pupils to share different ways of subtracting a single-digit number from a two-digit number mentally.
- List methods alongside their counterparts for addition.
- Draw two large circles. Write a selection of two-digit numbers in one and the numbers 5 to 9 in the other.
- Ask pupils to select a number from each circle to add or subtract. They should select different combinations of numbers until they have tried all the listed methods.
- Remind pupils that their choice of numbers is likely to influence their choice of method. Alongside each method, record when it might be the best method, for example counting on and back may be better for adding or subtracting single-digit numbers less than 5; compensation is a good strategy when addends or subtrahends are close to a decade number.

- Pupils are now ready to complete Question 1 in the Practice Book. Ask them to look at the numbers in the calculation and consider which method to use before calculating.

Same-day intervention

- Give pupils a ten-frame and ten double-sided counters (or ten counters in one colour and ten in another).

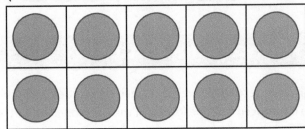

- Ask pupils to work in pairs, turning over (or swapping) one counter at a time so that they can see and record all the number bonds for 10.
- Give pupils a calculation such as 26 + 7 = ☐. Ask them to use their number bonds for 10 to find out how many they need to add to 26 to make 30. Agree that it must be 4, because 4 + 6 = 10.
- Partition the 7 into 4 and 3, recording the calculation as 26 + 4 + 3 = ☐. Underline 26 and 4, then record the number sentence again as 30 + 3 = ☐. Check that pupils recognise that 26 + 7 = 33.
- Repeat with another addition using counters in the ten-frame to model the number bond for the addend to bridge through 10, then move on to subtraction using number bonds for 10. Explore 43 − 8 and 51 − 8 by subtracting 10 then adding 2 back on, because 8 + 2 = 10. Explore subtracting 6, 7 or 9 in the same way.

Same-day enrichment

- Show pupils the following layouts for mixed calculations:  ☐☐ + ☐ − ☐ = ☐☐ and ☐☐ − ☐ + ☐ = ☐☐.
- Give pupils a set of digit cards each and ask them to estimate how many calculations, using the two layouts alternately, they can do in one minute.
- Time pupils for one minute to see if their estimates were correct.
- Challenge pupils to repeat to see if they can do one more in one minute.
- You could run this activity as a competition between pupils or challenge pupils to beat their own best score.

Question 2

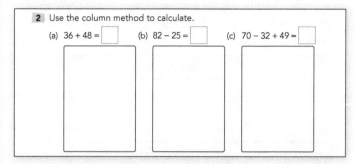

2 Use the column method to calculate.

(a) 36 + 48 = ☐   (b) 82 − 25 = ☐   (c) 70 − 32 + 49 = ☐

What learning will pupils have achieved at the conclusion of Question 2?

- The column method for addition and subtraction will have been practised.

Activities for whole-class instruction

- Show pupils an addition number sentence, written horizontally, for example 27 + 45 = ☐.

- Ask pupils how to set this out using the column method. Confirm the layout and work through the calculation together. If necessary, remind pupils how to add the ones and exchange 10 ones for 1 ten, then add that ten to the other tens in the calculation.

- Repeat for a subtraction number sentence such as 73 − 38 = ☐. As with addition, revise exchanging 1 ten for 10 ones and how to record it if necessary.

- Remind pupils that in Unit 1.1, they looked at some mixed addition and subtraction calculations. Show pupils the number sentence 96 − 63 + 41 = ☐ and ask them how to set this out using the column method. Check that pupils recognise that they must first subtract 63 from 96, then add 41 to the difference.

- Pupils are now ready to complete Question 2 in the Practice Book.

Same-day intervention

- Revisit the column method for addition and subtraction. Initially, choose numbers so that there is no exchange or regrouping, for example:

$$23 \qquad 47$$
$$+\,36 \qquad -\,26$$

- Use base 10 blocks to support the calculation.

- Focus on addition first. Keep 23 the same and work through a series of additions in column format, adding 37, 38 and 39. Exchange 10 ones for 1 ten within each addition. Ask pupils to work independently for the last addition and to explain why they need to exchange 10 ones for 1 ten.

- Move on to subtraction. Again, keep 47 the same and subtract 27, then 28 and finally 29. Ask pupils to work independently for the last subtraction and to explain why they need to exchange 10 ones for 1 ten.

Same-day enrichment

- Refer pupils to the Question 4 in Unit 1.1. In the previous unit, they were asked to complete mixed addition and subtraction calculations but no space was left to use the column method. In Question 2, space has been given to use the column method.

- Ask pupils if the calculations in Unit 1.1 Question 4 were easier than the mixed calculations in Question 2. Ask pupils to explain their thinking. Share ideas, for example the Unit 1.1 calculations rarely involved crossing a ten.

- Finally, ask pupils to give two examples of mixed calculations, one that can most easily be done mentally and another done most easily using the column method. Ask pupils to share their reasons for their choices.

Question 3

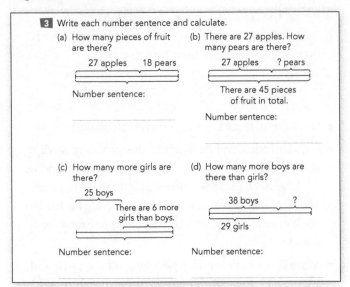

3 Write each number sentence and calculate.
  (a) How many pieces of fruit are there?

  27 apples   18 pears

  Number sentence:

  (b) There are 27 apples. How many pears are there?

  27 apples   ? pears

  There are 45 pieces of fruit in total.

  Number sentence:

  (c) How many more girls are there?

  25 boys

  There are 6 more girls than boys.

  Number sentence:

  (d) How many more boys are there than girls?

  38 boys   ?

  29 girls

  Number sentence:

What learning will pupils have achieved at the conclusion of Question 3?

- The bar model representation will have been expanded to include a line model version.
- Pupils will have revised how to use the bar and line models to support problem solving.

Activities for whole-class instruction

- Remind pupils of the bar model as a part/whole model.

whole	
part	part

- Explain that you are going to introduce them to another diagram for the part/whole model. Draw a line, marking the beginning, end and another point part way along the line. Use curly brackets (also called a brace) to highlight each section of the line.

- Ask pupils to help you label each bracket.

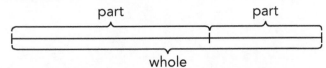

part   part

whole

- Ask: *What is the same and what is different about these two diagrams – the bar model and the line model?* Can pupils tell you that they show the same information but are drawn in a slightly different way? Explain that they will use both models in Year 3.

- Show pupils a word problem, for example there are 29 cars and 16 vans in the car park. How many vehicles are in the car park? Ask pupils to identify the parts so you can label the diagram.

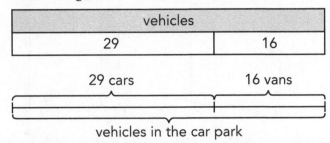

vehicles	
29	16

29 cars   16 vans

vehicles in the car park

- Ask pupils what number sentence they could use to find the number of vehicles in the car park. Agree $29 + 16 = \square$. Check that pupils recognise that they would generate the same number sentence using either diagram.

- Write another word problem with an unknown part. Label a new diagram and bar model and ask pupils to write the number sentence they would use to find the missing part.

- Move on to the comparison bar model. Show pupils a word problem, for example there are 29 cars and 16 vans in the car park. How many more cars than vans are there? Draw the matching bar model. Ask pupils to draw the matching line diagram and write the number sentence they would use to find out how many more cars than vans.

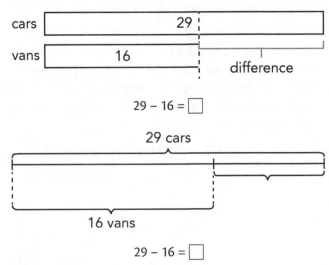

cars   29

vans   16

difference

$29 - 16 = \square$

29 cars

16 vans

$29 - 16 = \square$

- Pupils are now ready to complete Question 3 in the Practice Book.

Same-day intervention

- Give pupils a blank copy of the standard bar model.

- Ask pupils to tell you a simple number sentence, one they are sure is correct. Choose a number sentence such as 6 + 4 = 10 and ask pupils to label their bar model.

10	
6	4

- Remind pupils that instead of drawing a bar model, they can use a line to represent the part/whole model. Check frequently that pupils are drawing the line separating the parts in an appropriate position, that is, they should not always draw it in the middle of the bar, they should consider whether the parts are of roughly equal size; which is larger; whether it is very much larger.

- Ask pupils to draw a line 15 cm long and to mark where 10 cm is. Explain that lines do not always have to be this size, this is just an example. Show pupils how to draw the curly brackets to highlight the two parts and the whole, then ask pupils to label their parts and whole for the chosen number sentence.

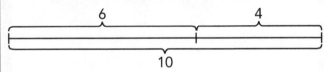

- Ask pupils to label 6 as apples, 4 as bananas and 10 as fruits altogether.

- Read out some word problems and ask pupils to give you their answer as a number sentence, for example: *There are ten fruits in the fruit bowl, four of them are bananas and the rest are apples. How many apples are there?* (10 – 4 = 6)

- Ask pupils to draw another line as before, this time labelling the whole as 24 and one part as 17, leaving the other part blank. Share ideas for a problem that could be solved using the bar model or line model.

- Label the numbers in the bar model or line model together, then record and solve the number sentence.

Same-day enrichment

- Give pupils a set of number cards, 11 to 40. Ask them to shuffle the cards and turn over the top two cards.

- Pupils can choose whether the two cards show two parts or one part and a whole.

- They then draw a line model and label it with the relevant numbers.

Question 4

> 4 Write each number sentence and calculate.
>
> (a) Grandma bought 56 eggs. After baking cakes, she had 19 eggs left. How many eggs did Grandma use?
>
> Number sentence: _____
>
> Answer: _____
>
> (b)
>
> *I have read 27 pages.*
>
> *I have read 14 more pages than you. How many pages have I read?*
>
> Erin   Sam
>
> Number sentence: _____
>
> Answer: _____

What learning will pupils have achieved at the conclusion of Question 4?

- Pupils will have been introduced to recording the answer to a word problem in a word sentence after completing the relevant number sentence.

Activities for whole-class instruction

- Remind pupils that they are often asked to solve a word problem. To solve it, they usually need to express the problem in a number sentence. To solve the problem, they think about what is happening in the 'story' so that the numbers make sense and they can write them as a number sentence.

- Tell pupils the following word problem: *There are 29 children in a class. 13 of them are boys. How many are girls?*

- Give pupils a few moments to draw a bar or line model to identify the knowns and unknowns. Then they will be able to write and solve the number sentence.

- Ask pupils what number sentence they recorded and solved. Share ideas, for example 29 – 13 = ☐ → 29 – 13 = 16 or 13 + ☐ = 29 → 13 + 16 = 29.

- Explain that many pupils think that solving the number sentence means they have solved the word problem, but they need to understand the problem before they can write the sentence and do the calculation. Ask: *What did you do that helped you understand what you needed to do? Was it the bar model or the number sentence that helped you understand?* Can pupils see that the number sentence doesn't help them understand the problem, it just helps them do the calculation? The bar model or line model is what helps pupils to see what they need to calculate – what the number sentence must be. Can pupils tell you this?

- Now ask pupils how they could record their solution to the word problem. Share some ideas, for example: *There are 16 girls in the class* or *The class has 16 girls.*

- Give pupils another straightforward word problem to solve. This time ask for the answer first, then the number sentence they used to solve it.

- Pupils are now ready to complete Question 4 in the Practice Book.

Same-day intervention

- Give pupils a number sentence, such as 13 + 15 = ☐, and ask them to solve the addition calculation mentally. Agree that the sum is 28.

- Explain that the answer to the number sentence is 28, but if the number sentence had been written to solve a word problem, then the answer to the word problem is likely to be, 'There are 28 _____ altogether.'

- Record the calculation 13 + 15 = 28 with a list of contexts below it, for example:

13	+	15 =	28
boys		girls	children in a class
oranges		apples	fruits in the fruit bowl
eggs used		eggs left	eggs bought
pages to read		pages read	pages in a book

- Remind pupils that each of the contexts could be the background story for a word problem.

- Say the word problems and answers for each background story together.

All say ... *There were 13 boys and 15 girls in a class. How many children in the class? There were 28 children in the class. There were 13 oranges and 15 apples in a fruit bowl. How many fruits in the fruit bowl? There were 28 fruits in the fruit bowl. And so on.*

Same-day enrichment

- Give pupils three answers to word problems, for example:

  - Sarah has read 37 pages.

  - There are 4 more boys than girls.

  - Joe received £7 change.

- Ask pupils to choose an answer to write a word problem for. They should then record and solve the appropriate number sentence.

- Compare problems for one of the answers. Have pupils written the same or similar problems? Or are they quite different? Ask pupils to identify how the problems are the same or different.

- Provided the answer given to pupils is the answer to their word problem, then the word problem is correct because there are many different word problems with the same answer.

Challenge and extension question

Question 5

5 Write each number sentence and calculate.

£47   £24   £38   £76

(a) How much cheaper is the toy plane than the toy car?

Number sentence: _____

(b) If you had £100 to spend, and you wanted to spend as much of it as possible, which toys could you buy? How much change would you get?

_____

_____

This question asks pupils to write the appropriate number sentence to solve a money problem. No indication is given of whether an addition or a subtraction calculation is needed – pupils are expected to decide this for themselves. While part (a) only has a single answer, part (b) is a two-step problem with a number of different solutions. Pupils could choose to tackle part (b) in different ways. If they choose to subtract from £100, it will be straightforward to check that they have not overspent. If they choose to add prices, they may forget that they have a limit of £100.

Unit 1.3
Addition and subtraction (2)

Conceptual context

This is the third of three units revising addition and subtraction of two-digit numbers. Challenge is extended to missing numbers and patterns in linked calculations. Pupils continue to explore a line model version of the bar model for part/whole relationships.

Learning pupils will have achieved at the end of the unit

- Knowledge of place value will have been reinforced, including in the context of money (Q1, Q4)
- Pupils will have revised partitioning to add and subtract mentally, including in the context of money (Q1, Q4)
- Pupils will have noticed and used patterns in numbers to solve a series of linked calculation (Q2)
- Pupils will have revisited how to use the solution from one calculation to solve another (Q2)
- Pupils will have extended their understanding of a line model as a representation of a part/whole relationship, including in different orientations (Q3)

Resources

place value arrow cards; mini whiteboards; number lines; bead strings; counters; cubes; base 10 apparatus; two-digit number cards; teddy bear or other small toys

Vocabulary

partition, pattern, adjust, addend, sum, minuend, subtrahend, difference, bar model, line model, rotated

Question 1

> **1** Calculate mentally.
>
> (a) 47 − 25 = ☐
> (b) 76 + 18 = ☐
> (c) 32 + 49 = ☐
>
> (d) 18 + 49 = ☐
> (e) 9 + 81 = ☐
> (f) 15 + 47 = ☐
>
> (g) 82 − 16 = ☐
> (h) 38 − 20 = ☐
> (i) 65 − 19 = ☐
>
> (j) 51 − 7 = ☐
> (k) 77 − 14 = ☐
> (l) 36 − 18 = ☐

What learning will pupils have achieved at the conclusion of Question 1?

- Knowledge of place value will have been reinforced.
- Pupils will have revised using partitioning to add and subtract mentally.

Activities for whole-class instruction

- Remind pupils that one of the most efficient ways to add or subtract a two-digit number mentally is to keep the first number whole but partition the second, then add or subtract the tens and add or subtract the ones. This method is particularly useful when you want to calculate as you read along the number sentence.
- Show pupils the example 56 + 37 = ☐. Explain that you are going to keep the 56 whole but partition 37 into 30 and 7. Say: *56 add 30 equals 86, 86 add 7 equals 93.* Explain that when it came to adding 7 you mentally added 4 then another 3. Ask: *Why did I choose to partition the 7 into 4 and 3?*

All say … 56 *add 30 equals 86, 86 add 7 = 93.*

- Pupils might also choose to add 6 and 7 and then add 13 to 80. Repeat for two more additions.
- Show pupils the example 56 − 37 = ☐. Say: *56 subtract 30 equals 26, 26 subtract 7 equals 19.* Explain that when it came to subtracting 7, you mentally subtracted 6 then another 1.

All say … 56 *subtract 30 equals 26, 26 subtract 7 equals 19.*

- Repeat for two more subtractions.
- Pupils are now ready to complete Question 1 in the Practice Book.

Same-day intervention

- Give pupils a set of place value arrow cards and call out some two-digit numbers for them to make and hold up. Ask pupils to record 26 on their whiteboards. Explain that this is the addend, the number they are going to add on to.
- Ask pupils to make 13 with place value cards. Write 26 + 13 = ☐.
- Ask pupils to separate 13 into 10 and 3 to add each part separately. Ask them to add 26 + 10 first. Provide a number line or bead string for support if necessary. Agree that the sum is 36. Now ask pupils to add 36 + 3. Agree that the sum is 39, 26 + 13 = 39. Repeat, adding numbers mentally to 26 that do not cross the tens boundary, then moving on to some that do.
- Now ask pupils to write 54 on their whiteboards. This is the minuend, the number they are going to subtract from.
- Ask pupils to make 12 with place value cards. Write 54 − 12 = ☐.
- Ask pupils to separate 12 into 10 and 2 to subtract each part separately. Ask them to calculate 54 − 10 first. Agree the difference is 44. Now ask pupils to calculate 44 − 2. Agree that the difference is 42, 54 − 12 = 42. Repeat, subtracting numbers mentally from 54 that do not cross the tens boundary, moving on to some that do. Place value counters might be needed for subtraction with regrouping.

Same-day enrichment

- Give pupils a set of two-digit number cards to shuffle and some counters.
- Working in pairs, pupils take it in turns to turn over the top two cards. Both pupils add the two numbers together and say the sum. If they both agree, they move on to the next two cards. If they don't agree, each pupil explains the steps of the addition, as in the whole-class activities, to find out who was correct. The pupil who was correct takes a counter.
- Repeat for subtraction, but first sort the cards into 'Pile A' and 'Pile B'. Pile A should contain the 20 or 30 highest numbers from the set and Pile B, the other cards. Pupils then take it in turns to turn over a card from each pile. Both pupils subtract the B card from the A card. If they both agree, they move on to the next two cards. If they don't, each pupil explains the

steps of the subtraction, as in the whole-class activities, to find out who was correct. The pupil who was correct takes a counter.

- The first pupil to collect ten counters is the mental maths champion, but their partner could challenge them to another round.

Question 2

> 2 Fill in the missing numbers.
>
> (a) $34 + 47 = \boxed{}$ (e) $93 - 29 = \boxed{}$ (i) $\boxed{} - 25 = 39$
>
> (b) $\boxed{} + 47 = 71$ (f) $93 - \boxed{} = 74$ (j) $\boxed{} - 35 = 39$
>
> (c) $\boxed{} + 47 = 61$ (g) $93 - \boxed{} = 54$ (k) $\boxed{} - 45 = 39$
>
> (d) $34 + \boxed{} = 51$ (h) $93 - \boxed{} = 34$ (l) $80 - \boxed{} = 39$

What learning will pupils have achieved at the conclusion of Question 2?

- Pupils will have noticed and used patterns in numbers to solve a series of linked calculation.
- Pupils will have revisited how to use the solution from one calculation to solve another.

Activities for whole-class instruction

- Remind pupils that when they have a list of calculations to complete, there might be a pattern in the calculations. They can then use the pattern to solve the next calculation by adjusting their answer to the previous calculation rather than recalculating.
- Show pupils the following list of calculations:

$85 - 18 = \boxed{}$

$85 - \boxed{} = 47$

$85 - \boxed{} = 27$

$85 - \boxed{} = 7$

- Ask: *Which is the easiest calculation?* Pupils will give different answers. Some might think the top one is easiest because both 'knowns' are on same side of the equals sign. Some might say the bottom one is easiest because all that is necessary is to subtract 7 from 85 to find the answer.
- Ask pupils to solve the first calculation, $85 - 18 = \boxed{}$. Agree that the difference is 67.
- Now ask pupils to look at the second calculation. The minuend is the same, but the difference is 20 smaller. Ask pupils what this tells them.

- Agree that the subtrahend must be increased by 20, 20 more has been subtracted so that 20 less is left. Ask pupils if the pattern continues. If so, can they use it to solve the rest of the calculations?
- Agree that the difference decreased by 20 in each new calculation, so the subtrahend had to increase by 20 since the minuend did not change.
- Now, ask pupils to start with the bottom calculation where the missing subtrahend is 78. Pupils should discuss with a partner what their next step would be if they started with the bottom calculation. Can they see they would still be able to find the other missing numbers?
- Explore a second list of calculations:

$\boxed{} - 16 = 74$

$\boxed{} - 26 = 74$

$\boxed{} - 36 = 74$

$\boxed{} - 46 = 74$

- Ask pupils to solve the calculation they find easiest and use the result to help them solve the rest. Agree that the first missing number is 90 and the rest are 100, 110 and 120. Each successive subtrahend increases by 10 but the difference stays the same, so the minuend must increase by 10 each time.
- Pupils are now ready to complete Question 2 in the Practice Book.

Same-day intervention

- Give pupils a calculation such as $53 + 31$ and ask them to make it using base 10 blocks, including the sum. Record the calculation.
- Build up a sequence of calculations by repeatedly removing 10 from the first addend and the sum, so that you generate the following list:

$53 + 31 = 84$

$43 + 31 = 74$

$33 + 31 = 64$

$23 + 31 = 54$

- Ask pupils to look at the list and describe the patterns they can see. The first addend is 10 less in the next calculation. Since they are adding less, the sum must be 10 less too.
- Give pupils the calculation $19 + 14$ and ask them to make it using base 10 blocks, including the sum. Record the calculation.

- Build up another sequence of calculations by repeatedly increasing the second addend by 10. Ask pupils to look at the list and describe the patterns they can see. The second addend is 10 more in the next calculation. Since they are adding 10 more, the sum must be 10 more too.

- Ask pupils to use the base 10 blocks to create and complete a number sentence. They should then create a series of number sentences by increasing or decreasing one of the addends and the total by the same value.

- Compare sequences by asking pupils to identify what patterns each pupil has used.

Same-day enrichment

- Give pupils a calculation such as 81 – 69 = 12. Explain that this is the third calculation in a series of four.

- Ask pupils to create the list of four calculations. Remind them that they can choose which number should stay the same and which number will change, and in what pattern.

- After giving pupils plenty of time to create their list, share lists. Explore what is the same and what is different about each list and the patterns used to create them.

Question 3

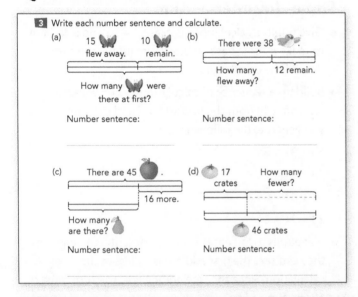

What learning will pupils have achieved at the conclusion of Question 3?

- Pupils will have extended their understanding of a line model as a representation of a part/whole relationship, including in different orientations.

Activities for whole-class instruction

- Remind pupils that they explored the line model in the previous unit. Ask them to explain how it is the same as and how it is different from the bar model.

- Show pupils two line models and ask them to look closely and explain to a partner what is the same and what is different about the two drawings.

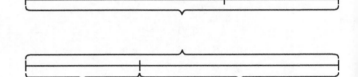

- Share and discuss pupils' comments. If necessary, explain that the two diagrams contain the same information but one shows the whole above the line and the other shows the whole below the line.

- Explain that when pupils draw a line model for a problem, at this stage in their learning, it will help if they use the space above the line to show the first piece of information in the number sentence and the space below the line for the rest of the information. So, sometimes their line models will have the long bracket at the top and sometimes at the bottom.

- Ask pupils which line model they would use for each problem:
  - There were 29 children in the class, 15 of them are boys. How many girls are there?
  - There are 15 boys and 14 girls in a class. How many children in the class?

- Since the first problem gives the total number of children in the class first, most pupils would choose to use the second diagram, because they can record that information at the top of the diagram.

- The second problem gives the value of the parts first. Most pupils would choose to use the first diagram, again because they can record that information at the top of the diagram.

- Repeat for the comparison line model. One line model should have the greater value at the top, the other the greater value at the bottom, as in Question 3. Show pupils the first comparison line model and give them a question such as: *Sumi has 37 stickers. She has 13 more than Tammy. How many stickers does Tammy have?*

- Ask pupils to draw and label a comparison line model to support them to find a solution.

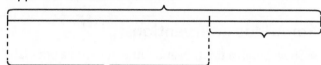

- Show pupils the second comparison line model and ask a question such as: *A shop is selling 17 boxes of oranges and 48 boxes of apples. How many fewer boxes of oranges does it have?* As before, ask pupils to draw and label a comparison bar model to support them to find a solution.

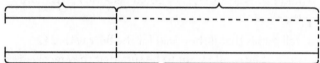

- Pupils are now ready to complete Question 3 in the Practice Book.

Same-day intervention

- Show pupils a line model, labelled as shown:

24 eggs in a tray

8 eggs used   How many eggs are left?

- Ask: *Where should I draw the mark on the line? In the middle? Why not? Where, roughly?*

- Ask pupils to write a number sentence to represent the problem. Agree $24 - 8 = \square$.

- Ask pupils to share their ideas for ways to solve the number sentence to find out how many eggs are left.

- Provide base 10 blocks, place value counters, cubes, bead strings and number lines for pupils to use if they wish.

- Ask pupils to explain their method, showing what they did with any apparatus. Agree that there are 16 eggs left.

- Show pupils another labelled line model. Ask them to explain what the problem is and how to find a solution.

12 bananas in   9 apples in
the basket   the basket

How many fruits in the basket?

- Ask: *Where should I draw the mark on the line? In the middle? Why not? Where, roughly?*

- Provide base 10 blocks, place value counters, cubes, bead strings and number lines for pupils to use if they wish.

- After agreeing that there are 21 pieces of fruit in the basket, explore the two line diagrams. Ask pupils to draw and label one of them the other way up, to show that it does not matter if the whole is on the bottom or the top of the diagram.

Same-day enrichment

- Show pupils a comparison line model:

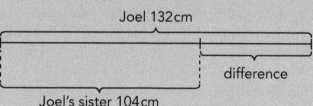

- Agree that the top of the line represents the larger quantity. The smaller quantity and the difference between them are shown below the line.

- Explain that this is a representation of a height chart, comparing the height of two children. Joel is 132 cm tall and his sister is 104 cm tall. Ask pupils how they could label the representation and use it to find out how much taller Joel is than his sister.

Joel 132 cm

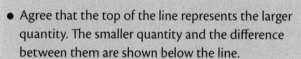

difference

Joel's sister 104 cm

$132 - 104 = 28$ Joel is 28 cm taller than his sister.

- If necessary, draw the comparison line model vertically (as above but rotated 90° anticlockwise) to help pupils to link the representation to a height chart.

- Challenge pupils to make up their own problems using a comparison line model.

Question 4

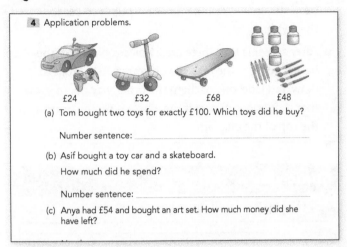

4 Application problems.

£24   £32   £68   £48

(a) Tom bought two toys for exactly £100. Which toys did he buy?

Number sentence: _____

(b) Asif bought a toy car and a skateboard.
    How much did he spend?

Number sentence: _____

(c) Anya had £54 and bought an art set. How much money did she have left?

What learning will pupils have achieved at the conclusion of Question 4?

- Knowledge of place value will have been reinforced, including in the context of money.
- Pupils will have revised partitioning to add and subtract mentally, including in the context of money.

Activities for whole-class instruction

- Remind pupils that we often carry out mental calculations in our everyday lives, for example when we go into a shop to buy something, we need to make sure we have enough money to pay.
- Explain that Rose wants to buy a game at £29 and a book at £19. She has £50. Does she have enough money?
- Ask pupils how they can find out. Agree that finding the total cost of the items she wants to buy would be a first step. £29 + £19 = £48
- Since she has £50, £50 − £48 = £2 tells us that she does have enough money and she will have £2 change. However, if she needs to keep £3 for her bus fare home, she does not have enough money to buy the game and the book.
- John wants to buy a different game at £38 and a teddy bear at £15. He also has £50. Does he have enough?
- Give pupils a few minutes to work out their answer.
- Agree that since £38 + £15 = £53, John does not have enough money to buy both things, though he can buy one of them.
- Check that pupils recognise that they have added and subtracted the amounts in exactly the same way as if they were numbers rather than amounts in pounds; the numbers work in the same way.

- Pupils are now ready to complete Question 4 in the Practice Book.

Same-day intervention

- Show pupils a teddy bear or other toy with a price label of £16.
- Ask pupils if they could buy the teddy bear if they had £10. Agree that they could not, because £10 is less than £16.
- Ask pupils if they could buy the teddy bear if they had £20. Agree that they could because £20 is more than £16.
- Tell pupils that if they paid for the bear with a £20 note, they would want to be sure that they received the correct change. Ask pupils what calculation they need to do to find out how much change they should receive.
- Agree the calculation £20 − £16 = £4; they will get £4 change.
- Ask pupils to find out how much change they would get if they bought toys costing £18, £13 or £11 with a £20 note, recording their answer in a number sentence. Provide base 10 blocks for pupils to use for support.

Same-day enrichment

- Give pupils ten two-digit numbers, selected at random.
- Explain that ten people each spent exactly £100 on two toys, but the cards tell you the cost of only one of the toys that each person bought. Ask pupils to find the price of the other toy that each person bought.
- Encourage pupils to record their solutions in a table for clarity:

Cost of first toy	Cost of second toy	Total spend
		£100
		£100
		£100

Challenge and extension question

Question 5

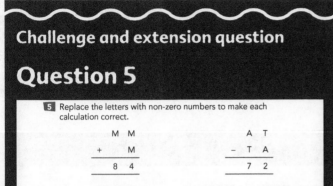

5 Replace the letters with non-zero numbers to make each calculation correct.

```
    M M              A T
  +   M            - T A
  -------          -------
    8 4              7 2
```

M = ☐   A = ☐   T = ☐

This question asks pupils to find values for unknown digits indicated by a letter within a column addition and a column subtraction. Pupils will need to complete the whole calculation to ensure that they are correct. An initial response to part (a) may be that M = 2. Although this works in the ones column, it will not allow successful completion of the calculation if used in the tens space.

Unit 1.4
Calculating smartly

Conceptual context

This unit introduces pupils to rounding and adjusting a calculation to make finding a solution straightforward. Deep understanding about addition and subtraction is necessary to enable pupils to calculate smartly. They will develop strategies based on sound number sense and on secure knowledge about relationships between addends and sum, and difference, subtrahend and minuend – all developed in Books 1 and 2.

Learning pupils will have achieved at the end of the unit

- Number bonds for 10 will have been revised, focusing on instant recall of pairs of numbers to make 10 (Q1)
- Number pairs for 10 will have been used to support mental calculation (Q1)
- Pupils will have been introduced to a new method for addition, 'rounding then adjusting – one up, one down' and subtraction, 'rounding then adjusting – both up or both down' (Q2)
- Pupils will have applied their knowledge of number bonds for 10 to decide which number to adjust and by how much (Q2)
- Pupils will have practised and refined rounding and adjusting to add and subtract mentally (Q3)
- The concept of the equals symbol as a balance will have been revisited and practised (Q4)

Resources

ten-frame; counters; 0–10 number track; squared paper; pan balance; base 10 apparatus; two-digit number cards; 0–100 number line; strips of red card; scissors

Vocabulary

number bond, digit, ten frame, multiple, rounding, adjusting, addend, sum, minuend, subtrahend, difference, balance

Question 1

> **1** Fill in the boxes.
>
> (a) 38 + ☐ = 40 (b) 60 + ☐ = 70 (c) ☐ + 54 = 60
>
> (d) ☐ + 71 = 80 (e) 45 + ☐ = 50 (f) 26 + ☐ = 30

What learning will pupils have achieved at the conclusion of Question 1?

- Number bonds for 10 will have been revised, focusing on instant recall of pairs of numbers to make 10.
- Number pairs for 10 will have been used to support mental calculation.

Activities for whole-class instruction

- Remind pupils that one of the most useful sets of number facts to help them calculate are the number bonds for 10.
- Call out random single-digit numbers for pupils to call back its bond to 10. Once pupils are confident, call out a two-digit number for pupils to call back its bond to the next multiple of 10.
- Show pupils a missing number calculation with the ones digit missing from the two-digit number, for example 5☐ + ☐ = 60. You should also check that pupils can do the reverse – call out the single-digit number and ask pupils to call back the relevant two-digit number.
- Pupils are now ready to complete Question 1 in the Practice Book.

Same-day intervention

- Give pupils a ten-frame and ten double-sided counters (or ten counters in one colour and ten in another).

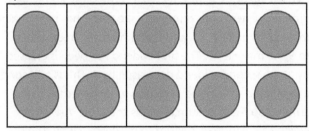

- Ask pupils to work in pairs, turning over (or swapping) one counter at a time so that they can see and record all the number bonds for 10.
- Now give pupils a 0 to 10 number track or ask them to make their own using squared paper.

- Ask pupils to draw a curved line from each number to its bond for 10, starting with 6 and 4, gradually moving along the number track to pair numbers.

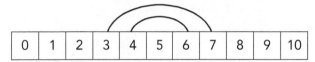

| 0 | 1 | 2 | 3 | 4 | 5 | 6 | 7 | 8 | 9 | 10 |

- Pupils continue until they have constructed a rainbow effect over the number track. Ask pupils which number does not have a bond in this arrangement. Agree that 5 does not have a partner, because there is only one 5 on a number track.
- Now give pupils a strip of 12 squares (using 2 cm squared paper or similar) to make their own number track with 2 fives. Give pupils scissors to cut up the track to create a set of 'number cards'.
- Ask pairs of pupils to jumble up their partner's numbers, then race to see who can find all the number bonds for 10 first.
- Give pupils lots of opportunities to practise recalling these six pairs of numbers so that they can begin to notice and exploit them in calculations.

Same-day enrichment

- Ask pupils to list all number bonds to 10 and use the list to help them work out the sum of the numbers 0 to 10. Ask: *How did the list help?*
- Ask: *How can you use what you have learned to find the sum of all the multiples of 10 from 10 to 100?*
- Share methods and agree solutions.

Question 2

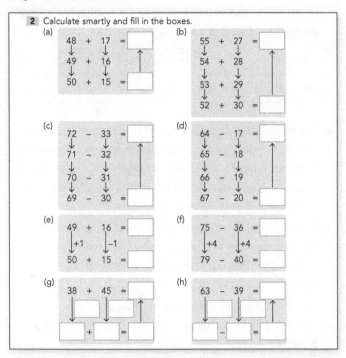

2 Calculate smartly and fill in the boxes.

(a)
48 + 17 =
49 + 16
50 + 15 =

(b)
55 + 27 =
54 + 28
53 + 29
52 + 30 =

(c)
72 − 33 =
71 − 32
70 − 31
69 − 30 =

(d)
64 − 17 =
65 − 18
66 − 19
67 − 20 =

(e)
49 + 16 =
  +1   −1
50 + 15 =

(f)
75 − 36 =
  +4   +4
79 − 40 =

(g)
38 + 45 =

   +   =

(h)
63 − 39 =

   −   =

What learning will pupils have achieved at the conclusion of Question 2?

- Pupils will have been introduced to a new method for addition, 'rounding then adjusting – one up, one down' and subtracting, rounding then adjusting – both up or both down.

- Pupils will have applied their knowledge of number bonds for 10 to decide which number to adjust and by how much.

Activities for whole-class instruction

- Explain that adding to or subtracting from a multiple of 10 is usually a straightforward calculation. However, many calculations do not have a multiple of 10 as their first addend or minuend. Explain that this can be changed by a method called 'rounding then adjusting – one up, one down'.

- Record: 39 + 17 = 56
         40 + 16 = 56

- Ask pupils to explain what has changed between the first and the second calculations. Agree that, first, 39 was rounded to 40. This is the 'rounding' in the name of the method. Then, because 39 has been increased by 1 (it has gone up by 1), the other number must go down, be decreased by the same value, so 17 has been reduced by 1 to 16. This ensures that the total is still the same, 56.

- Tell pupils they are going to explore some calculations to see why they can be adjusted like this.

- Show pupils the calculation 59 + 26 = ☐. Ask them to make 59 with base 10 blocks or place value counters, then make 26 and find the sum. Check that pupils have exchanged 10 ones for 1 ten to show the sum, 85.

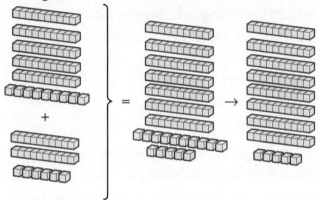

- Now show pupils the calculation 60 + 25 = ☐. Ask them to make 60 with base 10 blocks or place value counters, then make 25 and find the sum.

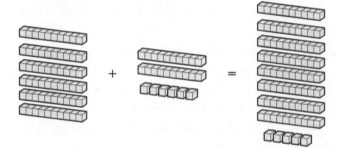

- Write the two calculations in a list:

  59 + 26 = 85

  60 + 25 = 85

- Agree that the sum is 85 in both cases. Compare the two calculations, exploring what is the same and what is different. Check that pupils can see that 1 has been taken from 26 to make it 25, and that 1 has then been added to 59 to make it 60. The first addend has been increased by 1, the second addend decreased by 1. This is adjusting, one addend up, one addend down, each by the same amount so that the sum remains the same. Ask pupils which calculation was easier to model and why. They should recognise that the second approach was more straightforward. Agree that the second calculation is also more straightforward to do mentally.

- Explain that they can adjust the two addends by any amount, as long as they increase one addend and decrease the other addend by the same amount. They could do this by adjusting by one repeatedly, or by adjusting by the required number immediately. Give pupils three addition calculations to change in the same way, for example 38 + 15, 49 + 14 and 28 + 27.

- Explore changing a subtraction calculation in a similar way. Ask pupils to model 34 − 15 = ☐ using base 10 blocks or place value counters. Pupils make 34 then exchange 1 ten for 10 ones to subtract 15, finding the difference of 19.

- Give pupils an A5 piece of card and a 0–100 number line and ask them to cut the card into a strip that fits between 15 and 34 on their number line.

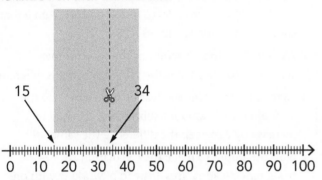

- Ask: *What number sentences does this number line and 'difference card' represent?* Pupils should see addition and subtraction sentences. Focus on:

34 − 15 = 19

- Tell pupils to focus on 34 then slide their card down the number line to 33. Ask what they notice. The other side of the card has moved to 14, showing that 33 − 14 = 19. Check that pupils recognise they have adjusted both the minuend and the subtrahend downwards by 1, maintaining a difference of 19.

- Ask pupils to make a 'difference card' to solve 48 − ☐ = 36 and 42 − ☐ = 36 on the number line.

- Check that pupils recognise that when they decrease the minuend, they must also decrease the subtrahend by the same value or the difference will not remain the same. Ask pupils to predict what they need to do to the subtrahend if they add to the minuend to make the calculation more straightforward. Check with examples such as 59 − 17 = 42 and 60 − ☐ = 42. Encourage pupils to make a new difference card for support.

- Give pupils three subtraction calculations to change in the same way, for example 51 − 24, 42 − 26 and 62 − 37.

- Pupils are now ready to complete Question 2 in the Practice Book.

Same-day intervention

- Give pupils the addition 27 + 15 = ☐ and ask them to solve it using base 10 apparatus (blocks or counters or both), keeping their apparatus laid out on the table.

- Agree that 27 + 15 = 42. Ask pupils to move a one cube or counter from the 15 to the 27 and to say what the addition is now: 28 + 14 = 42. Ask pupils to check that the total is still 42. Agree that it is, because they have moved one cube from one number to the other but they have not added or taken away any cubes.

- Repeat, moving another cube to see 29 + 13 = 42 and 30 + 12 = 42. Ask pupils to exchange the 10 ones for 1 ten to make the value clearer.

- Talk through how they have completed a similar calculation each time, they have simply rearranged the numbers a little. The total is the same, and the addition 30 + 12 is easier to carry out mentally.

- Now ask pupils to make the calculation 28 + 24 using base 10 blocks. Instead of working with one cube or counter at a time, ask pupils if they could move two cubes/counters at the same time to make this an easier addition.

- Agree that two cubes/counters could be moved from 24 to 28 to make the addition calculation 30 + 22. Encourage pupils to exchange the 10 ones for 1 ten to clearly see 30, or to line up the 10 ones like a ten stick.

- Ask pupils to find the sum. Agree that 30 + 22 = 52.

- Repeat with some similar calculations to develop pupils' confidence.

Same-day enrichment

- Point out that pupils have adjusted the numbers in a calculation to make it more straightforward, depending on which number is closest to a multiple of 10.

- Ask: *What would you do if the two numbers in a calculation were both the same distance from a multiple of 10, for example 42 − 28? How would you proceed?*

- Share pupils' ideas about how they would choose which way to adjust the numbers. The discussion will give pupils opportunities to practise relevant vocabulary and language skills.

Question 3

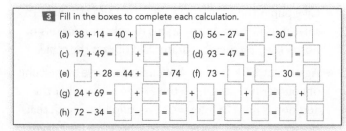

3 Fill in the boxes to complete each calculation.

(a) $38 + 14 = 40 + \square = \square$ (b) $56 - 27 = \square - 30 = \square$

(c) $17 + 49 = \square + \square = \square$ (d) $93 - 47 = \square - \square = \square$

(e) $\square + 28 = 44 + \square = 74$ (f) $73 - \square = \square - 30 = \square$

(g) $24 + 69 = \square + \square = \square + \square = \square + \square = \square + \square$

(h) $72 - 34 = \square - \square = \square - \square = \square - \square = \square - \square$

What learning will pupils have achieved at the conclusion of Question 3?

- Pupils will have practised and refined 'rounding then adjusting one up, one down' and 'rounding then adjusting both up or both down' to add and subtract mentally.

Activities for whole-class instruction

- Explore the addition calculation stream:

  $28 + 15 = \square$

  $29 + 14 = \square$

  $30 + 13 = \square$

- Explain that as pupils get used to doing this, they will be able to spot straightaway that they need to add 2 to round 28 to 30 and then subtract 2 from 15. They could immediately record $28 + 15 = 30 + 13 = 43$.

- Ask: *What could you do to* $26 + 38 = \square$ *to make it an easier calculation?* Agree that 38 is closer to a multiple of 10 than 26, so it is easier to increase the second addend, 38, by 2 to 40 and decrease 26 by 2. Emphasise that they can choose to round either of the addends in an addition – they should choose whichever is closest to a multiple of 10.

- Explore the following subtraction calculation stream:

  $42 - 16 = \square$

  $41 - 15 = \square$

  $40 - 14 = \square$

- Ask pupils how they could record this showing a single subtraction of 2. Agree that $42 - 16 = 40 - 14 = 26$.

- Ask pupils what they could do to $44 - 28$ to make it an easier calculation. Agree that 28 is closer to a multiple of 10 than 44, so it is easier to increase 28 and 44 by 2: $44 - 28 = 46 - 30 = 16$.

- Explain that this shows that they can choose to focus on either the minuend or the subtrahend in a subtraction. Remind pupils that they change both, but one is likely to be an easier number to change to a multiple of 10 than the other.

- Pupils are now ready to complete Question 3 in the Practice Book.

Same-day intervention

- Give pupils the subtraction $48 - 26 = \square$ and ask them to solve it using base 10 blocks or place value counters keeping their apparatus laid out on the table.

- Agree that $48 - 26 = 22$. Ask pupils to add two cubes (or counters) to 48 to make it 50. Remind pupils that they subtracted 26 from 48 to get to 22. Ask them to use their apparatus to find out how much they would need to subtract from 50 to be left with 22. Agree that they need to subtract 28, $50 - 28 = 22$.

- Ask pupils to look carefully at the two number sentences to see what is the same and what is different.

- Agree that the minuend was increased by 2, so the subtrahend, the amount subtracted, also had to increase by 2 to keep the difference the same.

- Now ask pupils to make the calculation $28 + 24 = \square$ using base 10 apparatus. Instead of working with one cube/counter at a time, ask pupils if they could move two cubes/counters at the same time to make this an easier addition.

- Agree that two cubes/counters could be moved from 24 to 28 to make the addition calculation $30 + 22$. Encourage pupils to exchange the 10 ones for 1 ten to clearly see 30, or to line up the 10 ones like a ten stick.

- Ask pupils to find the sum. Agree that $30 + 22 = 52$.

- Repeat with some similar calculations to develop pupils' confidence.

Same-day enrichment

- Give pupils a set of two-digit number cards. Pupils should shuffle the cards and turn over the top two cards.

- Pupils choose whether to add or subtract and find a solution. They then adjust the numbers to make one of them a multiple of 10, to make the calculation easier.

- After completing six calculations, ask pupils what was the largest adjustment they made.

- Ask pupils to predict what the greatest adjustment anyone is likely to make. Agree that it is likely to be 5, otherwise the number would be closer to a different multiple of 10.

Question 4

> **4** Draw lines to match the calculations that have the same answer.
>
> 55 + 28 ●   ● 60 – 13
>
> 13 + 48 ●   ● 53 + 30
>
> 62 – 15 ●   ● 78 – 40
>
> 74 – 36 ●   ● 10 + 51

What learning will pupils have achieved at the conclusion of Question 4?

- The equals sign as a balance will have been revisited and practised.
- The importance of the equals sign in 'rounding then adjusting' to add or subtract will have been explored.

Activities for whole-class instruction

- Remind pupils that two number sentences balance when they both have the same value. When they adjusted the number sentences in Questions 2 and 3, they adjusted both numbers by the same value, increasing one and decreasing the other for addition and increasing or decreasing both by the same amount for subtraction, so that the number sentences balanced.
- Ask pupils to look back at Question 3 (a), 38 + 14 = 40 + 12 = 52. This could be written without the total as a balance, 38 + 14 = 40 + 12.
- Ask pupils to record three different pairs of calculations from Question 3 as balances, with the equals symbol between the number sentences to show they have equal value.
- Pupils are now ready to complete Question 4 in the Practice Book.

Same-day intervention

- Remind pupils that the equals sign is a balance. The expressions on either side of the equals sign have the same value, for example 4 + 1 = 3 + 2, because both sides of the equals sign have the same total of 5.
- Give pairs of pupils a pan balance and ask them to place 29 in base 10 blocks in each pan and check that the scales balance.
- Ask pupils to remove the base 10 blocks from one pan. They partition the 29 in any way they choose, for example 14 + 15, and replace in the pan.
- Ask pupils to remove the base 10 blocks from the second pan and partition the 29 in any way they choose, for example 11 + 18, and replace in the pan.

- Remind pupils that they can record the two number sentences as a balance, because both are equal to 29, 14 + 15 = 11 + 18.
- Ask pupils to choose a number to place in both pans using base 10 blocks, then to record the balance using two different number statements.

Same-day enrichment

- Give pupils the following table to complete:

Calculation	=	<	>
37 + 28			
61 – 16			
49 + 47			
74 – 38			
45 + 39			
52 – 37			

- For each calculation, pupils need to record another equivalent calculation, one that would be less than the given calculation and another that would be greater than the given calculation.
- Ask pupils to compare tables. Have they changed the initial calculation in the same or different ways?

Challenge and extension question

Question 5

> **5** Ethan was adding two numbers. He mistook the digit 7 for 5 in the ones place of one of the addends and the digit 4 for 6 in its tens place. This gave him an answer of 92.
>
> The correct answer should be ☐.

This question asks pupils to work out the effect on the total of adding two two-digit numbers when digits are misread. Pupils will need to read the text carefully, then recognise the effect of misreading the numbers to find the correct answer.

Unit 1.5
What number should be in the box?

Conceptual context

This is the final addition and subtraction revision unit in Chapter 1. To support pupils to be able to reason about how they might find a missing number in an addition or subtraction problem, this unit revisits the relationship between addition and subtraction and fact families.

Learning pupils will have achieved at the end of the unit

- The inverse relationship between addition and subtraction will have been revisited and revised (Q1)
- Pupils will have used the inverse relationship between addition and subtraction to check a calculation and find a missing number (Q1, Q2)
- Pupils will have been reminded that they need to calculate, not simply shuffle the numbers when using the inverse relationship to check a calculation (Q1)
- Pupils will have revisited and revised using a bar model and a line model to explore part/whole relationships (Q2)
- Generating a family of facts from a part/whole model will have been revised and extended (Q2)
- A single bracket will have been used to express a part/whole relationship (Q3)
- Pupils will have made connections between different representations by interpreting questions represented pictorially and representing them as number sentences (Q3)

Resources

calculators; assorted dice; counters; base 10 blocks; sticky notes; mini whiteboards

Vocabulary

addition, subtraction, inverse, missing number, fact family, calculate, bar model, line model, bracket, known, unknown, sum, difference

Question 1

> **1** Fill in the boxes.
>
> (a) 52 + ☐ = 81   (b) 47 − ☐ = 9   (c) ☐ − 25 = 39
>
> (d) 81 − 52 = ☐   (e) 47 − 9 = ☐   (f) 39 + 25 = ☐
>
> (g) 33 + ☐ = 67   (h) 56 − ☐ = 28   (i) ☐ − 24 = 44
>
> (j) ☐ + 18 = 18   (k) ☐ − 37 = 63   (l) 67 − ☐ = 28
>
> (m) 64 − ☐ = 34   (n) ☐ − 73 = 19   (o) 48 + ☐ = 82

What learning will pupils have achieved at the conclusion of Question 1?

- The inverse relationship between addition and subtraction will have been revisited and revised.
- Pupils will have used the inverse relationship between addition and subtraction to check a calculation and find a missing number.
- Pupils will have been reminded that they need to calculate, not simply shuffle the numbers when using the inverse relationship to check a calculation.

Activities for whole-class instruction

- Remind pupils that addition and subtraction are the inverse of each other. Ask them to explain what that means.
- Share ideas. Suggestions are likely to be along the lines that subtraction 'undoes' addition (and *vice versa*), you can check an addition (or subtraction) using the inverse calculation and use an inverse calculation to find a missing number.
- Tell pupils they are going to use their knowledge about the inverse relationship between addition and subtraction to check whether an addition sentence is correct. Give pupils the addition calculation 58 + 27 = 85. Ask pupils what the inverse calculation is. Agree that 85 − 27 = 58 and 85 − 58 = 27 are both calculations that are the inverse of 58 + 27 = 85.
- Explain that rewriting the calculation as a subtraction by simply rewriting the numbers is not enough. They must actually calculate 85 − 27 or 85 − 58. Ask: *How will we know if the addition calculation is correct?* Can pupils tell you that the subtraction calculation must be correct – then you will know that the addition is correct as well?
- Ask pupils to choose one of the inverse calculations and quickly find a solution. 85 − 27 = 58 and 85 − 58 = 27, revealing that 58 + 27 = 85 is correct.
- Repeat with another example such as 27 + 35 = 62. Pupils should use an inverse calculation to help them find out that 62 − 35 = 27 and 62 − 27 = 35 are both correct, so the original addition calculation is correct.

- Next, repeat again, starting with a subtraction so that pupils use addition to check.
- Then, start with a subtraction that is *not* correct; the additions that are used to check will reveal that an error has been made because the additions will both reach a solution other than the minuend in the subtraction sentence.
- Pupils are now ready to complete Question 1 in the Practice Book.

Same-day intervention

- Give pairs of pupils a simple calculator, a 1–6 dice and some counters.
- Show pupils how to play a secret number game. Player 1 chooses any two-digit number and enters it on the calculator. Player 2 rolls a dice and says add or subtract. Player 1 then adds or subtracts the dice number from their number and shows Player 2 the result.
- Player 2 must carry out the inverse operation to identify the secret number. Players then swap roles.
- For the first turn each, allow Player 2 to use the calculator to find the secret number. After that, they should calculate mentally or use the written column method.
- Remind players that if they told their partner to add, they must subtract (the inverse operation) the same number to find the secret number. If they told their partner to subtract, they must add (the inverse operation) the same number to find the secret number.
- Players should take a counter if they correctly identify the secret number. The first player to collect five counters is the winner.

Same-day enrichment

- You will need a range of dice and some counters.
- Show pupils how to play a secret number game. Player 1 chooses any two-digit number. Player 2 rolls a dice and says add or subtract. Player 1 then adds or subtracts the dice number from their number and gives the result, saying, 'The number is ☐, what is my secret number?'
- Player 2 must carry out the inverse operation to identify the secret number. Players then swap roles.
- Give pupils a range of different dice to make the game more interesting.

- Players should take a counter if they correctly identify the secret number. The first player to collect ten counters is the winner.
- Ask players to think about which numbers they should select to make the calculation easier or harder for their partner.

Question 2

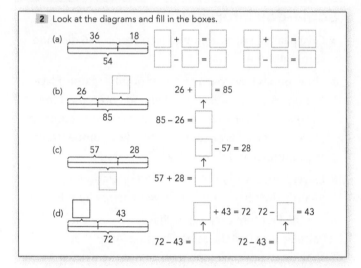

2 Look at the diagrams and fill in the boxes.

(a) 36 | 18 / 54   □ + □ = □   □ + □ = □   □ − □ = □   □ − □ = □

(b) 26 | □ / 85   26 + □ = 85 ↑   85 − 26 = □

(c) 57 | 28 / □   □ − 57 = 28 ↑   57 + 28 = □

(d) □ | 43 / 72   □ + 43 = 72 ↑   72 − □ = 43 ↑   72 − 43 = □   72 − 43 = □

What learning will pupils have achieved at the conclusion of Question 2?

- Pupils will have revisited and revised using a bar model and a line model to explore part/whole relationships.
- Generating a family of facts from a part/whole model will have been revised and extended.
- Pupils will have used another fact from the fact family to find an unknown or missing number.

Activities for whole-class instruction

- Give pupils three numbers in a part/whole relationship, for example 33, 67 and 90. Ask them to draw both a bar model and a line model with brackets to show this part/whole relationship.
- Tell pupils that they are going to write as many number sentences as they can about the information in these models. Ask: *How many sentences do think there will be?* Explain that you will think about the sentences together and they might be surprised at how many they can write from just one family of three numbers. Remind them that the sum or difference can be at the beginning or the end of the calculation, what is important is that both sides of the equals sign are equal in value.

- List the calculations together:

  33 + 67 = 90

  67 + 33 = 90

  90 − 67 = 33

  90 − 33 = 67

  90 = 33 + 67

  90 = 67 + 33

  33 = 90 − 67

  67 = 90 − 33

- Agree that, together, the class wrote eight number sentences about just one family of three numbers. Ask: *Do you think you could do the same thing for another family of three numbers? Try 23, 39 and 62. Start by drawing a bar model or line model.*

  23 + 39 = 62

  39 + 23 = 62

  62 − 39 = 23

  62 − 23 = 39

  62 = 23 + 39

  62 = 39 + 23

  23 = 62 − 39

  39 = 62 − 23

- Remind pupils that being able to quickly generate the family of facts allows them to manipulate the numbers to make it easier to find a missing number.
- Ask pupils to imagine that the value 39 was unknown in this set of calculations and on the bar and line model.
- Give pupils a few minutes to talk to their partner about which number sentence they would use to help them find the missing number.
- Go down the list and ask pupils to put their hand up if they would choose that number sentence.
- Explain that most pupils chose 62 − 23 = □ because the unknown is at the end of the number sentence. The unknown is also the only number on one side of the equals sign, so it is clear from the other side of the equals sign what to do to work it out. Although □ = 62 − 23 is the same number sentence, this is likely to be a less popular choice because we do not see calculations written in this order as often as we do those with the single number at the other end of the calculation.
- Give pupils two numbers, 17 and 43. Explain that 17 is a part and 43 is the whole. Ask them to draw either a bar or line model and record the number sentence used to find the unknown.

- After a few minutes, agree that the missing number is 26 and check which number sentence most pupils used. Was it 43 − 17 = ☐?
- Pupils are now ready to complete Question 2 in the Practice Book.

Same-day intervention

- Give pupils base 10 blocks and ask them to make the addition number sentence 13 + 15 = ☐, then solve it using more blocks. Give pupils sticky notes to record + and = signs on so that they can lay out their base 10 blocks to make the number sentence. Ask pupils to record the number sentence on a whiteboard. Agree that 13 + 15 = 28.
- Show pupils how to move the sum and the equals sign to the beginning of the number sentence so it now says 28 = 13 + 15. Ask pupils to check that the number sentence is correct and record it on their whiteboard.
- Ask: *How could you use the same numbers to make two more addition number sentences?* If necessary, show them how to swap the 13 and 15 to make two more number sentences, 28 = 15 + 13 and 15 + 13 = 28, recording these on their whiteboards.
- Ask pupils to draw a bar model or line model to illustrate the part/whole relationship shown by the family of addition number sentences.
- Remind pupils that there are also four subtraction number sentences in the same family. Invite pupils to rearrange their addition calculation into a subtraction calculation, using the same base 10 blocks. Give pupils another sticky note to record the subtraction sign so that they can make a subtraction number sentence.
- If necessary, show pupils how to move their blocks to make 28 − 15 = 13, 13 = 28 − 15, 28 − 13 = 15 and 15 = 28 − 13 and ask pupils to record them on their whiteboards.
- Ask pupils to check that their chosen part/whole diagram also illustrates their family of subtraction number sentences. Confirm that it does, because all the number sentences are just different ways of showing the part/whole relationship between the three numbers.
- Ask pupils to draw a bar model or a line model with 39 as the whole and 13 as one of the parts. Ask them to use their base 10 blocks to help them to find the unknown part and then to record eight number sentences about this part/whole model. If some pupils have not found all the number sentences, share ideas so that the group can find all eight.

Same-day enrichment

- Give pupils a blank line model, or ask them to draw their own.

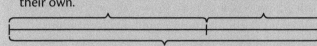

- Ask pupils to choose any three letters of the alphabet and use them to label each of the parts and the whole with a different letter.
- Now ask pupils to record the addition and subtraction fact family for their line model.
- For example, if a, b and c were chosen, with a as the whole, then $a = b + c$, $a = c + b$, $a − c = b$ and $a − b = c$. Each calculation could also be written with the sum or difference first.
- Pupils should cover their line model and challenge a partner to tell them which letter represents the whole and which letters represent the parts.
- Pupils can then compare their fact families with other pupils. Ask them if it matters which letters they chose.

Question 3

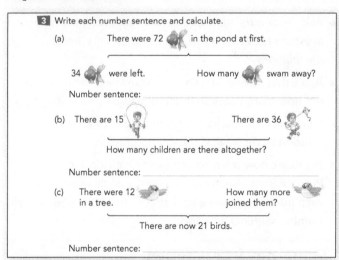

3 Write each number sentence and calculate.

(a) There were 72 🐟 in the pond at first.

34 🐟 were left. How many 🐟 swam away?

Number sentence: _____

(b) There are 15 🏃 There are 36 🏃

How many children are there altogether?

Number sentence: _____

(c) There were 12 🐦 in a tree. How many more 🐦 joined them?

There are now 21 birds.

Number sentence: _____

What learning will pupils have achieved at the conclusion of Question 3?

- A single bracket will have been used to express a part/whole relationship (Q3)
- Pupils will have made connections between different representations by interpreting questions represented pictorially and representing them as number sentences.

Activities for whole-class instruction

- Remind pupils that number problems usually tell a story. It is possible to tell a story – and at the same ask questions about it – using a combination of words, pictures and diagrams.

- A long bracket shows that the whole of something is contained within it, so a label for the whole bracket, where the point is drawn, tells us about the whole. Sometimes the point is above and sometimes below.

whole

part   part

part   part

whole

- At both ends of the bracket, information about the parts can be shown.

- Show pupils the following problem:

There are 64 pages in a book.

48 have been read.   How many pages are there left to read?

- Ask: *What is the number sentence that asks the same question? How would you find the solution? What is the solution?*

- Agree that solving $64 - 48 = \square$ would help them to find out how many pages were left to read, $64 - 48 = 16$ (pages).

- Point out that you have written (pages) after your solution to show what the solution means.

- Pupils are now ready to complete Question 2 in the Practice Book. Ask them to record a single word after each number sentence so that everyone knows what their number sentence was about.

Same-day intervention

- Give pupils a simple, familiar problem such as: *There are 16 girls and 17 boys in a class. How many children in the class?*

- Draw a large bracket

and ask pupils to help you show how the problem could have been recorded in this format, for example:

How many children?

16 girls   17 boys

- Ask pupils if they have enough information to solve the problem. Ask them to record a number sentence for the problem and solve it: $16 + 17 = \square$, $16 + 17 = 33$ (children).

- Explain that you have written (children) to show what the problem was about.

- Change the problem to 16 birds in a tree and 17 more birds arriving. Ask pupils to record a number sentence for the problem and solve it: $16 + 17 = \square$, $16 + 17 = 33$ (birds). Confirm that the solution is the same, all that has changed is what it is about.

- Draw another large bracket. Agree a problem focus with pupils and label the parts and whole. This time, leave one part unknown and support pupils to write a number sentence for the problem and solve it.

Same-day enrichment

- Give pupils two number sentences, for example $47 + 26 = 73$ and $92 - 67 = 25$.

- Ask them to write two problems, one for each sentence.

- Pupils can choose what is unknown in each problem and number sentence.

- Share problems and check that one of the given number sentences expresses the solution to the problem.

Challenge and extension question

Question 4

> **4** Look at each number sentence. Then choose > or < to write in the circle below.
> (a) ■ + 17 = 26 + ●   (b) ■ – 9 = ● – 15   (c) ■ + 8 = ● – 8
>   ■◯●   ■◯●   ■◯●

Pupils are asked to reason about number statements including symbols. Since some of the numbers are unknown, pupils may find it easier substitute values for those unknowns. Each calculation includes an equals sign so pupils will be able to substitute values to make the calculation balance. They can then use their chosen values to help them compare pairs of unknowns. Some pupils may be able to reason about the relative values of each pair of unknowns without giving them specific values.

Unit 1.6
Let's revise multiplication

Conceptual context

This is the first of two units aimed at revising multiplication and division in preparation for further development in Chapter 2. Pupils will revise multiplication tables and explore the relationship between multiplication and division.

Learning pupils will have achieved at the end of the unit

- Pupils will have revisited and revised the multiplication tables for 2, 4, 5, 8 and 10 (Q1)
- Understanding of multiplication tables will have begun to develop for 3, 6, 7 and 9 (Q1)
- Multiplication and division fact families will have been revisited, expressed and represented in different ways, developing fluency (Q1, Q2, Q3)

Resources

cubes; counters; mini whiteboards

Vocabulary

multiplication, multiplication table, multiplicand, multiplier, factor, product, array, dividend, divisor, quotient, fact family

Question 1

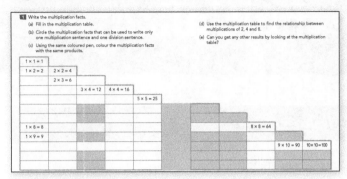

What learning will pupils have achieved at the conclusion of Question 1?

- Pupils will have revisited and revised the multiplication tables for 2, 4, 5, 8 and 10.
- Understanding of multiplication tables will have begun to develop for 3, 6, 7 and 9.
- Relationships between different multiplication tables will have been revisited and extended.

Activities for whole-class instruction

- Chant the 4 times and 8 times tables together.
- Skip-count together in multiples of 5 from 0 to at least 100. Move on to counting in multiples of 10 to at least 100.
- Draw up a grid as below and complete together

×	1	2	3	4	5	6	7	8	9	10	11	12
5	5	10										
10	10	20										

- Ask pupils to work in pairs to describe the two sequences of numbers, focusing on what is the same and what is different.
- Agree that both counts included multiples of 10. Each of these numbers had 0 in the ones place. Only half the numbers spoken, when counting in multiples of 5, had 0 in the ones place – the other half had 5 in the ones place. All numbers spoken were multiples of 5. Pupils may have other ideas.
- Choose a product, for example 30. Ask pupils to find the product in the grid and record the matching multiplication sentences, $5 \times 6 = 30$ and $10 \times 3 = 30$. Remind pupils that they can also write $5 \times 6 = 10 \times 3$ since both multiplication sentences have the same product; they are equivalent calculations.
- Ask pupils what they notice about the equivalent calculations. Can they see that on the right side of the equals sign the first number is double what it is on the left side; the second number is half what it is on the left; and the product remains the same?
- Ask pupils to use the grid to help them find at least three more pairs of equivalent calculations. Ask them to check if one number is doubled on the other side of the equals sign, while the other is halved.
- Ask pupils to look at the column headed by 2. Ask: *If 2 is the quotient, what are the two calculations indicated in that column?* Pupils might benefit from a reminder about the meaning of division-related vocabulary 'quotient', 'dividend' and 'divisor'. Agree that equivalent division calculations in the column in which the quotient is 2 are $10 \div 5 = 2$ and $20 \div 10 = 2$.
- Remind pupils that these can be written as $10 \div 5 = 20 \div 10$.
- Ask pupils to find at least three more pairs of equivalent division calculations.
- Pupils are now ready to complete Question 1 in the Practice Book.

Same-day intervention

- Ask pupils to count together in twos, from 0 to 30. Use cubes in twos to support the count.
- Working together, list the multiplication table for 2, beginning with $2 \times 1 = 2$ and continuing to $2 \times 12 = 24$.
- Ask pupils to look at each fact in the multiplication table for 2 and write its double alongside it, for example:

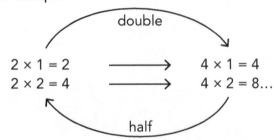

- Double cubes joined in twos to make cubes joined in fours to confirm that double 2 is 4.
- Continue doubling each fact, with cubes for support if necessary, until pupils have also written the multiplication table for 4.
- Ask pupils to imagine that they started with the multiplication table for 4. Ask them how they could use this to find the multiplication table for 2. Agree that this is half the multiplication table for 4, because 2 is half of 4.

Same-day enrichment

- Ask pupils to explain why the multiplication table in the Practice Book is not a square. Can they explain why the columns are taller on the left? (With each new column, some of the facts have already been shown in the previous columns so are not repeated.)

- Ask pupils to find the products they have coloured 3 times. Ask: *What are the multiplication facts for each product? How are they related?* (Each multiplication fact contains two factors of the product.)

- Ask pupils to identify which factors of each of the three products are missing from the multiplication facts. (12: all factors present; 24: 1 and 24 missing; 36: 1, 2, 18 and 36 missing.)

Question 2

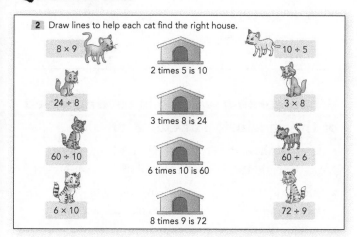

2 Draw lines to help each cat find the right house.

8 × 9 | 2 times 5 is 10 | 10 ÷ 5
24 ÷ 8 | 3 times 8 is 24 | 3 × 8
60 ÷ 10 | 6 times 10 is 60 | 60 ÷ 6
6 × 10 | 8 times 9 is 72 | 72 ÷ 9

What learning will pupils have achieved at the conclusion of Question 2?

- Multiplication and division fact families will have been revisited, expressed and represented in different ways.

- The relationship between multiplication and division will have been explored through linking a multiplication or division fact to another fact from the same fact family.

Activities for whole-class instruction

- Show pupils the following array and ask them to write two multiplication and two division facts for it.

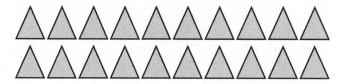

All say ... $10 \times 2 = 20$, $2 \times 10 = 20$, $20 \div 2 = 10$, $20 \div 10 = 2$

- Now give pupils sets of three numbers, such as 5, 7, 35; 3, 4, 12 and 6, 8, 48. Ask pupils to draw a representation of each set and record two multiplication and two division facts alongside it. Invite some pupils to share their array or representation with the rest of the class. Can pupils say two multiplication and two division facts for each representation?

- Check that the class number sentences are the same as pupils' number sentences.

- Pupils are now ready to complete Question 2 in the Practice Book.

Same-day intervention

- Ask pupils to arrange 12 counters in an array on a mini whiteboard, with three rows of four identical shapes, for example:

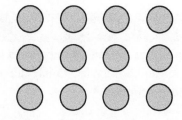

- Check that all the arrays are similar.

- Choosing one array, draw a ring around each row and record the multiplication sentence $3 \times 4 = 12$. Check that pupils can see 3 rows, each with 4 counters.

All say ... $3 \times 4 = 12$

- Choosing a second array, draw a ring around each column and record the multiplication sentence $4 \times 3 = 12$. Check that pupils can see 4 columns, each with 3 counters.

All say ... $4 \times 3 = 12$

- Choosing a third array, draw a ring around the whole array and write $12 \div 4 = $. Ask pupils how many groups of 4 they can see within the whole 12 counters. Agree that there are 3 groups, indicating each row. Complete the division sentence,

All say ... $12 \div 4 = 3$

- Choosing a fourth array, draw a ring around the whole array and write $12 \div 3 = $. Ask pupils how many groups of 3 they can see within the whole 12 counters. Agree that there are 4 groups, indicating each column. Complete the division sentence,

 All say ... $12 \div 3 = 4$

- Look at all four number sentences and ask pupils which numbers have been used each time. Agree that 3, 4 and 12 are used in every sentence, so these numbers link together to make a family of facts.

- Give pupils the three numbers 3, 5 and 15. Ask them to make a matching array on their mini whiteboards and record two multiplication and two division facts alongside their array.

- Look at everyone's facts and confirm that everyone has written the same four facts.

Same-day enrichment

- Remind pupils that the three numbers in a multiplication and division fact family can be shown in a triangle, for example:

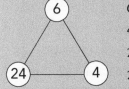

$6 \times 4 = 24$
$4 \times 6 = 24$
$24 \div 6 = 4$
$24 \div 4 = 6$

- Ask pupils to draw and label the triangle in the centre of their mini whiteboard or on a large piece of paper. Challenge pupils to add a further triangle at each vertex, and then continue to do the same for the new triangles. For example, the vertex containing 4 could be one vertex in the triangle 2, 2, 4 while 6 could be one vertex in the triangle 2, 3, 6 and 24 could be one vertex in the triangle 3, 8, 24.

- Compare the triangle 'maps'. Have pupils used the same sets of numbers? Which numbers are generally present in the outer ring of triangles? (0, 1)

Question 3

3 Use the multiplication facts to fill in the boxes.

(a) The product is 20.

☐ × ☐ = 20   ☐ × ☐ = 20
☐ × ☐ = 20   ☐ × ☐ = 20

(b) The product is 24.

☐ × ☐ = 24   ☐ × ☐ = 24
☐ × ☐ = 24   ☐ × ☐ = 24

(c) The product is 30.

☐ × ☐ = 30   ☐ × ☐ = 30
☐ × ☐ = 30   ☐ × ☐ = 30

(d) The quotient is 3.

☐ ÷ ☐ = 3   ☐ ÷ ☐ = 3
☐ ÷ ☐ = 3   ☐ ÷ ☐ = 3

(e) The quotient is 5.

☐ ÷ ☐ = 5   ☐ ÷ ☐ = 5
☐ ÷ ☐ = 5   ☐ ÷ ☐ = 5

(f) The quotient is 8.

☐ ÷ ☐ = 8   ☐ ÷ ☐ = 8
☐ ÷ ☐ = 8   ☐ ÷ ☐ = 8

What learning will pupils have achieved at the conclusion of Question 3?

- Pupils will have revisited and used the language of multiplication and division to explore relationships.

- Fluency in multiplication and division facts will have been developed through exploring products and quotients.

Activities for whole-class instruction

- Ask: *What does product mean?* Agree that the product is the result of multiplying two numbers together. Agree two multiplication sentences, factor × factor = product and group size × number of groups = product.

- Draw a spider diagram with a central product such as 12. Ask pupils to suggest different factor pairs to multiply to make 12 and record these in the outer circles.

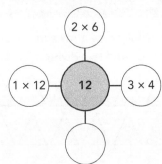

- Give pupils products such as 36, 48 and 60 to create their own spider diagrams.

- List factor pairs for each product. Have pupils used them all?

- Ask pupils to make arrays of counters with four rows or four columns – can they label the dividend, divisor and quotient if the quotient is 4?

- Draw a spider diagram with a central quotient such as 4. Ask pupils to suggest different dividends and divisors and to record these in the outer circles. Challenge pupils to find all the dividends and divisors with a quotient of 4 using arrays or from the multiplication tables that they know. Pupils who realise that they need the matching division facts for the multiplication table for 4 are demonstrating a clear understanding of the relationship between multiplication and division.

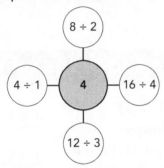

- Pupils are now ready to complete Question 3 in the Practice Book.

Same-day intervention

- Give pupils 12 counters and ask them to make an array. Check that pupils recall what an array is and that every row must have the same number of counters and every column must have the same number of counters.

- After pupils have explored and recorded some multiplication sentences, ask them to work systematically with you to make sure that they have found them all.

- Begin with one row of 12, $1 \times 12 = 12$. Ask pupils to turn their array around so that it now has 12 rows and record the multiplication sentence that represents the array now: $12 \times 1 = 12$. Can pupils see that they have a pair of equivalent calculations? ($1 \times 12 = 12 \times 1$)

- Use the same 12 counters to make two equal rows, $2 \times 6 = 12$, then turn it around. How many rows does it have? Represent this array as a multiplication sentence $6 \times 2 = 12$. Can pupils see that they have another pair of equivalent calculations, $2 \times 6 = 6 \times 2$?

- Continue to make arrays with a product of 12 until all possibilities have been found, recording equivalent calculations.

- Pupils should list all the multiplication sentences for 12.

Same-day enrichment

- Ask pupils if the quotient is always, sometimes or never less than the dividend and divisor. They should give some examples to support their answer.

- The quotient is sometimes less than the dividend and divisor, for example it is less in the division sentence $6 \div 3 = 2$, but not in the sentence $6 \div 2 = 3$.

- Ask pupils what happens when we divide a number by 1? What can they say about the quotient then? The quotient is the same as the dividend, the number that was divided by 1.

Challenge and extension question

Question 4

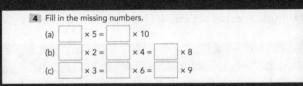

This question asks pupils to link multiplication facts with the same product. Initially, pupils link two facts. The first pair of facts link multiplying by 5 and 10. Since no other factors are supplied and the product is not specified, there are several options available. Some pupils could be asked to find all the possible solutions. The second and third linked multiplication facts both have three linked facts. The first one links multiplying by 2, 4 and 8 while the second one links multiplying by 3, 6 and 9. Pupils looked at the relationship between multiplications by 2, 4 and 8 in Question 1 of the Practice Book so are likely to find the first of these linked multiplications straightforward. Pupils will need to identify a common product first before they can find the relevant factors. Pupils may assume that the relationship between 3, 6 and 9 will be the same as that between 2, 4 and 8 so they will need to check their initial thoughts carefully.

Unit 1.7
Games of multiplication and division

Conceptual context

This is the second of two units aimed at revising multiplication and division in preparation for further development in Chapter 2. The focus is on multiplication and division as grouping and scaling. Expressing any 'left overs' after grouping is also explored. Pupils also focus on solving problems with multiplication and division.

Learning pupils will have achieved at the end of the unit

- Pupils will have revised and revisited multiplication and division as grouping (Q1, Q2)
- Left overs after grouping in multiplication will have been expressed as an addition or subtraction to the multiplication sentence and as a remainder (recorded with an r) at the end of the division sentence (Q1, Q2)
- Multiplication as scaling will have been revised and expressed in multiplication sentences (Q3)

Resources

counters; cubes; four multipacks of eight biscuits or packaging from these; pictures of a car and a minibus, or toys; sunflower seeds; circles of brown paper; animal dice; 1–6 dice

Vocabulary

groups, array, multiplication, addition, division, quotient, left over, remainder

Question 1

> **1** Look at the diagrams and fill in the boxes. Write the number sentence underneath each one.
>
> (a) ☆☆☆
> ◇◇◇◇◇◇◇◇◇
>
> There are ☐ times as many ☐ as ☐.
>
> Number sentence: _____
>
> (b) (△△) (△△) (△△)
> (△△) (△△)
>
> There are ☐ groups of ☐ in ☐.
>
> Number sentence: _____
>
> (c) How many footballs are there in total? Group the footballs in two different ways and write the number sentences.
>
> ⚽⚽⚽⚽⚽ ⚽⚽⚽⚽⚽⚽
> ⚽⚽⚽⚽⚽ ⚽⚽⚽⚽⚽⚽
> ⚽⚽⚽⚽⚽ ⚽⚽⚽⚽⚽
>
> Number sentence: Number sentence:
> ☐ × ☐ + ☐ = ☐   ☐ × ☐ + ☐ = ☐
>
> Number sentence: Number sentence:
> ☐ × ☐ − ☐ = ☐   ☐ × ☐ − ☐ = ☐

What learning will pupils have achieved at the conclusion of Question 1?

- Pupils will have revised and revisited multiplication and division as grouping.
- Left overs after grouping in multiplication will have been expressed as an addition or subtraction to the multiplication sentence.

Activities for whole-class instruction

- Explain that when we put a set of objects into groups, there may be some left over.
- Give pupils 11 cubes and ask them to put the cubes in groups of 2. Check that pupils have 5 groups of 2 and 1 single cube. Show pupils that they could write this as 5 × 2 + 1 = 11.
- Now ask pupils to put their cubes into groups of 3 and record the results in the same way. (3 × 3 + 2 = 11)
- Record the sentence ☐ × ☐ + ☐ = 11. Ask pupils to work in pairs to find as many different ways as they can to complete the sentence.
- Check that pupils recognise that 11 cannot be placed into equal groups, other than groups of 1 and 11.
- Show pupils the sentence ☐ × ☐ − ☐ = 11 and ask them how this is the same as and different from the sentence ☐ × ☐ + ☐ = 11. Agree that instead of adding the left overs, something is being subtracted but the total is still 11.

- Show pupils the sentence 5 × 2 + 1 = 11 and ask them to suggest how they could change it to the form ☐ × ☐ − ☐ = 11. Check that pupils recognise that since they need to subtract, the product of the multiplication must be more than 11. 6 × 2 = 12, so 6 × 2 − 1 = 11.
- Ask pupils to work in pairs to change all their ☐ × ☐ + ☐ = 11 sentences into ☐ × ☐ − ☐ = 11.
- Pupils are now ready to complete Question 1 in the Practice Book.

Same-day intervention

- Give pupils seven counters each. Ask them to arrange their counters in groups of 2, then to align the counters in an incomplete array as in Question 1c in the Practice Book:

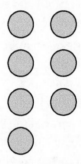

- Check that pupils can see 3 groups of 2 and record this as a multiplication sentence, 3 × 2 = 6.
- Remind pupils that they have 7 counters and model recording 3 × 2 + 1 = 7. Ask pupils to point to the 3 × 2 counters and then the + 1 counter.
- Now ask pupils to put the 7 counters into groups of 3 and align them in an incomplete array:

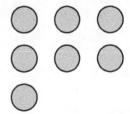

- Ask pupils to record a multiplication sentence for the first 2 rows, 2 × 3 = 6. Agree that they have 2 groups of 3 and point out that they also have one more counter. Ask pupils how they could record this. Agree 2 × 3 + 1 = 7.
- Repeat with a group of 4 counters and 3 more, 1 × 4 + 3 = 7. Ask pupils to work individually to record placing the counters in a group of 5, then 6. Check pupils have recorded 1 × 5 + 2 = 7 and 1 × 6 + 1 = 7.

- Confirm that they could make 1 group of 7, but there would be none left over, so they would not need to use the sentence □ × □ + □ = 7 to record this.
- Ask pupils to explain why their counters cannot be put into groups of 8, 9 or 10 counters. Agree that they only have 7 counters, so they do not have enough to make groups of 8, 9 or 10.
- Repeat with an incomplete 3 × 5 array with the 15th item missing.

Same-day enrichment

- Challenge pupils to find other numbers below 20 where the quantity cannot be put into groups of equal size other than groups of 1 and the number itself. Some pupils will find cubes or counters useful for support.
- Ask pupils to choose one number to find all the possible number sentences □ × □ + □ = △, where △ has the same value.
- Pupils who are unsure of how to get started should look at Question 1 part c in the Practice Book and find all the other possibilities for 17. Pupils who work consistently from 1 to 20, checking each number in turn, and those who eliminate the products of known multiplication facts are demonstrating an even greater depth of understanding.

Question 2

2 Write each number sentence and find the answer.

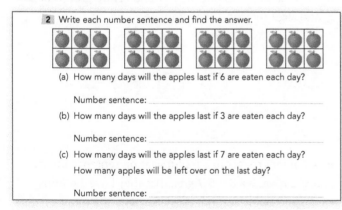

(a) How many days will the apples last if 6 are eaten each day?

Number sentence: _____

(b) How many days will the apples last if 3 are eaten each day?

Number sentence: _____

(c) How many days will the apples last if 7 are eaten each day? How many apples will be left over on the last day?

Number sentence: _____

What learning will pupils have achieved at the conclusion of Question 2?

- Pupils will have further revised and revisited multiplication and division as grouping.
- Left overs after grouping in division will have been expressed as a remainder (recorded with an r) at the end of the division sentence.

Activities for whole-class instruction

- Show pupils four packs of eight snack biscuits. Use multipacks of biscuits (full or empty) or sticks of eight cubes to represent the multipacks. Pupils should work in pairs to find out how many biscuits there are altogether.
- Share different ways of finding out, including counting, doubling and multiplying by 4 or 8. Agree that recalling the multiplication facts 4 × 8 = 32 or 8 × 4 = 32 would be the quickest way to find out. Saying the multiplication table for 8 to 8 × 4 would be a 'smart' way to find out if you could not recall the fact.
- Ask pupils how many days the biscuits would last if 8 were eaten each day. Ask pupils to express this in a division sentence. (32 ÷ 8 = 4)
- Ask similar questions to explore putting the 32 biscuits into groups of different sizes, expressing the results as division sentences, with and without remainders, for example: *What if 4 were eaten each day?* (32 ÷ 4 = 8) *What if 6 were eaten each day?* (32 ÷ 6 = 5 with 2 left over)
- Talk about the meaning of the word 'remainder'. Show pupils how to record the left overs as a remainder, using the letter 'r' for remainder after the quotient.
- Pupils are now ready to complete Question 2 in the Practice Book.

Same-day intervention

- Give pupils 15 sunflower seeds and three brown circles to represent pots of soil.
- Count the seeds together and ask pupils to set out the three brown circles in a row. Now ask them to 'plant' the same number of seeds in each circle of soil.

ⓘ Pupils who share the seeds by placing one at a time in each circle are demonstrating that they are more confident with division as sharing. Pupils who try a small group on each circle or recall the multiplication fact 5 × 3 = 15 and use that to put 5 seeds on each circle are demonstrating that they understand the grouping aspect of division.

- Confirm that each pot has 5 seeds. Record the matching division statement, 15 ÷ 3 = 5.

- Explain that one pot of soil is broken and cannot be used. Collect a circle from each pupil and ask them to plant the same number of seeds in each circle of soil again.

- Confirm that each pot has 7 seeds and there is 1 seed left over. Remind pupils that they had 15 seeds and record the matching division sentence 15 ÷ 2 = 7 r1.

- Ask pupils what each number represents, ensuring that they recognise the meaning of each part of the number sentence, including the remainder of 1.

- Repeat with a different number of seeds and pots, asking pupils to record the matching number sentences, including any remainders.

Same-day enrichment

- Remind pupils that when they recorded incomplete groups in multiplication, they used the number sentence □ × □ + □ = △. When they recorded incomplete groups in division, they used the number sentence △ ÷ □ = □ r □.

- Show pupils how they can use both number sentences to make a fact family of two multiplication and two division sentences, for example 9 × 2 + 1 = 19, 2 × 9 + 1 = 19, 19 ÷ 2 = 9 r1, 19 ÷ 9 = 2 r1.

- Ask pupils to find two more fact families for 19.

- Once pupils have completed their fact families, ask what is the same and what is different about these fact families. Pupils should notice that there are four numbers in each fact family instead of three and the value of the left over is the same within each family, whether this is expressed as + □ or r □.

Question 3

> **3** Draw the number of shapes as indicated and then fill in the number sentences below.
>
> In the first row, draw 2 ☆.
>
> In the second row, draw 4 △.
>
> In the third row, there are 4 times as many ◯ as ☆ in the first row.
>
> (a) There are [ ] times as many ◯ as △.
>
>   [ ] × [ ] = [ ]
>
> (b) 5 times the number of ◯ is [ ].
>
>   [ ] × [ ] = [ ]

What learning will pupils have achieved at the conclusion of Question 3?

- Multiplication as scaling will have been expressed in multiplication sentences.

Activities for whole-class instruction

- Show pupils a picture of a car and a minibus, or use toy vehicles. Ask: *If there are two people in the car and twice as many people in the minibus, how many people are in the minibus?* Agree that there must be four people in the minibus, 2 × 2 = 4.

- Ask pupils to explain what is meant by 'twice as many'. Agree that whatever you have must be multiplied by 2 to find out what twice as many is. Call out some numbers for pupils to call back twice as many.

- Ask: *If there are three people in the car and 5 times as many people in the minibus, how many people are in the minibus?* Agree that there must be 15 people in the minibus, 5 × 3 = 15.

- Ask pupils to explain what is meant by '5 times as many'. Agree that whatever you have must be multiplied by 5. Call out some numbers or quantities for pupils to call back 5 times as many or as much, for example 3 cars, 8 triangles, 100, 20 people and 50 should produce responses of 15 cars, 40 triangles, 500, 100 people and 250.

- Pupils are now ready to complete Question 3 in the Practice Book.

Same-day intervention

- Show pupils two triangles in a row. Ask them to make another row with twice as many triangles in.

- Check that pupils have four triangles. Do pupils understand that they need to replace each single triangle with two triangles? Show pupils that they have multiplied by 2, or doubled, to find twice as many, so they can record what they did in a multiplication sentence, $2 \times 2 = 4$.

- Now show pupils three triangles and ask them to make another row with twice as many triangles in.

- Check that pupils have made a row of six triangles, confirming that each triangle has been replaced by two triangles. Ask pupils what the matching number sentence is, $3 \times 2 = 6$.

- Ask pupils what '3 times as many' might mean. Agree that each single item must be replaced by three items to make 3 times as many. Make a row of three triangles and ask pupils to make another row with 3 times as many triangles. Confirm that the matching number sentence is $3 \times 3 = 9$.

- Repeat with a more examples as necessary.

Same-day enrichment

- Give pupils a 1–6 dice and an animal dice, for example:

- Play the game (in groups of 2–4) as follows:

1. Roll the animal dice. Agree the number of legs that the animal has.	If the dice showed a horse, pupils should draw a set of 4 legs or 4 lines to represent legs.
2. Roll the number dice to tell you how many times to draw the set of legs.	If pupils had thrown 3 on the number dice and had drawn a set of 4 legs, they should draw 3 sets of 4 legs.
3. Record the matching number sentence.	Record $3 \times 4 = 12$
4. Repeat, keeping count of the total number of legs 'found' so far.	

- The winner is the first to reach a total of 50 legs.

Challenge and extension question

Question 4

> **4** All the pupils in a class are assembled on the sports field. They form exactly 5 rows of the same number of pupils. Tom is in the second row. He is in the fourth place from the left. He is also in the fourth place from the right. How many pupils are there in the class?

This question asks pupils to identify a quantity by exploring how many equivalent groups can be made. The position of one child in a row relative to the ends of that row allows pupils to identify the length of the rows. They can then multiply the number of rows by the length of each row to find the total number of children in the class. Some pupils will find it helpful to draw the row that Tom is in to help clarify how many children are in a row.

Chapter 1 test (Practice Book 3A, pages 21–23)

Test question number	Relevant unit	Relevant questions within unit
1	1.1	1
	1.2	1
	1.3	1
	1.6	1(a)
2	1.1	2, 3, 4
	1.2	1
3	1.1	5
	1.3	2
	1.4	1
	1.5	1, 2
	1.6	1, 2, 3, Challenge and extension
4a	1.3	3(d)
	1.5	3(a), 3(c)
4b	1.3	3(b)
4c	1.3	3(a)
	1.5	3(b)
4d	1.2	3(a)
	1.3	3(c)
	1.5	3(b)
4e	1.2	3(d)
	1.3	3(c)
	1.5	3(c)
4f	1.7	3(a), 3(b)
4g	1.7	2(a), 2(b)
5	1.3	Challenge and extension
	1.4	Challenge and extension

Chapter 2
Multiplication and division (II)

Chapter overview

Area of mathematics	National Curriculum statutory requirements for Key Stage 2	Shanghai Maths Project reference
Number – Multiplication and division	Year 3 Programme of study: Pupils should be taught to: ■ recall and use multiplication and division facts for the 3, 4 and 8 multiplication tables	Year 3, Units 2.2, 2.10, 2.11. 2.12, 2.13
	■ write and calculate mathematical statements for multiplication and division using the multiplication tables that they know, including for two-digit numbers times one-digit numbers, using mental and progressing to formal written methods	Year 3, Units 2.1, 2.2, 2.3, 2.4, 2.5, 2.6, 2.7, 2.8, 2.10, 2.11, 2.12, 2.13
	■ solve problems, including missing number problems, involving multiplication and division, including positive integer scaling problems and correspondence problems in which *n* objects are connected to *m* objects.	Year 3, Units 2.1, 2.2, 2.3, 2.4, 2.5, 2.7, 2.8, 2.9 2.10, 2.11, 2.12, 2.13
Number – Multiplication and division	Year 4 Programme of study: Pupils should be taught to: ■ recall multiplication and division facts for multiplication tables up to 12×12	Year 3, Units 2.1, 2.2, 2.3, 2.4, 2.5, 2.6, 2.7, 2.8, 2.9, 2.10, 2.11, 2.12, 2.13
	■ use place value, known and derived facts to multiply and divide mentally, including: multiplying by 0 and 1; dividing by 1; multiplying together three numbers	Year 3, Units 2.10, 2.11, 2.12, 2.13
	■ recognise and use factor pairs and commutativity in mental calculations	Year 3, Unit 2.12

Unit 2.1
Multiplying and dividing by 7

Conceptual context

Pupils have had experience of concrete and pictorial representations to develop their conceptual understanding of multiplication. So far they have learned about multiplying and dividing by 10, 5, 2, 4 and 8. In this unit, they will focus on both multiplying and dividing by 7. It is important that pupils understand how the two operations of multiplication and division are linked early in their experiences of multiplication. Often pupils can recall a multiplication fact, but they also need conceptual understanding to be able to make relationships and apply their knowledge to a range of contexts. Throughout this unit, pupils should be encouraged to answer in full sentences using the correct mathematical vocabulary.

Pupils' mathematical thinking should be developed by ensuring pupils are talking about the process as well as the solution.

Learning pupils will have achieved at the end of the unit

- Pupils will have used manipulatives and images to develop understanding of multiplying and dividing by 7 (Q1)
- Pupils will have developed a deeper understanding of inverse (Q1)
- Flexibility in mental calculation will have been developed through practice in deriving related multiplication and division facts (Q2)
- Pupils will have practised recall of the 7 multiplication table to improve fluency (Q3)
- Pupils will have applied their knowledge of known multiplication and division facts to solve problems (Q4)

Resources

counting stick; mini whiteboards; interlocking cubes; number cards; counters; **Resource 3.2.1a** Related multiplication and division facts representations; **Resource 3.2.1b** Domino cards

Vocabulary

multiplication, multiply, times, factor, product, divide, division, divided by, inverse

Question 1

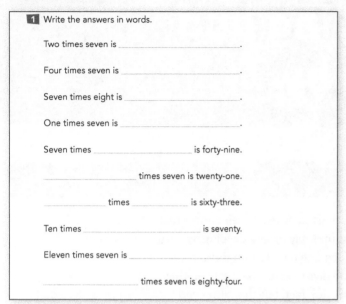

1　Write the answers in words.

Two times seven is _____ .

Four times seven is _____ .

Seven times eight is _____ .

One times seven is _____ .

Seven times _____ is forty-nine.

_____ times seven is twenty-one.

_____ times _____ is sixty-three.

Ten times _____ is seventy.

Eleven times seven is _____ .

_____ times seven is eighty-four.

What learning will pupils have achieved at the conclusion of Question 1?

- Pupils will have practised recalling multiplication facts where 7 is the product and where 7 is either the first or second factor in a multiplication sentence.

- Pupils will have developed a deeper understanding of inverse.

Activities for whole-class instruction

- Ask pupils to take seven counters. Ask: *How many groups of 7 do you have?* (one) Ask pupils to take some more counters so that they have 2 groups of 7. Ask: *How many counters do you have altogether?* (14)

 All say …　2 groups of 7 is 14. 2 times 7 is 14.

- Pupils should take another group of seven counters. Ask: *How many groups of 7 do you now have?* (3) Without counting, can they say how many counters they will have altogether? (21) Ask: *How do you know? Convince me.* Can pupils explain that they just need to add 7 to the 14 they already had? Do they have other explanations?

 All say …　3 times 7 is 21.

- With their counters, ask pupils to make an array that represents 3 groups of 7. Can they add to their array so that there are now 4 groups of 7?

 All say …　4 times 7 is 28.

- Repeat for 5 groups, 6 groups, 7 groups, 8 groups, 9 groups, 10 groups, 11 groups and 12 groups.

- Show pupils the following multiplication sentences. Can they complete them with a partner?

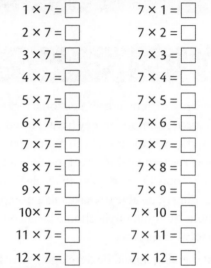

1 × 7 = ☐	7 × 1 = ☐
2 × 7 = ☐	7 × 2 = ☐
3 × 7 = ☐	7 × 3 = ☐
4 × 7 = ☐	7 × 4 = ☐
5 × 7 = ☐	7 × 5 = ☐
6 × 7 = ☐	7 × 6 = ☐
7 × 7 = ☐	7 × 7 = ☐
8 × 7 = ☐	7 × 8 = ☐
9 × 7 = ☐	7 × 9 = ☐
10× 7 = ☐	7 × 10 = ☐
11 × 7 = ☐	7 × 11 = ☐
12 × 7 = ☐	7 × 12 = ☐

 All say …　1 times 7 is 7, 2 times 7 is 14…. 12 times 7 is 84.

- Ask: *What's the same? What's different? What do you notice about the 7 multiplication table? Can you see any patterns?*

- Show pupils the array below. Using their mini whiteboards, can they write a multiplication sentence?

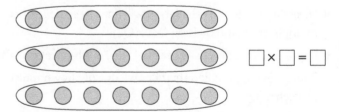

☐ × ☐ = ☐

- Ask: *How do you know your multiplication sentence is correct? What does the array represent? Could you write a different multiplication sentence?*

(i) This image shows 3 groups of 7 (3 × 7). If the columns were ringed, it would show 7 groups of 3. If the rings are removed, no groups are indicated so the array would show both 3 × 7 and 7 × 3.

- Show pupils the array below. Using their mini whiteboards, can they write two multiplication sentences?

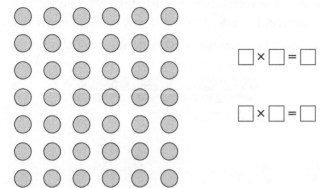

$$\square \times \square = \square$$

$$\square \times \square = \square$$

- It is important that pupils are able to recall relevant multiplication facts using different styles of questions. Often pupils are asked for the product of a multiplication question or the quotient of a division question, but by varying where the missing value is their thinking becomes more flexible.

- Write the sentence $7 \times 7 = 49$ on the board.

 Seven times seven is forty-nine.

- Write '$\square$ times seven is thirty-five' on the board. Ask pupils to talk to a partner about the missing factor. What did they do to calculate the answer? How do they know they are correct?

(i) We want to encourage pupils to share their strategies when calculating, but we must ensure that we encourage the strategy that is most efficient. Therefore, repeated addition is no longer to be encouraged since it is not the most efficient strategy. Instead, pupils should be recalling or deriving multiplication facts. Use manipulatives to create arrays frequently so that pupils have secure experience on which to build understanding. Support pupils to visualise arrays to use as mental images when thinking about multiplication calculations and problems.

- Pupils are now ready to complete Question 1 in the Practice Book.

Same-day intervention

- Working in pairs, using interlocking cubes, ask pupils to build 12 towers of seven interlocking cubes. Ask: *Without counting, how many cubes do you have altogether? How do you know?* Pupils should be able to refer to the number of equal groups that they have.

- Using these 'groups' of 7, can pupils answer: 'Four times seven is $\square$?' Ask: *Can you rearrange your cubes to show 'seven times four is twenty-eight'? Did you need any more cubes? Why not?*

- Ask pupils to rebuild their groups of 7 and leave them in front of them to help them answer the following questions:

  $\square$ *times seven is fourteen.*

  Seven times $\square$ *is thirty-five*

  Five times $\square$ *is thirty-five*

  Seven times $\square$ *is forty-two*

  One times seven is $\square$

- Encourage pupils to explain how they know their answer is correct. What did they do to help them work it out?

Same-day enrichment

- Ask pupils to look at the first two sentences in Question 1 of the Practice Book.

- Say: *Peter said, 'I used my knowledge of $2 \times 7 = 14$ to help me calculate 4×7.'* Can pupils explain why Peter said this? Did they do the same? Ask: *What other facts could you use to calculate unknown facts?* Ask if pupils can look at the other sentences and explain how they calculated the missing value.

Question 2

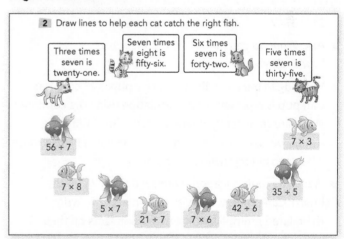

What learning will pupils have achieved at the conclusion of Question 2?

- Flexibility in mental calculation will have been developed by practice in deriving related multiplication and division facts.

Activities for whole-class instruction

- Give pupils the number cards 7, 28, 0, 77 and 8. Pupils should select one of the cards to complete number sentences with missing values.

  Four times seven is ☐

  Seven times ☐ *is fifty-six*

  ☐ *times seven is zero*

  Eleven times seven is ☐

  ☐ *times three is twenty-one*

- Working in mixed ability pairs, ask pupils to talk to their partner about the number missing from the sentence. Pupils should hold up their card when they have found the answer, but it is important that they are given time to think and discuss first.

- When pupils are answering each question, they should say the full number sentence.

- To develop pupils' ability to recall and use multiplication and division facts for the 7 times table, pupils need to understand the equivalence between them to develop greater fluency.

- Write the following sentences on the board:

  $3 \times 7 = 21$

  $7 \times 3 = 21$

  $21 \div 7 = 3$

  $21 \div 3 = 7$

- Ask: *What do you notice? What's the same? What's different?*

- Working in mixed ability pairs, give pupils 21 counters. Can pupils represent each calculation using the counters? What are they doing differently each time? Why? Pupils should be able to refer to their knowledge of equal groups when explaining their representations.

- Write the multiplication sentence $6 \times 7 = 21$ on the board. Can pupils work with their partner to write the related multiplication and division facts on their mini whiteboards?

- Pupils are now ready to complete Question 2 in the Practice Book.

Same-day intervention

- Visual representations will help to develop pupils' conceptual understanding of how the facts are related.

- Show pupils **Resource 3.2.1a** Related multiplication and division facts representations.

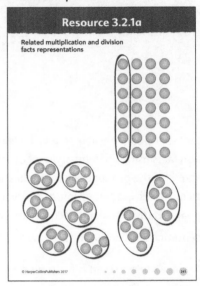

- Ask them what they think each image represents. Why?

- Give pupils the multiplication sentence $5 \times 7 = 35$. Can they use counters to represent this and the related multiplication and division facts?

- Repeat with $56 \div 8 = 7$

Same-day enrichment

- Give pupils three facts. Ask: *Which fact is not related to the others? How do you know?*

7×4	$21 \div 7$	$28 \div 7$
6×7	$49 \div 7$	7×6
$21 \div 3$	3×7	3×8

Question 3

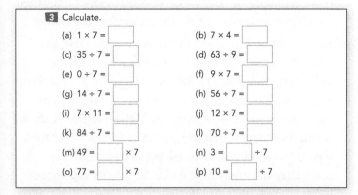

3 Calculate.

(a) $1 \times 7 =$ ☐

(b) $7 \times 4 =$ ☐

(c) $35 \div 7 =$ ☐

(d) $63 \div 9 =$ ☐

(e) $0 \div 7 =$ ☐

(f) $9 \times 7 =$ ☐

(g) $14 \div 7 =$ ☐

(h) $56 \div 7 =$ ☐

(i) $7 \times 11 =$ ☐

(j) $12 \times 7 =$ ☐

(k) $84 \div 7 =$ ☐

(l) $70 \div 7 =$ ☐

(m) $49 =$ ☐ $\times 7$

(n) $3 =$ ☐ $\div 7$

(o) $77 =$ ☐ $\times 7$

(p) $10 =$ ☐ $\div 7$

What learning will pupils have achieved at the conclusion of Question 3?

- Pupils will have practised their recall of the 7 times table to improve fluency.

Activities for whole-class instruction

- Opportunities to develop pupils' fluency should be provided regularly and an important component of this is to provide varied practice. It is important to change the structure of multiplication and division sentences so that the = symbol does not always come at the end.

- Using a counting stick will help to improve pupils' fluency when recalling multiplication facts. Using tools such as a counting stick will also provide pupils with an image to visualise later.

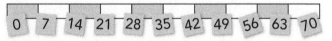

| 0 | 7 | 14 | 21 | 28 | 35 | 42 | 49 | 56 | 63 | 70 |

All say... *0 times 7 is 0, 1 times 7 is 7, 2 times 7 is 14 … 10 times 7 is 70.*

- It is important that pupils answer in full sentences.

ⓘ Many counting sticks in school only allow you to recall facts to 10 times …; however, pupils should also continue to recall 11 times and 12 times … You should refer to these as an imaginary extension of the counting stick.

- Begin to remove the sticky notes gradually when pupils show greater confidence. Can they still recall the fact when asked?

- Attach the number 35 onto the counting stick at the correct position. Ask pupils to discuss in pairs and then record on their mini whiteboards what the division fact would be. ($35 \div 5 = 7$) How do they know?

- Point to where '70' would be on the counting stick. Ask: *What would the division fact be? What would the multiplication fact be?*

- You may wish to split the class into smaller groups for this activity. Give pairs of pupils number cards with the multiples of seven from 0 to 84 and number cards 0 to 12 below this. Ask pupils to lay them in front of them on their desks so that they correspond as below:

| 0 | 7 | 14 | 21 | 28 | 35 | 42 | 49 | 56 | 63 | 70 | 77 | 84 |

| 0 | 1 | 2 | 3 | 4 | 5 | 6 | 7 | 8 | 9 | 10 | 11 | 12 |

- Ask pupils a series of multiplication and division questions, asking them to select the answer as quickly as possible and raise it above their heads.

- Pupils are now ready to complete Question 3 in the Practice Book.

Same-day intervention

- Give small groups of pupils a set of domino cards with multiplication and division facts from **Resource 3.2.1b** Domino cards.

Resource 3.2.1b

Domino cards

0 × 7	7	7 × 7	9
14 ÷ 7	21	7 × 8	6
7 × 3	7	63 ÷ 7	84
28 ÷ 4	2	11 × 7	0
5 × 7	49	7 × 12	35
42 ÷ 7	56		

242   © HarperCollinsPublishers 2017

- Can pupils work together to create a domino run, matching the multiplication/division question with the answer? Pupils may want to use a mini whiteboard to support them when calculating each answer.

Same-day enrichment

- Getting pupils to explain their thinking requires a deep level of understanding and for some pupils can pose a challenge. Asking pupils to explain how and why they have chosen a particular method develops their fluency. Pupils must demonstrate some flexibility in their approach and this is best done when pupils are able to reflect on their approach and adapt.

- Can pupils explain what multiplication facts they used to calculate the following?

  3×7

  9×7

  2×7

  5×7

- Ask: *Did you only recall multiplication facts from the 7 times table?*

Question 4

> 4 Application problems.
>
>   (a) 14 sweets were shared equally by 7 children. How many sweets did each child get?
>
>   Answer: _____
>
>   (b) In one week, Zoe used 3 pieces of paper every day. How many pieces of paper did she use in the whole week?
>
>   Answer: _____
>
>   (c) In Maths lessons for Class 3A, the teacher divides the pupils into 7 groups. There are 5 pupils in each group. How many pupils are there altogether?
>
>   Answer: _____
>
>   (d) Class 3B has 6 more pupils than Class 3A. How many pupils are there in Class 3B?
>
>   Answer: _____

What learning will pupils have achieved at the conclusion of Question 4?

- Pupils will have applied their knowledge of known multiplication and division facts to solve problems.

Activities for whole-class instruction

- Pupils need opportunities to solve multiplication and division problems presented in real-world contexts to appreciate the relevance of mathematics to the real world. Developing pupils' problem-solving skills makes them more competent mathematicians and being able to solve

a range of problems is an essential part of mathematics. If required, pupils can use concrete equipment or draw their own representations to help them calculate the answer.

- Ask seven pupils to come to the front of the class. Say: *I have 21 pencils that I need to share equally with the volunteers. How many pencils should I give to each of them?* Allow pupils to discuss with a partner (the seven pupils can also discuss). As the language in the question is 'share', pupils may suggest that the pencils are physically shared. If you start sharing pencils, ask pupils to predict how many they will have at the end. Ask: *How do you know? What multiplication fact did you use?* At the end, ask: *Were you correct? How many pencils would each volunteer get if I started with 28? How do you know?*

- Ask the seven pupils to sit back down and ask the class what the important information was that you gave to help pupils work it out. How did they know what operation to use? Could they have done it differently?

- Write the following problem on the board: *Mrs Eaton eats 6 apples each week. How many apples does she eat over 7 weeks?*

- Ask: *What is the important information in this problem? Tell your partner which operation you might need to use to calculate the answer to the problem.*

- At this stage, you do not want pupils to give you the answer as it is important that they understand the thought process required to solve a problem.

- Using mini whiteboards, ask pupils to draw something to represent this problem. Share their representations with the class. Are they correct? It is important that pupils understand that they can use representations to support their thinking when problem solving if required. Using these images, can they now write a number sentence and calculate the answer?

- **Look out for** … pupils drawing detailed pictures to help them calculate the answer. Explain to pupils that their drawings only need to symbolise what is in the problem and the importance should be on the mathematics. For example, rough circles can be drawn instead of apples.

- Pupils are now ready to complete Question 4 in the Practice Book.

Same-day intervention

- Use concrete equipment to help pupils solve problems.

Same-day enrichment

- Can pupils think of their own multiplying and dividing by 7 problems for a partner to solve?

- Give pupils an image. Can they create a word problem using this image? For example:

Challenge and extension question

Question 5

> 5 Use the numbers below to make number sentences. Who can make the most number sentences? Operations can be used on both sides of the equation.
>
> 14, 42, 6, 7, 2, 4, 28, 35, 5

Giving pupils questions with more than one solution encourages them to be systematic and organised. To answer this question, pupils will need to apply their knowledge of multiplication and division, identifying which of the numbers would be factors and products in a multiplication sentence and which of the numbers could be the dividend, divisor and quotient in division sentences.

Ask: *How do you know you have found all the possibilities?*

Unit 2.2
Multiplying and dividing by 3

Conceptual context

This unit further develops pupils' understanding of multiplying and dividing by 3. It allows them to practise and apply knowledge in a range of contexts. Pupils revisit the language 'dividend', 'divisor' and 'quotient' from Book 2. Pupils will need to have a secure understanding of inverse operations and understand how multiplication and division are linked.

Although the questions within this unit do not provide images for pupils to refer to, it would be appropriate to allow pupils to use concrete equipment and pictorial representations to continue to secure and develop their conceptual understanding. This is particularly true when pupils are applying knowledge to more abstract ideas.

At this stage, pupils need to be able not only to recall multiplication and division facts but also to use them to solve problems. They should make reasoned decisions about the mathematics that that they use.

Learning pupils will have achieved at the end of the unit

- Pupils will have practised writing related multiplication and division facts (Q1)
- Pupils will have practised their recall of the 7 times table to improve fluency (Q2)
- Pupils will have further developed their understanding of division as the inverse of multiplication by finding the missing dividend, divisor and quotient (Q3)
- Pupils will have interpreted and calculated multiplication and division calculations that are presented through word problems and are different structurally (Q4)
- Pupils will have applied their knowledge of known multiplication and division facts to solve problems (Q5)

Resources

mini whiteboards; interlocking cubes; number cards; symbol cards; counters; 100 square; **Resource 3.2.2a** Dividend, divisor, quotient; **Resource 3.2.2b** Multiply and divide by 3; **Resource 3.2.2c** Multiply and divide by 3 match up

Vocabulary

multiplication, multiply, times, factor, product, divide, division, divided by, inverse, dividend, divisor, quotient

Question 1

> **1** Complete the multiplication facts. Then write multiplication and division sentences.
>
> (a) Three times five is _____.
>
> _____   _____
>
> _____   _____
>
> (b) _____ times _____ is thirty-three.
>
> _____   _____
>
> _____   _____
>
> (c) Three times _____ is eighteen.
>
> _____   _____
>
> _____   _____

What learning will pupils have achieved at the conclusion of Question 1?

- Pupils will have practised writing related multiplication and division facts in words, varying the missing element (factor or product) that must be inserted.

Activities for whole-class instruction

- Show pupils the images (or use concrete objects) of cookies on a plate. Ask pupils to discuss what calculation the images 'could' represent. Ask: *Can you think of both multiplication and division sentences?*

- Write these sentences on the board:

$$8 \times 3 = 24 \qquad 3 \times 8 = 24$$
$$24 \div 3 = 8 \qquad 24 \div 8 = 3$$

- Using mini whiteboards, can pupils work in pairs to represent these multiplication and division sentences? Can they explain what they have drawn and why?

- Give pupils a multiplication or division sentence. Can they move around the classroom to find the related facts? (You may have to consider your class size for this activity. You may only want to give three facts and get pupils to write the missing fourth, for example.)

$15 \div 3 = 5$	$15 \div 5 = 3$	$3 \times 5 = 15$	$5 \times 3 = 15$

- When pupils have found their group, take a card from another group. Ask them to write the related facts on their whiteboard. Ask: *How did you do it? What knowledge did you use?*

- Select two of the 'families' of multiplication and division sentences to be written in words.

- Pupils are now ready to complete Question 1 in the Practice Book.

Same-day intervention

- Give pupils the number and symbol cards: three, seven, twenty one, ×, ÷ and =. Can pupils use these cards to write as many calculations as they can? They can only use each card once. Ask: *How many sentences do you think you will be able to write? Why?*

- Support pupils' conceptual understanding by using cubes/counters to represent each calculation. Ask: *What's the same? What's different?*

Same-day enrichment

- Working in pairs, can pupils write word problems for the following multiplication and division sentences?

  three × nine = twenty-seven

  nine × three = twenty-seven

  twenty-seven ÷ three = nine

  twenty-seven ÷ nine = three

Question 2

> **2** Calculate.
>
> (a) $6 \times 3 = \square$ (b) $3 \times 8 = \square$ (c) $3 \times 9 = \square$
>
> (d) $11 \times 3 = \square$ (e) $7 \times 3 = \square$ (f) $3 \times 12 = \square$
>
> (g) $30 \div 3 = \square$ (h) $5 \times 3 = \square$ (i) $36 \div 3 = \square$
>
> (j) $2 \times 9 = \square \times 3$ (k) $27 \div 9 = 3 \times \square$ (l) $4 \times \square = 2 \times 6$
>
> (m) $3 \times \square = 12$ (n) $\square \times 3 = 12 + 3$ (o) $\square + 3 = 12 \times 3$

What learning will pupils have achieved at the conclusion of Question 2?

- Pupils will have practised their recall of the 3 times table to improve fluency.

Activities for whole-class instruction

- To demonstrate fluency, pupils need not only to recall a fact but also to understand why they are doing what they are doing and know when it is appropriate to use each fact. Encourage pupils to make decisions about the multiplication and division facts that they use and explain why.

- Write $3 \times 7 = \square \times \square$ on the board. Ask pupils to discuss in pairs what the missing factors are. Can they write the equivalent calculation on their mini whiteboards? ($3 \times 7 = 7 \times 3$) Pupils are applying their understanding of the commutative property of multiplication and understanding of equivalence.

- Write $3 \times 4 = 36 \div \square$ on the board. Ask pupils to discuss in pairs what the missing numbers are. Can they complete the equivalent calculation on their mini whiteboards? What did they do first? It is Important that pupils are given opportunities to communicate their understanding with one another.

- Pupils are now ready to complete Question 2 in the Practice Book.

Same-day intervention

- Give pupils a 100 square. Ask them to shade the products of the 3 times table. Ask: *How do you know you have them all? What did you do? What do you notice about the numbers you have shaded?* Using the 100 square, can pupils tell you what $36 \div 3$ is? Ask: *How many groups of 3 are there in 36?* Can they now write a multiplication sentence to show this fact? Say: *Convince me you are correct.*

- Write the following calculations on the board. Can pupils work with a partner to calculate the answers?

  $3 \times 9 = \square$

  $15 \div 3 = \square$

  $3 \times 10 = \square \times 5$

Same-day enrichment

- Show pupils the 3, 6 and 9 times tables. Can pupils identify any patterns? Are there any relationships?

- Pupils could use a multiplication square to help them spot patterns and make connections. Can they explain them?

Question 3

3 Complete the table.							
Dividend	3		18	27		30	0
Divisor		7	6		3		3
Quotient	1	3		9	12	3	

What learning will pupils have achieved at the conclusion of Question 3?

- Pupils will have further developed their understanding of division being the inverse of multiplication by finding the missing dividend, divisor and quotient.

Activities for whole-class instruction

(i) Pupils were introduced to the terms 'dividend', 'divisor' and 'quotient' in Book 2. Language is important in helping pupils to express mathematical relationships.

$$30 \quad \div \quad 3 \quad = \quad 10$$
dividend   divisor   quotient

(i) In mathematics teaching in Shanghai, the processes of sharing and grouping are not differentiated – they are both considered simply as dividing. Therefore, pupils will experience dividing by 4 as making groups of 4 and as sharing into 4 groups in the same set of questions. Both are referred to as dividing; the result is, of course, the same.

$12 \div 4 = 3$

There are 3 groups of 4 in 12.

If you share 12 things equally between 4 people (that is, divide them into 4 groups) they will have 3 things each.

- Ask pupils to explain what $12 \div 4 = 3$ means. Can they represent this using concrete apparatus?

- Give pupils this division sentence on a strip of paper.

12	÷	4	=	3

- Ask pupils to cut out each number. Can they put them under the following headings?

Dividend	Divisor	Quotient

 12 is the dividend. 4 is the divisor. 3 is the quotient.

- Write the headings on the board. Give pupils three blank pieces of square paper. Can they think of a division sentence to use multiplication and division facts from the 3 times table? Ask pupils to come to the board and to stick their numbers under the headings. As a class, check they are correct.

- Write 18 under the heading 'dividend'. Write 6 under the heading 'quotient'. Ask pupils to talk to their partner about what the missing divisor is. How do they know? Ask: *The dividend is 0, the divisor is 3. What is the quotient?*

- Pupils are now ready to complete Question 3 in the Practice Book.

Same-day intervention

- Write 30 ÷ 3 = 10 on the board. Ask: *Which number in the calculation is the dividend/divisor/quotient?*

- Pupils should discuss in pairs. How do they know? Ask: *Can you write a sentence where the dividend is 15 and the divisor is 3? Can you write a sentence where the quotient is 9?* Pupils will be correct if they give either a sharing or grouping example.

- Initially, pupils will find it easier to find the missing dividend/divisor/quotient with the calculation written horizontally. Ask pupils to complete the table in **Resource 3.2.2a** Dividend, divisor, quotient.

Resource 3.2.2a

Dividend, divisor, quotient

Dividend	Divisor	Quotient
6	3	
33	3	
15		5
	10	3
27	9	

© HarperCollinsPublishers 2017

- Ask: *How did you find the missing number?* When they have completed the table, ask: *What do you notice? Is the divisor always 3? Why not?*

Same-day enrichment

- This activity challenges pupils in a similar way as Question 3 in the Practice Book as it requires pupils to apply their knowledge of both multiplication and division. Ask pupils if they can use the given information to complete the unknown facts in the multiplication table. Ask: *Where did you start? Why? Which was the hardest to work out? Why?*

×		6
3	9	
		48

- Ask pupils to create their own 'unknown facts table' for a partner to complete.

Question 4

> 4 Write a number sentence for each question.
> (a) What is the product of 2 threes?
>
>   Number sentence: _____
> (b) How many threes should be subtracted from 15 so the result is 0?
>
>   Number sentence: _____
> (c) What is 7 times 3?
>
>   Number sentence: _____

What learning will pupils have achieved at the conclusion of Question 4?

- Pupils will have interpreted and calculated multiplication and division calculations that are presented through word problems and are different structurally.

Activities for whole-class instruction

- Pupils need to have a secure understanding of different expressions and their equivalence to be able to understand the relationships between multiplication and division. Questions that demand more than just the product or quotient and use a range of mathematical language require higher order thinking from pupils.

- Give pupils a copy of the grid on **Resource 3.2.2b**
  Multiply and divide by 3.

Resource 3.2.2b

Multiply and divide by 3

9	6	27
36	15	3

© HarperCollins Publishers 2017

- Ask: *What do you notice about all the numbers on the grid?*
- Explain that you are going to read some questions (below), the answers to which they will find on their resource sheet. Pupils use a counter to cover the answer to each question. As you read each question, it may be appropriate to provide pupils with apparatus or to draw models/images on the board to support, for example pupils may need to draw a number line for the question that requires them to use repeated subtraction. Pupils may also want to ask questions about particular language used.

What is the product of 3 and 5?
What is 9 times 3?
How many threes should be subtracted from 9 so the answer is 0?
How many groups of 5 are there in 30?
What is 12 multiplied by 3?

- Ask pupils to write on their mini whiteboards the only number that is left uncovered and hold it up. (9)
- To show a secure understanding of a question given in words, pupils should be able to interpret it to write a number sentence. Give pupils strips of paper with the questions you asked. Ask: *Can you write the number sentence for each on your mini whiteboards?*
- Pupils are now ready to complete Question 4 in the Practice Book.

Same-day intervention

- To provide further support, give pupils **Resource 3.2.2c**
  Multiply and divide by 3 match up.

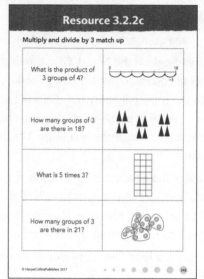

- This activity requires pupils to interpret the question to find the related image. Can they then use the image to calculate the answer?

Same-day enrichment

- Show pupils the diagram for $12 \times 5 = 60$.
  The related division facts are:
  $$60 \div 12 = 5$$
  $$60 \div 5 = 12$$

$60 = 5 \times 12$ The product is 60.

$12 \times 5 = 60$ The factors are 12 and 5.

There are 12 groups of 5.   There are 5 groups of 12.

- Give pupils $8 \times 3 = 24$
- How many facts can they write?

Question 5

> **5** Application problems.
> (a) Joe and his parents visit a museum. The admission ticket is £8 per person. How much do they have to pay?
>
> Answer: _____
>
> Joe pays with a £50 note. How much change should he get?
>
> Answer: _____
>
> (b) 6 ducks are on a river. There are 4 times as many ducks on the bank as on the river. How many ducks are on the bank?
>
> Answer: _____
>
> How many ducks are there altogether?
>
> Answer: _____

What learning will pupils have achieved at the conclusion of Question 5?

- Pupils will have applied their knowledge of known multiplication and division facts to solve problems.

Activities for whole-class instruction

- Placing the mathematics into real-life contexts helps pupils to develop understanding of the relationships when expressed through words.

- Write the problem on the board: *Three children are going to the zoo. Each ticket costs £6. How much will it cost for all three children to go to the zoo?*

- Bring three pupils to the front of the class. Give each child a 'ticket' to the zoo with £6 written on the front (scenario can be changed as appropriate). Ask pupils to write a calculation on their mini whiteboards that will help them work out the answer. At this stage, you do not want the answer.

- Share some pupils' calculations. Are they the same? Ask: *Why have you chosen this calculation?* Ask pupils to calculate the answer. (£18) Check with a partner, do they have the same answer?

- **Look out for** … pupils who focus only on the numbers within the question and 'guess' the operation required. It is important that pupils also focus on the language used and understand that they must interpret that correctly.

- Write the following problem on the board: *Peter paid £3 for a book. He bought three more for his friends. How much did Peter pay?* Ask pupils if they can represent this problem using pictures on their whiteboard. Share some of the pupils' boards with the rest of the class. Ask pupils to explain their representations of the problem.

- Ask: *If Peter paid for the books with a £20 note, how much change should he get?* Ask pupils to work with a partner to calculate the answer. Get pairs to check their answer with another pair. Did they solve the problem in the same way? Which way was most efficient?

- Encourage pupils to use bar models to represent the problem and support as required. For example:

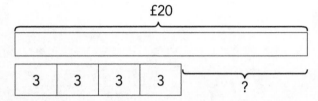

- Pupils are now ready to complete Question 5 in the Practice Book.

Same-day intervention

- To support pupils with solving problems they need to be given the opportunity to talk about the mathematics and understand the language used. It would be appropriate for pupils to use equipment and/ or draw pictures to represent the problem.

- Write the following problem on the board: *Cupcakes were sold in packs of 3. Julie bought 6 packs. How many cupcakes did she have altogether?*

- Ask pupils to use counters to represent the problem. Can they solve the problem using the counters?

- Repeat using similar problems.

Same-day enrichment

- Ask pupils to write their own multiplication and division problems.

Challenge and extension questions

Question 6

6 A bag of sweets can be divided equally into 3 groups. It can also be divided equally into 6 groups. What could be the smallest number of sweets in the bag?

Answer: _____

This problem has more than one solution and pupils may initially start by using trial and error. This question requires pupils to recall multiplication facts for the 3 and 6 times tables but they also need to apply other mathematical skills, such as working systematically and organising their work.

Question 7

7 A monkey picked 24 peaches. She gave all the peaches to her two baby monkeys. The older baby monkey got 2 times as many peaches as the younger one.

(a) How many peaches did the younger baby monkey get?

Answer: _____

(b) How many peaches did the older monkey get?

Answer: _____

Pupils need to apply their knowledge of multiplying by scaling up and also understand the inverse. To support pupils, they may find it useful to use the bar model.

24	Peaches

	Youngest baby monkey

		Eldest baby monkey

Unit 2.3
Multiplying and dividing by 6

Conceptual context

This unit focuses on multiplying and dividing by 6 and pupils are also required to recall other known facts. Pupils will be given the opportunity in this unit to apply their knowledge of the commutative property of multiplication. To ensure fluency, pupils should be making choices about which multiplication/division fact(s) to use; it is not enough for pupils simply to memorise facts; they need to understand why they are doing what they are doing and know when it is appropriate to apply their knowledge.

Pupils should be beginning to notice relationships between multiplication tables.

Learning pupils will have achieved at the end of the unit

- Pupils will have practised 'translating' arrays into multiplication sentences in numerals and words (Q1)
- Knowledge of multiplication and division facts will have been applied to compare values using <, > and = (Q2)
- Pupils will have used their multiplication knowledge to find alternative factor pairs (Q3)
- Pupils will have practised their recall of multiplication and division facts to improve fluency (Q4)
- Pupils will have applied their knowledge of known multiplication and division facts to solve problems (Q5)

Resources

mini whiteboards; interlocking cubes; squared paper; number cards; counters; **Resource 3.2.3a** Making connections; **Resource 3.2.3b** Multiplication squares

Vocabulary

multiplication, multiply, times, factor, product, divide, division, divided by, inverse, dividend, divisor, quotient

Question 1

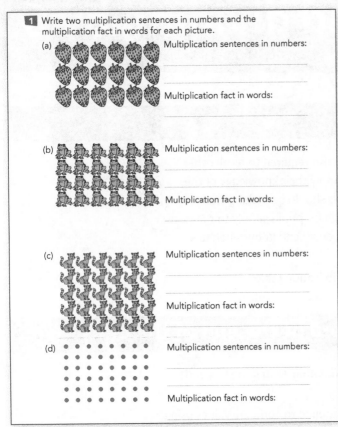

1 Write two multiplication sentences in numbers and the multiplication fact in words for each picture.

(a) Multiplication sentences in numbers:

Multiplication fact in words:

(b) Multiplication sentences in numbers:

Multiplication fact in words:

(c) Multiplication sentences in numbers:

Multiplication fact in words:

(d) Multiplication sentences in numbers:

Multiplication fact in words:

What learning will pupils have achieved at the conclusion of Question 1?

- Pupils will have practised 'translating' arrays (in which one of the factors is 6) into multiplication sentences in numerals and words.

Activities for whole-class instruction

- Write 2 × 6 = on the board. Can pupils draw an array to represent this? Can they change their array to represent 3 × 6? Ask: *What did you do? Why?*

All say ...   *3 times 6 is 18.*

- Ask pupils to write the multiplication fact in words on their mini whiteboards. Some pupils may write 'three times six is eighteen', others may write 'six times three is eighteen'. Both sentences are correct.

- Write 'Seven times six is forty-two'. Read together. Ask pupils to close their eyes and visualise the array. What are they 'seeing'? Ask them to draw it on their mini whiteboards. Ask pupils to write the number sentence, and another.

- Give pupils various representations of multiplication facts (similar to those at the top of the next column). With a partner, can they take it in turns to say the multiplication fact?

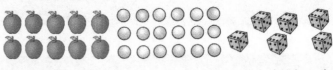

- Can pupils write these in words on their mini whiteboards?
- Pupils are now ready to complete Question 1 in the Practice Book.

Same-day intervention

- Write 4 × 6 = 24.

All say ...   *4 times 6 is 24. 4 multiplied by 6 equals 24.*

- Ask pupils to make the array using counters.

Same-day enrichment

- Give pupils squared paper and ask them to draw an array of 42 squares. Ask: *Can you write a multiplication sentence in numbers and words that you think the array represents? Why have you chosen those factors? Are they the same or different as someone else's?*
- Draw all the arrays for 42. List all the multiplication sentences where 42 is the product.

Question 2

2 Write >, < or = in each ◯.

(a) 3 × 6 ◯ 9 × 2   (b) 6 × 12 ◯ 70

(c) 6 × 9 ◯ 5 × 9 + 9   (d) 60 ÷ 10 ◯ 66 ÷ 6

What learning will pupils have achieved at the conclusion of Question 2?

- Knowledge of multiplication and division facts will have been applied to compare values using <, > and =.
- Use of reasoning will have enabled pupils to see relationships between operations.

Activities for whole-class instruction

- Ask pupils to look at the images below. Ask: *What do you notice?*

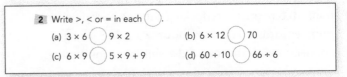

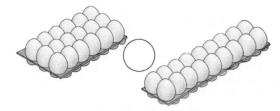

- In mixed ability pairs, pupils decide whether the circle should contain a <, > or =. Pupils should be familiar with these symbols from Book 2. Pupils explain their answers.

- Using the images, pupils now write a number sentence. ($3 \times 6 = 2 \times 9$)

- Write $6 \times 5 = 36$ on the board. Ask: *Is this true or false?* Allow pupils to discuss. Ask: *How do you know? What could you change to make this sentence correct? Is there only one way?*

  - $6 \times 5 < 36$

  - $6 \times 5 = 30$

  - $6 \times 6 = 36$

- Write: $18 \div 6 \bigcirc 24 \div 3$ on the board. Ask pupils to write the correct symbol on their mini whiteboards. Say: *Convince me that you are correct.*

- Write $4 \times 6 + 6 + 6 \bigcirc 6 \times 6$. Ask: *What do you notice?* Ask pupils to talk to a partner about what they did to calculate the answer. What facts did they use? Ask pupils to represent this using interlocking cubes.

- Can pupils write $4 \times 6 + 6 + 6$ as a multiplication sentence?

- Pupils are now ready to complete Question 2 in the Practice Book.

  (i) When there is more than one operation, pupils may need to be reminded to calculate in the order that they appear, reading from left to right. At this stage, you do not teach about order of operations as they will learn about it later. If they ask, tell them that we always multiply and divide before we add and subtract and that they will learn more about it later.

Same-day intervention

- Continue to use pictorial representations and/or concrete apparatus to support pupils.

- Give pupils cards with >, < and = on. Write $3 \times 6 \bigcirc 10 \times 2$ on the board. Ask pupils to use interlocking cubes to represent the multiplication or division sentence. Ask: *What do you notice? Which symbol should they chose? Why?* Get pupils to explain their thinking.

- Repeat with $4 \times 6 + 6 \bigcirc 11 \times 3$

Same-day enrichment

- Ask pupils to look at the calculations. Ask: *What do you notice? How else could you write these calculations? Can you write some of your own?*

  $4 \times 6 = 2 \times 6 + 6 + 6$

  $9 \times 6 = 5 \times 6 + 6 + 6 + 6 + 6$

  $3 \times 6 + 3 \times 6 = 6 \times 6$

Question 3

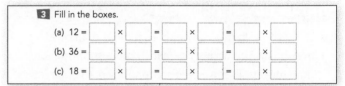

3 Fill in the boxes.
(a) $12 = \boxed{} \times \boxed{} = \boxed{} \times \boxed{} = \boxed{} \times \boxed{}$
(b) $36 = \boxed{} \times \boxed{} = \boxed{} \times \boxed{} = \boxed{} \times \boxed{}$
(c) $18 = \boxed{} \times \boxed{} = \boxed{} \times \boxed{} = \boxed{} \times \boxed{}$

What learning will pupils have achieved at the conclusion of Question 3?

- Pupils will have used their multiplication knowledge to find alternative factor pairs.

Activities for whole-class instruction

- Write the number 48 on the board. In mixed ability pairs, ask: *How many factors can you find? How do you know you have found them all?* Ask pupils to share the factors that they have found. Record them as multiplication sentences:

  2×24

  4×12

  6×8

- Ask: *What is the product of these multiplications? Is the value of each one the same? How could we write all three multiplications in one sentence?* ($48 = 2 \times 24 = 4 \times 12 = 6 \times 8$)

- Write the number 32 on the board. Pupils work with a partner to write their own multiplication number sentence.

  (Look out for) … pupils completing the calculation by applying the commutative property of multiplication. This is of course correct, but encourage pupils to find other factor pairs.

- Pupils are now ready to complete Question 3 in the Practice Book.

Same-day intervention

- Tell pupils to take 24 counters. Ask: *How many different arrays can you make using all of the counters?* Ask pupils to record the multiplication sentences on their whiteboards. Ask: *Did the product change?*

- Model writing an equivalent calculation using the = symbol.

  $24 = 2 \times 12 = \square \times \square$

- Repeat for other multiples of 6.

Same-day enrichment

- Share this rich task where there is more than one possibility. This activity demands more sophisticated thinking. Pupils may need some support to organise their work and to work systematically.

- Say: *During a storm, animals took shelter in a farmyard barn. The farmer counted 42 legs from under the door. Ask: What animals could have been in the barn? How do you know you have found all possibilities? Remember to include insects and spiders.*

Question 4

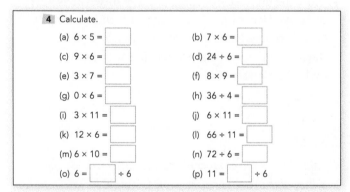

4	Calculate.		
(a)	$6 \times 5 =$	(b)	$7 \times 6 =$
(c)	$9 \times 6 =$	(d)	$24 \div 6 =$
(e)	$3 \times 7 =$	(f)	$8 \times 9 =$
(g)	$0 \times 6 =$	(h)	$36 \div 4 =$
(i)	$3 \times 11 =$	(j)	$6 \times 11 =$
(k)	$12 \times 6 =$	(l)	$66 \div 11 =$
(m)	$6 \times 10 =$	(n)	$72 \div 6 =$
(o)	$6 = \square \div 6$	(p)	$11 = \square \div 6$

What learning will pupils have achieved at the conclusion of Question 4?

- Pupils will have practised their recall of multiplication and division facts to improve fluency.

Activities for whole-class instruction

- When answering multiplication and division questions, pupils should be asking themselves questions about the most efficient method. When calculating mentally, they will need to draw upon other relevant knowledge and will need to know what is relevant. Pupils should now be able to make connections.

- Write $\boxed{\begin{array}{l} 6 \times 7 = 42 \\ 3 \times 7 = 21 \end{array}}$ on the board.

Ask: *What do you notice?*

- Write $\boxed{\begin{array}{l} 6 \times 9 = 54 \\ 3 \times 9 = ? \end{array}}$ on the board.

Ask pupils to calculate the answer. Ask: *What did you do?* Pupils should be able to explain that the product must be half of 54. Can they explain why? Show pupils **Resource 3.2.3a** Making connections.

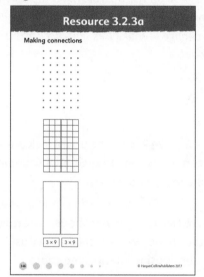

- Ask: *What do these images show? What is the same about them? What is different?*

(i) Pupils have already begun to learn that arrays can be represented on squared paper and so will find counting the squares on each side of the array a natural extension to what they already understand. The next step is to imagine the squares within the outline of the third image and to begin to understand that it is not necessary to 'see' each individual square – that it is enough to know how many are there and that as long as the diagram is drawn and labelled correctly, the individual squares do not need to be drawn. This links, of course, with representing area, but in this context also prepares pupils to understand multiplication of numbers that are not whole numbers in the future. At this stage, you are just introducing the 'area model' for multiplication; gently expanding pupils' concept of multiplication by exposing them to the image, with some understanding.

- Write $\boxed{\begin{array}{l} 36 \div 6 = 6 \\ 36 \div 3 = ? \end{array}}$ on the board.

Ask pupils to calculate the answer. Ask: *What did you do?* Pupils should be able to explain that the quotient is double 6. Can they explain why? Can they draw an image to represent this?

- Write 36 ÷ 9 = ? on the board. Ask pupils to calculate the answer. Ask: *What did you do?* Pupils should be able to say that the quotient is 4. Can they explain why?

- Pupils are now ready to complete Question 4 in the Practice Book.

Same-day intervention

- Give pupils some cards with multiplication and division facts and the answer (see below). Instruct pupils to place the cards face down. Take it turns to turn over two cards. Ask: *Are they equivalent? Can you prove it using squared paper?*

6 × 6	60 ÷ 10	4 × 6	36
6	18 ÷ 6	24	3

Same-day enrichment

- Show pupils **Resource 3.2.3b** Multiplication squares.

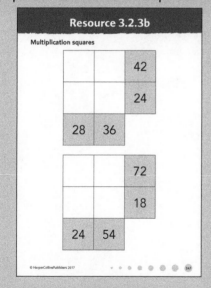

Resource 3.2.3b

Multiplication squares

- Explain to pupils that the greyed out squares contain the products of the two numbers before/above it. Ask pupils to complete the squares.

Question 5

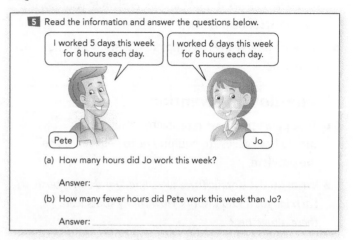

5 Read the information and answer the questions below.

I worked 5 days this week for 8 hours each day.

I worked 6 days this week for 8 hours each day.

Pete

Jo

(a) How many hours did Jo work this week?

Answer: _____

(b) How many fewer hours did Pete work this week than Jo?

Answer: _____

What learning will pupils have achieved at the conclusion of Question 5?

- Pupils will have applied their knowledge of known multiplication and division facts to solve problems.

Activities for whole-class instruction

- By giving pupils real-life problems, they are able to see how and when they will need to draw upon their knowledge.

- Share the statements below. Say: *Chris bought the most muffins. True or false?*

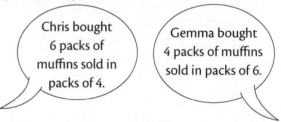

Chris bought 6 packs of muffins sold in packs of 4.

Gemma bought 4 packs of muffins sold in packs of 6.

- Ask pupils to draw something on their mini whiteboards to represent the muffins that Chris bought and the muffins that Gemma bought. Ask: *What's the same? What's different?* Share some of the pupils' representations. By sharing work, pupils may learn a 'better way' of doing something.

- Share the statements below. Say: *Sarah drove 2 miles more than Gary. True or false?*

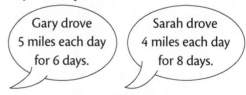

Gary drove 5 miles each day for 6 days.

Sarah drove 4 miles each day for 8 days.

- Ask pupils to draw something on their mini whiteboards to represent the miles that Gary drove and the miles that

Sarah drove. Ask: *What's the same, what's different? What did you do to work it out? Can you explain to your partner why your statement is true?*

- Pupils are now ready to complete Question 5 in the Practice Book.

Same-day intervention

- To support pupils with problem solving, they should have access to equipment to help them 'see' the problem.

- Write on the board: *There were 5 ladybirds on a bush. Each ladybird had 6 spots. How many spots were there altogether?*

- Read the problem together and discuss. Ask: *What do you need to do to solve the problem?* Ask pupils to use counters to help them solve the problem.

Same-day enrichment

- Challenge pupils to write their own problems for a partner to complete. Explain that the problem must require their partner to apply their knowledge of the 6 times table facts.

Challenge and extension questions

Question 6

6 Given ● + ● + ● + ● = ▲ + ■ and ■ = ▲ + ▲,
   if ● = 6, then ▲ = ☐ .

This question challenges pupils' algebraic thinking. Pupils should apply their knowledge of equivalences and multiplication and division facts. Some pupils may start this problem by using trial and error, but this is not an efficient method and pupils should be given time to consider their approach to the problem. What do they need to do first? What information do they need to know?

Question 7

7 Simon starts reading his book at page 1. He reads 6 pages each day.
   He reads the book for 4 days.
   On the fifth day, Simon starts reading from page ☐ .

If needed, pupils should be encouraged to represent this problem pictorially first. To add further challenge, explain to pupils that pages 1 and 2 include a contents page also.

Unit 2.4
Multiplying and dividing by 9

Conceptual context

This unit focuses on multiplying and dividing by 9. Pupils have already covered multiplying and dividing by 7, 3 and 6 in previous units. They have had opportunities to notice patterns and relationships between different multiplication tables and need to continue to do this for the 9 times table. It is important that pupils use their knowledge of known facts such at the 10 times table to help them.

Once pupils are confident with recalling multiplication facts, they should be able to apply their knowledge to tasks that are more open-ended and to solve problems.

Learning pupils will have achieved at the end of the unit

- Pupils will have practised writing related multiplication and division facts (Q1)
- Pupils will have used their reasoning skills to derive known multiplication facts (Q2)
- Using their knowledge of inverse operations, pupils will have practised writing multiplication and division sentences (Q3)
- Pupils will have practised applying their knowledge of multiplying and dividing by 9 to solve problems (Q4)

Resources

counting stick; mini whiteboards; interlocking cubes; number cards; counters; ribbon, **Resource 3.2.4** Problem matching and solving; 9 cm pieces of ribbon (or paper)

Vocabulary

multiplication, multiply, times, factor, product, divide, division, divided by, inverse, dividend, divisor, quotient

Question 1

> **1** Recall the multiplication facts, and write the answers in words and then in numbers.
>
> (a) One times nine is _____.
>
> $1 \times 9 =$ _____ $9 \times 1 =$ _____
>
> (b) Three times nine is _____.
>
>   _____
>
> (c) Four times nine is _____.
>
>   _____
>
> (d) Eleven times nine is _____.
>
>   _____
>
> (e) _____ is fifty-four.
>
>   _____
>
> (f) _____ is eighteen.
>
>   _____

What learning will pupils have achieved at the conclusion of Question 1?

- Pupils will have deepened their understanding of the relationship between multiplication and division when they practise writing related multiplication and division facts.

Activities for whole-class instruction

- Pupils will know that multiplication is commutative and therefore should be able to recall the best known fact to give the product.

- Working in pairs, ask pupils to write the 9 times table. Pupils must write the full multiplication sentence and recording them in a list will help them to spot patterns.

    $1 \times 9 = 9$
    $2 \times 9 = 18 \dots$

- Ask pupils what they notice about all the numbers in the 9 times table.

- Working in pairs, Pupil A should use counters to represent one of the facts as an array. Can Pupil B write two multiplication sentences that represent the same multiplication fact and explain their decisions?

- Say: *Carly says she can calculate 9 × 8 easily because she knows 10 × 8 = 80. Lewis says he can calculate 6 × 9 easily because he knows 6 × 10 = 60.* Ask: *How is Carly and Lewis' knowledge of the 10 times table helping them to multiply by 9?*

- Write 4×9, 9×4 and $3 \times 9 + 9$ on the board. Ask: *What's the same? What's different?* Pupils should apply their reasoning skills to explain how $3 \times 9 + 9$ is the same as 4×9.

- Pupils are now ready to complete Question 1 in the Practice Book.

Same-day intervention

- Ask pupils to represent the multiplication facts below using counters:

    $5 \times 9 = \square$
    $\square = 9 \times 2$
    $7 \times 9 = \square$
    $\square = 9 \times 3$

- Show pupils the multiplication sentences below. Ask: *What do you notice?* Enourage pupils to spot patterns and talk about patterns within and across the multiplication tables. Ask: *How might this help you when you are calculating?*

$1 \times 10 = 10$	$1 \times 9 = 9$
$2 \times 10 = 20$	$2 \times 9 = 18$
$3 \times 10 = 30$	$3 \times 9 = 27$
$4 \times 10 = 40$	$4 \times 9 = 36$
$5 \times 10 = 50$	$5 \times 9 = 45$
$6 \times 10 = 60$	$6 \times 9 = 54$
$7 \times 10 = 70$	$7 \times 9 = 63$
$8 \times 10 = 80$	$8 \times 9 = 72$
$9 \times 10 = 90$	$9 \times 9 = 81$
$10 \times 10 = 100$	$10 \times 9 = 90$

Same-day enrichment

- Write on the board: 'The sum of the digits of a multiple of 9 is always 9.'

- Get pupils to work together to discuss the vocabulary in the statement and understand it. Can they say whether the statement is always, sometimes or never true?

Question 2

> **2** What is the greatest number you can write in each box?
>
> (a) $2 \times \boxed{} < 19$ (b) $10 \times \boxed{} < 25$ (c) $\boxed{} \times 9 < 23$
>
> (d) $5 \times \boxed{} < 24$ (e) $\boxed{} \times 5 < 54$ (f) $\boxed{} \times 6 < 36$
>
> (g) $8 \times \boxed{} < 20$ (h) $\boxed{} \times 9 < 55$ (i) $\boxed{} \times 9 < 100$

What learning will pupils have achieved at the conclusion of Question 2?

- Pupils will have used their reasoning skills to derive known multiplication facts.

Activities for whole-class instruction

- Write the facts below on the board. Ask pupils to work with a partner to decide if they are true or false. How do they know?

    $5 \times 6 < 32$

    $2 \times 9 < 16$

    $10 \times 4 < 63$

- Ask pupils to explain how they know.

- Write $9 \times \square < 30$ on the board. Ask: *What could the missing number be? What is the greatest number that could fill the missing space? How do you know?*

- Repeat for $\square \times 9 < 71$

    Look out for … pupils calculating the answer by counting in multiples of 9 from 0. This will help pupils to find the missing number, but it is more efficient for pupils to start from a known fact, for example $5 \times 9 = 45$.

- Pupils are now ready to complete Question 2 in the Practice Book.

Same-day intervention

- Start by skip-counting in nines using a counting stick; this should help pupils to have the multiples of 9 in their mind. Can pupils use interlocking cubes to make 10 'sticks' of 9?

- Write $\square \times 9 < 46$ on the board. Ask pupils to find the greatest number that will fill the gap. Encourage pupils to use the sticks of '9' to help them. Agree that 5 sticks of 10 would be 50 so 5 sticks of 9 will be less than 50. Ask: *Could the answer be 6 sticks of 9?* Can pupils 'see' that 6 sticks of 9 will be 6 less than 60, which is more than 50? So the answer cannot be 4×9.

- Repeat with $9 \times \square < 30$. Agree that 3 sticks of 10 would be 30 so 3 sticks of 9 will be less than 30. Ask: *Could the answer be 4 sticks of 9?* Can pupils 'see' that 4 sticks of 9 will be 4 less than 40, which is more than 30? So the answer cannot be 4×9.

Same-day enrichment

- Ask pupils to find all possibilities to fill each space.

    $7 \times 9 > \square \times 6$

    $\square \times 9 < 12 \times 4$

- Ask pupils to write their own for a partner to complete.

Question 3

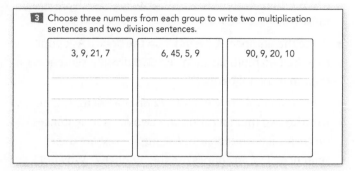

3 Choose three numbers from each group to write two multiplication sentences and two division sentences.

3, 9, 21, 7	6, 45, 5, 9	90, 9, 20, 10

What learning will pupils have achieved at the conclusion of Question 3?

- Using their knowledge of inverse operations, pupils will have practised writing multiplication and division sentences.

Activities for whole-class instruction

(i) A 'function machine' is a diagram that takes an input, applies a rule, such as multiplication, and delivers the answer as an output. Pupils will need to determine the input, the output or whatever operation takes place inside the machine. Pupils should be familiar with function machines from Key Stage 1.

- Draw a simple function machine on the board. Ask pupils to calculate what went into the machine.

- Repeat for outputs of 18 and 72.
- Repeat for ×6, ÷3 and so on.
- Show pupils the table:

input	output
25	5
60	12
45	9

Ask: *What is the rule? How do you know?*

- On a piece of paper, pupils should write a rule (a symbol followed by 3, 6 or 9) and draw a table like the one above, including three inputs and outputs that follow their rule.

Ask a pupil to come to the front of the class and complete a new input/output table with their values. The class must work out the rule.

- Write the numbers 9, 27 and 3 on the board. Ask pupils to write two multiplication and two division sentences on their mini whiteboards.

- Write the numbers 8, 9, 7 and 72 on the board. Ask pupils to write two multiplication and two division sentences on their mini whiteboards. Ask pupils to share their sentences. Did any pupils use the number 7? Why not?

- Pupils are now ready to complete Question 3 in the Practice Book.

Same-day intervention

- Continue to use the concept of a function machine with pupils but instead of using numerals, ask pupils to represent the values using interlocking cubes. Use the diagram below. Ask: *What happens to the cubes to calculate what comes out of the function machine?* (÷5)

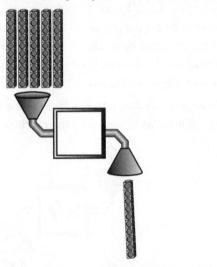

Same-day enrichment

- Ask pupils to solve the following:

  I think of a number. I multiply my number by 6. The product is 54. What was my number?

  I think of a number. I divide by number by 9. My answer is a whole number less than 6. What could my number be?

- When pupils have finished, ask them to write their own for a partner to solve.

Question 4

4 Application problems.

(a) A rabbit collected 54 carrots and gave all of them equally to 6 baby rabbits. How many carrots did each baby rabbit get?

Answer: _____

(b) Suraj, Max and Erin went to Anya's home to celebrate her birthday. Anya's family prepared food including chocolate, cake and fruit.

The four children shared 36 chocolate bars equally. How many bars did each child get?

Answer: _____

(c) There were 9 bananas in a bunch. There were 3 bunches. How many bananas were there altogether?

Answer: _____

(d) There were 9 pieces of cake in each box. How many boxes were there for 18 pieces of cake?

Answer: _____

What learning will pupils have achieved at the conclusion of Question 4?

- Pupils will have practised applying their knowledge of multiplying and dividing by 9 to solve problems.

Activities for whole-class instruction

- Pupils should be encouraged to visualise images when solving problems. It may be appropriate for some pupils to use manipulatives or draw pictures to help them make sense of problems.

- Share the following problem with the pupils: *Ribbon was sold in 9 cm lengths. Kate needed 54 cm. How many pieces of ribbon did she need to buy?*

- Give pairs of pupils a 9 cm piece of ribbon (or paper). Ask: *What would you do to solve the problem? Can you write this as a number sentence?*

- Show pupils this number line. Ask: *Does this represent a solution to the problem? Was your answer correct?*

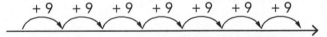

- Show pupils the image below:

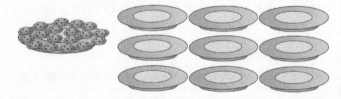

- Ask: *What do you see in the picture?*

- Ask pupils to talk to their partner about what the problem might be. Can they write a problem that involves these biscuits and plates on their whiteboards?

- Allow time for pupils to construct a problem. Agree that it will be something like: 'Jill had baked 18 cookies. She was sharing them equally between 9 people. How many cookies did each person get?'

- This image prompts the process of sharing between the plates or people so the number sentences would be $18 \div 9 = 2$. Model how to solve the problem by sharing the cookies onto the plates. Ask pupils to record this as a division sentence. ($18 \div 9 = 2$)

- Repeat for the following image, giving pupils time to talk about what the image represents first. ($5 \times 9 = 45$)

- Pupils are now ready to complete Question 4 in the Practice Book.

Same-day intervention

- Give pupils **Resource 3.2.4** Problem matching and solving.

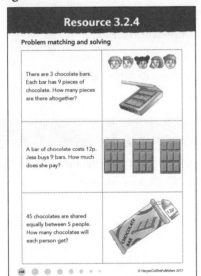

- Pupils should match the problem with the correct image. When they have matched them up, they use the image to help them write a number sentence to solve the problem. If needed, pupils can use counters or other equipment to help them calculate the answer, but it is important that pupils are not using the equipment to count but to represent the structure of the multiplication or division problem.

- Ask pupils to write their own multiplication and division problems for the number sentences below:

$4 \times 6 = 24$

$30 \div 3 = 10$

$45 \div 9 = 5$

- Ask them to think of their own for a partner to solve. They may want to challenge themselves and their partner further by writing a multi-step problem.

Challenge and extension questions

Question 5

5 Complete the number patterns.

(a)

9 5 18 10 27 15 ☐ ☐

(b)

1 3 7 15 31 ☐ ☐

It would be helpful for pupils to complete the number patterns in pairs. This will allow them to talk about what they notice. Pupils will have to apply their knowledge of other known multiplication facts. Can pupils prove that they have continued the pattern correctly? Ask pupils to write their own patterns for a partner to solve.

Question 6

6 (a) It took Molly 18 seconds to run from the first floor to the third floor. How many seconds did it take Molly to run from one floor to the next floor on average?

Answer: _____

(b) How many seconds would it take Molly to run from the first floor to the eighth floor if she could keep up the same speed?

Answer: _____

Pupils are required to apply their knowledge of multiplication and division to solve the problem. As this problem involves time, pupils may find it more difficult to visualise or draw an image to help them. They may benefit from using a number line as a tool to support them if needed.

Unit 2.5
Relationships between multiplications of 3, 6 and 9

Conceptual context

Pupils should now be fluent at recalling multiplication and division facts for 3, 6 and 9. They will have noticed connections between the 3, 6 and 9 multiplication tables and this unit will focus on those relationships so that pupils understand them more deeply. This will equip them to think flexibly, 'navigating' within their knowledge of multiplication facts and reasoning about how one fact helps to derive another: 'I know that $3 \times 4 = 12$ so I also know that $6 \times 4 = 24$ because when the number of fours is doubled, the product will also double.'

Learning pupils will have achieved at the end of the unit

- Using arrays, pupils will have explored the similarities in structure of the 3, 6 and 9 multiplication tables (Q1)
- Pupils will have explored equivalences in the 3, 6 and 9 multiplication tables (Q2)
- When interpreting mathematical vocabulary, pupils will have calculated multiplication and division questions for the 3, 6 and 9 times tables (Q3)
- Pupils will have practised applying their knowledge of relationships between multiplications of 3, 6 and 9 to solve problems (Q4)

Resources

counting stick; mini whiteboards; interlocking cubes; counters; plastic plant pots; beans/seeds; hoops and bean bags; **Resource 3.2.5** Scoring points

Vocabulary

multiplication, multiply, multiple, times, factor, product, divide, division, divided by, inverse

Question 1

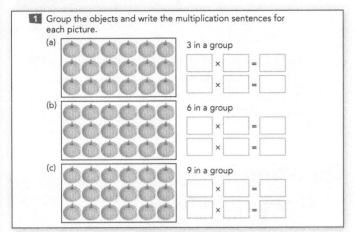

> **1** Group the objects and write the multiplication sentences for each picture.
> (a) 3 in a group
> ☐ × ☐ = ☐
> ☐ × ☐ = ☐
> (b) 6 in a group
> ☐ × ☐ = ☐
> ☐ × ☐ = ☐
> (c) 9 in a group
> ☐ × ☐ = ☐
> ☐ × ☐ = ☐

What learning will pupils have achieved at the conclusion of Question 1?

- Using arrays, pupils will have explored the similarities in structure of the 3, 6 and 9 multiplication tables.

Activities for whole-class instruction

- Start by skip-counting with pupils in threes, sixes and nines using a counting stick.

- Write the number line on the board. Ask pupils to come up to the board and complete the missing numbers. Ask: *What do you notice? Are any of the multiples of 3, 6 or 9 the same? Why?*

3	6		12			21	24		30	

	12			30		42				66	

9		27	36				72		90	

- Give pupils 18 counters. Ask: *How many groups of 3 can you make?* Ask pupils how many groups of 6 they can make using the same counters. Ask: *What did you do to make the groups of 6? Did you count six counters each time?* (Pupils should have combined 2 groups of 3.) Now ask pupils to make groups of 9. Ask: *How many groups of 9 are there? What can you tell me about the number 18?* (It is a multiple of 3, 6 and 9.)

- Together, write the multiplication sentences on the board. Encourage pupils to apply their knowledge of the commutative property of multiplication.

- Pupils are now ready to complete Question 1 in the Practice Book.

- Repeat the activity with the following array:

- Ask: *How many groups of 3 are there? How many groups of 6 are there? How many groups of 9 are there?*

Same-day enrichment

- Ask: *How many multiplication sentences in the 3, 6 and 9 multiplication tables can you write with a product of 36?*

Question 2

> **2** Find the missing numbers.
> (a) 18 = 3 × ☐ = 6 × ☐ = 9 × ☐
> (b) 36 = ☐ × 4 = ☐ × 6 = 3 × ☐
> (c) 54 = ☐ × 9 = ☐ × 6 = ☐ × 3

What learning will pupils have achieved at the conclusion of Question 2?

- Pupils will have explored equivalences in the 3, 6 and 9 multiplication tables.

Activities for whole-class instruction

- Write 3 × 8 and 6 × 4 on the board. Ask: *What do you notice?* (The product for both multiplications is 24.) Ask: *How could you record this as a number sentence?*

- Write 24 = 3 × 8 = 6 × 4. Remind pupils that this is called an equivalent calculation – all calculations have the same value – the equals symbols show us this.

 24 has the same value as 3 times 8, which has the same value as 6 times 4.

- Show pupils the array below. Ask: *What does the array represent?* (3 × 10 or 10 × 3)

- Model how th e array could be restructured to represent 6 × 5 or 5 × 6. Discuss.

- Ask pupils to complete the following:

30 = ☐ × 3 = ☐ × 6.

  30 has the same value as 10 × 3, which has the same value as 5 × 6.

- Ask pupils to write their own.

- Pupils are now ready to complete Question 2 in the Practice Book.

Same-day intervention

- Give pupils three 3 × 18 arrays. Ask pupils what the product is. Can they count in threes?

- Ask pupils to work in pairs, can they cut two of the arrays to change them to represent ☐ × 6 = 54 and ☐ × 9 = 54.

- Pupils should eventually have the following three arrays for 54. Ask: *What can you tell me about 54?*

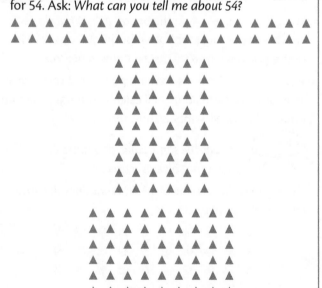

Same-day enrichment

- Present this scenario: In space, there are three alien planets. On one planet, the aliens all have 3 legs. On the second planet, the aliens all have 6 legs and on the third planet, the aliens all have 9 legs. There are the same number of legs on each planet. How many aliens could there be on each planet?

Question 3

> 3 Write a number sentence for each question.
> (a) What is the sum of adding 7 sixes together?
>
> Number sentence: _____
> (b) How many fours are there in 36?
>
> Number sentence: _____
> (c) How many threes should be subtracted from 24 so the result is 0? How many sixes?
>
> Number sentence: _____
>
> Number sentence: _____

What learning will pupils have achieved at the conclusion of Question 3?

- When interpreting mathematical vocabulary, pupils will have calculated multiplication and division questions for the 3, 6 and 9 times tables.

Activities for whole-class instruction

- Draw this bar model on the board:

3	3	3	3	3	3
18					

- Ask: *What number sentences can you write?* (3 × 6 = 18, 18 ÷ 3 = 6)

- Add the following bar to the model.

6		6		6	
3	3	3	3	3	3
18					

- Ask: *What do you notice? Use the bar to add further number sentences on your whiteboard.* (3 × 6 = 18, 18 ÷ 6 = 3)

9			9		
6		6		6	
3	3	3	3	3	3
18					

- Ask: *What do you notice? Use the bar to add further number sentences on your whiteboard.* (2 × 9 = 18, 18 ÷ 9 = 2)
- Ask pupils to draw their own bar model on their whiteboards to help them answer the following questions:

  What is the sum of adding 4 threes together?

  How many nines are there in 45?

  How many sixes should be subtracted from 36 so the result is zero?

- Pupils are now ready to complete Question 3 in the Practice Book.

Same-day intervention

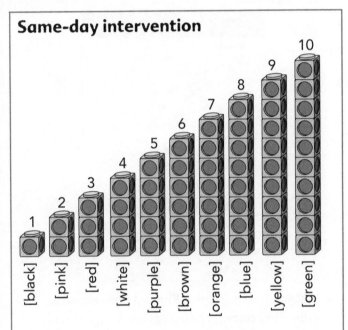

- Ask pupils to look at the image (above). Ask: *What is the value of the green cubes?* (10) *How do you know? What is the value of the brown cubes?* (6)
- Ask: *Using sticks of red cubes, can you make a bar of 15?*

  [red]

- Ask: *How many threes are there in 15? How do you know?*
- Ask pupils to make a stick of 30 brown cubes. Ask: *How many sixes should be subtracted from 30 so the result is zero?*
- Repeat, getting pupils to add and subtract groups of 3, 6 and 9. Once pupils are confident with the representation of the concrete materials, they should draw a bar to help instead.

- Present this scenario: Penny had 54p in her purse. She wanted to buy some cookies. She could buy a bag filled with 3p cookies, a bag filled with 6p cookies or a bag filled with 9p cookies. How many cookies would there be in each bag?

Question 4

> 4 Application problems.
> (a) There are 9 roses. There are twice as many tulips as roses. How many tulips are there?
>
> Answer: _____
>
> (b) If a bouquet of flowers is made up of 9 flowers, how many bouquets can be made up with these tulips and roses?
>
> Answer: _____
>
> (c) There are 6 bunches of balloons and there are 6 balloons in each bunch. The balloons are to be shared equally among 3 children. How many balloons will each child get?
>
>
>
> Answer: _____

What learning will pupils have achieved at the conclusion of Question 4?

- Pupils will have practised applying their knowledge of relationships between multiplications of 3, 6 and 9 to solve problems.

Activities for whole-class instruction

- Ask pupils to make eight towers of three interlocking cubes. Ask: *How many cubes are there altogether.* (24) *What do the cubes represent?* (8 × 3, 24 ÷ 3)
- Ask pupils what they would need to do to change the cubes so that they represent 4 × 6 or 24 ÷ 6. Allow pupils time to explore in pairs. Ask them to take the last group of three cubes and distribute them among the first three groups, then take the seventh group of three and distribute each cube among the next three groups. Ask pupils how many groups there are now. What is the group size? Can they write the multiplication and division sentence on their whiteboards?
- Pupils can then see that two groups have become 6 ones that got distributed between six groups that were left after two groups got broken up. Pupils need to really

believe that it is simply all about redistributing the same 'things' when they work with relationships between their multiplication tables.

- Some pupils should also notice that they need to increase the group size to 6. Therefore, they will reduce the number of groups by half. Ask pupils to discuss this. Ask: *Has the number of cubes changed?* (no)

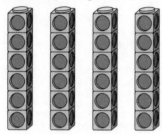

- On the board, draw eight pots of flowers, each with three flowers. Give pairs of pupils counters and ask them to represent the pots of flowers with counters. They should make eight piles or groups of eight counters. Draw a similar representation on the board. Number the groups 1–8.

- Tell pupils to take group 8 and share the counters among groups 1–3, then take group 7 and share its counters among groups 4–6. Ask: *What do you have now?* They should have:

- Say: *Now there are 6 groups of 4, instead of 8 groups of 3 because 6 × 4 = 8 × 3 = 24.*

- On the board, draw nine sets of keyrings each with three keys, like the image below.

- Tell pupils that the keys need to be changed so that there are nine keys on each keyring. How many bunches of keys will there now be? Ask pupils to explain what they did to work it out. Ask them to write a number sentence. (27 ÷ 9 = 3)

- Give pupils 36 seeds/beans and nine small plastic plant pots (counters could be used to represent the seeds). Say:

Jack has a packet of seeds that contains 36 seeds. He plants them into six plant pots; each plant pot has the same number of seeds. How many seeds are there in each plant pot? If Jack had nine plant pots, how many seeds could he put in each one if he wanted the same number in each pot?

- Pupils are now ready to complete Question 4 in the Practice Book.

Same-day intervention

- Show pupils the image from **Resource 3.2.5** Scoring points.

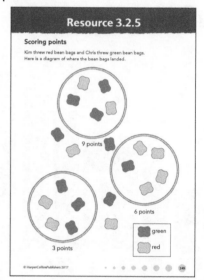

- Explain to the pupils that Kim and Chris played a game. They threw 10 bean bags into hoops to score points. The closest hoop scored them 3 points per bean bag that landed inside the hoop, the next hoop scored them 6 points per bean bag that landed inside the hoop and the final hoop scored them 9 points per bean bag that landed inside the hoop.

- Ask: *How many points did Kim score? Write the multiplication sentences. How many points did Chris score? Write the multiplication sentences.*

- Pupils can also play the game to help them understand the context and to apply their knowledge to keep their own score. It is better if pupils calculate their score at the end of the game. This will encourage pupils to use their multiplication facts rather than repeated addition.

Same-day enrichment

- Write on the board:

$6 \times 6 = 36$	$7 \times 3 = 21$
How does this help you to calculate the following?	How does this help you to calculate the following?
$60 \times 6 =$	$70 \times 3 =$
$30 \times 6 =$	$7 \times 30 =$
$34 \times 6 =$	$700 \times 3 =$

- Can pupils write any more of their own facts?

Challenge and extension question

Question 5

5 There is a bag of marbles. They can be counted in both threes and sixes without any left over. What could be the smallest number of marbles in the bag?

Answer: _____

By the end of this unit, pupils should be able to solve this problem by recalling a common multiple of both the 3 and 6 times tables.

Question 6

6 A goat is 3 times as heavy as a cat. The cat is 6 times as heavy as a squirrel. The squirrel is twice as heavy as a chick. How many times as heavy is the goat as the chick?

Answer: _____

Presenting the information below to pupils will provide some support if needed.

cat × 3 = dog

squirrel × 6 = cat

chick × 2 = squirrel

Unit 2.6
Multiplication grid

Conceptual context

In this unit, pupils will explore numbers in a multiplication grid. The structure of the grid draws upon and reinforces pupils' knowledge of commutativity. Pupils will find that, as they move through the columns, they are able to draw on known facts and the numbers of facts that they need to memorise decreases. Pupils will need opportunities to explore and understand the multiplication grid that they create.

Learning pupils will have achieved at the end of the unit

- Pupils will have explored numbers in the multiplication grid and have applied their knowledge of commutativity (Q1)
- Pupils will have practised writing related multiplication and division facts (Q2)
- When using pictorial representations, pupils will have practised writing related multiplication and division facts in different ways (Q3)

Resources

mini whiteboards; interlocking cubes; number cards; symbol cards; counters; 100 square; 1–6 dice; envelopes; **Resource 3.2.6a** Multiplication grids; **Resource 3.2.6b** Growing multiplication; **Resource 3.2.6c** Multiplication grid puzzle; **Resource 3.2.6d** Multiplication statements; **Resource 3.2.6e** Representations; **Resource 3.2.6f** Match up sentences; **Resource 3.2.6g** Missing numbers

Vocabulary

multiplication, multiply, times, factor, product, divide, division, divided by, inverse, dividend, divisor, quotient, multiplication grid, commutative,

Question 1

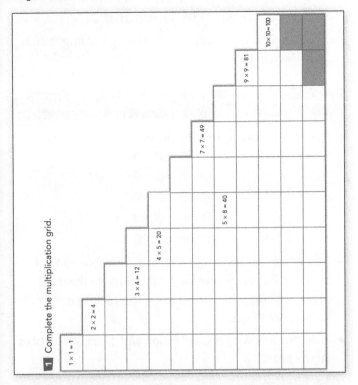

What learning will pupils have achieved at the conclusion of Question 1?

- Pupils will have explored numbers in the multiplication grid and applied their knowledge of commutativity.

Activities for whole-class instruction

- Give pupils a copy of **Resource 3.2.6a** Multiplication grids, which shows two different multiplication grids.

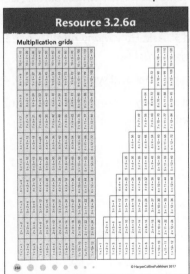

- Ask: *What's the same and what's different?* It is important pupils not only discuss the way the tables are structured but also recognise that the commutative property of multiplications means that the same multiplication facts do not appear twice in the second table.

- Ask pupils to talk about patterns within and between the times tables.
- Show pupils **Resource 3.2.6b** Growing multiplication.

Resource 3.2.6b
Growing multiplication

1	2	3	4	5	6	7	8	9	10
2	4	6	8	10	12	14	16	18	20
3	6	9	12	15	18	21	24	27	30
4	8	12	16	20	24	28	32	36	40
5	10	15	20	25	30	35	40	45	50
6	12	18	24	30	36	42	48	54	60
7	14	21	28	35	42	49	56	63	70
8	16	24	32	40	48	56	64	72	80
9	18	27	36	45	54	63	72	81	90
10	20	30	40	50	60	70	80	90	100

© HarperCollinsPublishers 2017   251

What do they notice? Ask them to look at the size and the shape of the boxes. (Pupils should recognise that the product within the table sits in a cell that represents the value of that number as an array.)

- Ask pupils to find numbers that are repeated. What do they notice?
- Ask pupils to find the numbers 4, 9, 16, 25, 36, 49, 64, 81 and 100. What do they notice?
- Pupils are now ready to complete Question 1 in the Practice Book.

Same-day intervention

- Give pupils a copy of **Resource sheet 3.2.6c** Multiplication grid puzzle. In order to complete the multiplication grid, pupils should be using known facts to help them. Pupils should cut out the statements from **Resource 3.2.6d** Multiplication statements. Ask them to use the statements to help them to complete the multiplication table.

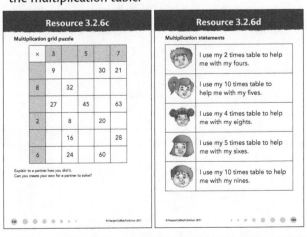

Same-day enrichment

- Give pupils **Resource 3.2.6c** Multiplication grid puzzle. Can they complete the multiplication table without support?

Resource 3.2.6c

Multiplication grid puzzle

×	3		5		7
	9			30	21
8		32			
	27		45		63
2		8	20		
		16			28
6		24	60		

Explain to a partner how you did it.
Can you create your own for a partner to solve?

252 © HarperCollinsPublishers 2017

Question 2

2 Complete the multiplication facts. Then write multiplication and division sentences.

(a) Four times seven is _____.

_____   _____

(b) Five times eight is _____.

_____   _____

(c) Six times eleven is _____.

_____   _____

(d) Three times _____ is twelve.

_____   _____

(e) Four times _____ is twenty-four.

_____   _____

(f) _____ times nine is thirty-six.

_____   _____

What learning will pupils have achieved at the conclusion of Question 2?

- Pupils will have practised writing related multiplication and division facts.

Activities for whole-class instruction

- Draw the following array on the board:

- Ask: *What does the array represent? Write four number sentences on your whiteboard.* Ask a pupil to the front of the class with their board to explain their four number sentences.

- Write the following multiplication and division sentences on the board:

$3 × 12 = \square$

$12 × 3 = \square$

$12 ÷ 3 = \square$

$\square ÷ 12 = 3$

- Using counters, can pupils represent this as an array? Ask: *What is the product?* (24)

- Using the same number of counters, can pupils make a different array? ($6 × 6$, $2 × 18$, $4 × 9$) Using their new array, ask them to record two multiplication sentences and two division sentences on their whiteboards.

- Give pairs of pupils the number and symbol cards 6, 7, 42, ×, ÷ and =. Ask: *How many number sentences can you make with the cards?* (4)

- Repeat with number and symbol cards 5, 9, 45, ×, ÷ and =.

- Pupils are now ready to complete Question 2 in the Practice Book.

Same-day intervention

- Give pupils a pictorial representation to help them identify the relationship between the multiplication and division facts; see below for some examples. Ask pupils to use the arrays to help them write two multiplication facts and two division facts.

Same-day enrichment

- Give pairs of pupils four dice and two envelopes. Player 1 rolls two dice and adds them together for the first number. Write this number on a piece of paper and place in envelope A. Player 2 then rolls their two dice and adds them together for the second number. Write this number on a piece of paper and place in envelope B. On their whiteboards, pupils must multiply both numbers together and then find the two inverse calculations.

- Can they refer to the numbers as letters instead? (A × B = ?; B × A = ?; ? ÷ A = B; ? ÷ B = A) Ask: *Do you enjoy describing the sentences using letters? Why? Why not?*

Question 3

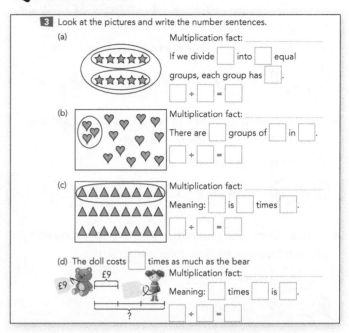

What learning will pupils have achieved at the conclusion of Question 3?

- When using pictorial representations, pupils will have practised writing related multiplication and division facts in different ways.

Activities for whole-class instruction

- Show pupils **Resource 3.2.6e** Representations. What do the pupils think the images represent? Give them time for discussion in pairs or small groups.

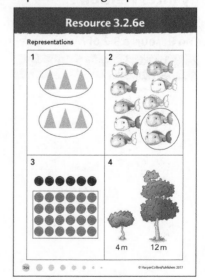

- Share pupils' responses. Encourage them to explain their thinking and give reasons for their answers.

● Write the sentence stems under each image. Ask pupils to talk to their partner. What numbers would go into the missing spaces to complete the sentence?

1. Dividing ☐ into ☐ equal groups, each group has ☐.

 Dividing 6 into 2 equal groups, each group has 3.

● Ask pupils, with a partner, to write a multiplication and division sentence on their mini whiteboards.

2. There are ☐ groups of ☐.

 There are 5 groups of 2.

● With a partner, ask pupils to write a multiplication and division sentence on their mini whiteboards.

3. ☐ is ☐ times ☐

 30 is 5 times 6.

● With a partner, ask pupils to write a multiplication and division sentence on their mini whiteboards.

4. ☐ times ☐ Is ☐

 4 times 3 is 12.

● Ask pupils, with a partner, to write a multiplication and division sentence on their mini whiteboards.

● Pupils are now ready to complete Question 3 in the Practice Book.

Same-day intervention

● Ask pupils to match the images with the correct sentences on **Resource 3.2.6f** Match up sentences.

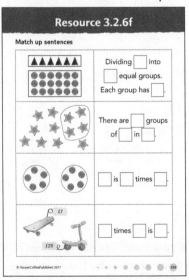

Same-day enrichment

● Show pupils the following image:

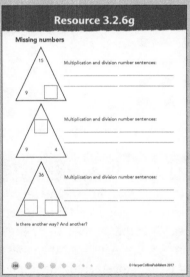

$$9 \times 5 = 45$$
$$5 \times 9 = 45$$
$$45 \div 5 = 9$$
$$45 \div 5 = 9$$

● Ask pupils what the relationship might be between the numbers.

● Give pupils **Resource 3.2.6g** Missing numbers. Can they complete the missing numbers and then write the multiplication and division sentences?

Challenge and extension question

Question 4

4 Write a suitable operation symbol in each ◯ so that the equation is correct.

(a) 2 ◯ 2 ◯ 2 ◯ 2 = 0   (b) 2 ◯ 2 ◯ 2 ◯ 2 = 1

(c) 2 ◯ 2 ◯ 2 ◯ 2 = 2   (d) 2 ◯ 2 ◯ 2 ◯ 2 = 3

This activity requires pupils to have a good understanding of all four operations and inverses as it will be necessary to 'do' and 'undo'.

The equations have more than one solution. Can pupils find all possibilities? How do they know they have them all?

Unit 2.7
Posing multiplication and division questions (1)

Conceptual context

In this unit and the next, pupils will need to apply their knowledge as they write multiplication and division number sentences. This unit focuses on questions presented in words, adding an extra layer of difficulty for pupils. Practice with word problems will increase pupils' depth of understanding.

Learning pupils will have achieved at the end of the unit

- Pupils will have used reasoning to interpret images, identifying the operation required, in order to write number sentences (Q1)
- Pupils will have practised reading and understanding problems written in words to write number sentences (Q2)

Resources

mini whiteboards; interlocking cubes; counters; **Resource 3.2.7a** Match up problems; **Resource 3.2.7b** What's the problem?

Vocabulary

multiplication, multiply, times, factor, product, divide, division, divided by, inverse, dividend, divisor, quotient

Question 1

1. Look at each picture. Pose a question and then write a number sentence. The first one has been done for you.

(a) How many sweets are there in total?

Number sentence:
5 × 4 = 20 (sweets)

(b) Question: _____

Number sentence: _____

(c) Question: _____

Number sentence: _____

(d) Question: _____

Number sentence: _____

What learning will pupils have achieved at the conclusion of Question 1?

- Pupils will have used reasoning to interpret images, identifying the operation required, in order to write number sentences.

Activities for whole-class instruction

- Give pupils three groups of two interlocking cubes. Tell them that you can no longer find the question that you had intended to ask. Ask: *What could the multiplication or division question have been?*

- Take pupils responses and record them on the board. For example: What is 6 divided by 3? How many cubes are there altogether? How many groups of 2 are there in 6?

- Ask pupils to write a number sentence for each question on the board. Are they the same or different?

- Draw the image below on the board. What do pupils think the image represents? Ask them to discuss with a partner.

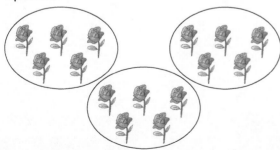

- Ask pupils which question they think relates to the image.

1. *How many roses are there in total?*

2. *If 15 roses were shared equally between 3 people, how many roses would each person get?*

- Pupils should realise that both questions could be correct. Ask: *Which question requires a division number sentence and which question requires a multiplication number sentence? Write them on your whiteboard.*

- Show pupils the image below or draw it on the board.

- In pairs, ask Pupil A to write a question and Pupil B to read the question and write the number sentence. Can Pupil B now write a different question using the image?

- Pupils are now ready to complete Question 1 in the Practice Book.

Same-day intervention

- Using counters, ask pupils to make one group of four, another group of four and another group of four. Ask: *How many groups of 4 are there?*

 There are 3 groups of 4.

- Ask: *Can you write this as a multiplication sentence?* (3 × 4 = 12).

- Ask pupils to take 12 counters each. Ask: *Can you make three equal groups? How many counters are in each group?*

 Dividing 12 into 3 equal groups, means there are 4 in each group.

- Ask: *Can you write this as a division sentence?* (12 ÷ 3 = 4)

- Repeat for 5 × 6 = 30 and 30 ÷ 5 = 6.

Same-day enrichment

- Say:
  - *There are 30 people travelling to the zoo by car. Each car can take five people. Will five cars be enough?*
  - *At the zoo, there are six monkeys in each cage. There are four cages. How many monkeys are there altogether?*
  - *During a picnic, 27 cupcakes are shared equally between three plates. How many cakes will go on each plate?*
- Ask: *How could you solve these three problems using:*
  - *objects?*
  - *pictures?*
  - *numbers?*
  - *words?*
  - *other ways?*

Question 2

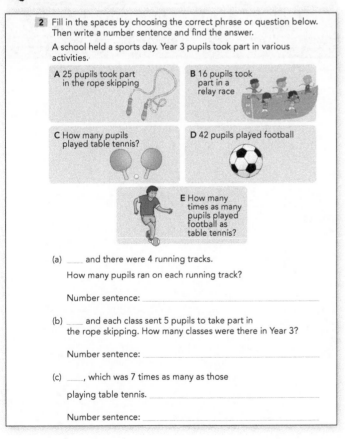

2 Fill in the spaces by choosing the correct phrase or question below. Then write a number sentence and find the answer.

A school held a sports day. Year 3 pupils took part in various activities.

A 25 pupils took part in the rope skipping

B 16 pupils took part in a relay race

C How many pupils played table tennis?

D 42 pupils played football

E How many times as many pupils played football as table tennis?

(a) _____ and there were 4 running tracks.
How many pupils ran on each running track?

Number sentence: _____

(b) _____ and each class sent 5 pupils to take part in the rope skipping. How many classes were there in Year 3?

Number sentence: _____

(c) _____, which was 7 times as many as those playing table tennis. _____

Number sentence: _____

What learning will pupils have achieved at the conclusion of Question 2?

- Pupils will have practised reading and understanding problems written in words to write number sentences.

Activities for whole-class instruction

- Write the following word problem on the board: In a cross-country competition, 5 children each ran 6 laps of the field. What is the total number of laps completed by all the children?
- Read the problem together. Ask pupils, in pairs, to identify what they think is the important information to help them solve the problem. Share pupils' responses with the class and ask them to explain why.
- Ask pupils to draw an image to represent the problem. Then ask them to write a number sentence to answer the question. ($5 \times 6 = 30$)
- Ask: *If each lap of the field was 12 m long, how far did each child run?* Read the problem together. Ask pupils, in pairs, to identify what they think is the important information to help them solve the problem.
- Ask pupils to draw an image to represent the problem. Share the image below. Ask: *How could this help you?*

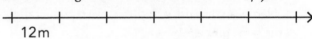

12 m

- Ask pupils to write a number sentence to answer the question. ($12 \times 6 = 72$)
- Pupils are now ready to complete Question 2 in the Practice Book.

Same-day intervention

● Give pupils **Resource 3.2.7a** Match up problems.
  Ask them to use counters to represent the problem
  and help them to match the problem to the correct
  number sentence.

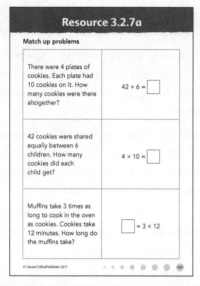

Resource 3.2.7a

Match up problems

There were 4 plates of cookies. Each plate had 10 cookies on it. How many cookies were there altogether?	42 ÷ 6 = ☐
42 cookies were shared equally between 6 children. How many cookies did each child get?	4 × 10 = ☐
Muffins take 3 times as long to cook in the oven as cookies. Cookies take 12 minutes. How long do the muffins take?	☐ = 3 × 12

© HarperCollinsPublishers 2017   257

Same-day enrichment

● Give pupils the images on **Resource 3.2.7b** What's the
  problem? Ask them to work with a partner to write a
  problem. Can they draw an alternative representation
  of the same problem?

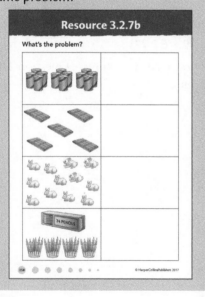

Resource 3.2.7b

What's the problem?

258   © HarperCollinsPublishers 2017

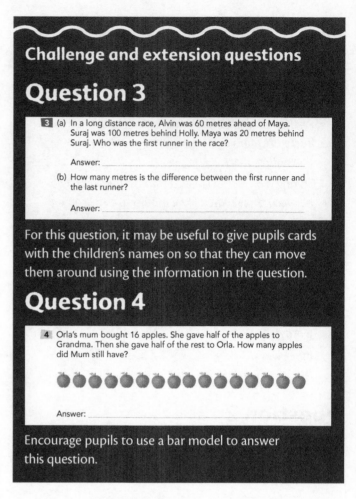

Challenge and extension questions

Question 3

3 (a) In a long distance race, Alvin was 60 metres ahead of Maya.
Suraj was 100 metres behind Holly. Maya was 20 metres behind
Suraj. Who was the first runner in the race?

Answer: _____

(b) How many metres is the difference between the first runner and
the last runner?

Answer: _____

For this question, it may be useful to give pupils cards
with the children's names on so that they can move
them around using the information in the question.

Question 4

4 Orla's mum bought 16 apples. She gave half of the apples to
Grandma. Then she gave half of the rest to Orla. How many apples
did Mum still have?

Answer: _____

Encourage pupils to use a bar model to answer
this question.

Unit 2.8
Posing multiplication and division questions (2)

Conceptual context

In this unit, pupils will continue to develop their understanding of the conditions in which multiplicative situations arise. These are circumstances where, in real-life, conditions are linked within a context in such a way that means something becomes multiplied or grouped or shared, scaled up or scaled down for a reason determined by the context.

Pupils' understanding of situations and conditions that give rise to reasons to multiply and divide can only develop when they are exposed to numerous opportunities to engage with such situations. This unit will therefore provide further practice with real-life problems involving multiplication and division so that pupils deepen their conceptual knowledge about how we use these mathematical processes every day in a variety of ways.

Learning pupils will have achieved at the end of the unit

- When presented with information in real-life contexts, pupils can recognise situations that require multiplication and division, can identify appropriate operations and construct number sentences (Q1)
- Pupils will have applied their knowledge of known multiplication and division facts to solve problems (Q2)

Resources

mini whiteboards; interlocking cubes and counters; **Resource 3.2.8a** Statement pairs; **Resource 3.2.8b** Linked facts

Vocabulary

multiplication, multiply, times, factor, product, divide, division, divided by, inverse, dividend, divisor, quotient

Question 1

1 Read the text below. Some facts can be linked to write questions that can be solved using multiplication and division. The first one has been done for you.

- These are the flowers in Avni's garden.
- There are 10 pots of roses and each pot contains 4 roses.
- There are 36 lilies.
- There are 9 pots of tulips with each pot containing 2 tulips, and there are 72 orchids planted in 9 pots.
- There are also 6 empty pots.

(a) Linked fact(s): _There are 10 pots of roses and each pot contains 4 roses._

Question: _How many roses are there in total?_

Number sentence: _10 × 4 = 40_

(b) Linked fact(s): _____

Question: _____

Number sentence: _____

(c) Linked fact(s): _____

Question: _____

Number sentence: _____

What learning will pupils have achieved at the conclusion of Question 1?

- When presented with information in real-life contexts, pupils can recognise situations that require multiplication and division, can identify appropriate operations and construct number sentences.

Activities for whole-class instruction

- Pupils should cut out the statements from **Resource 3.2.8a** Statement pairs.

Resource 3.2.8a

Statement pairs

There are 5 cars.	There are 7 lamp-posts on each street.
There are 90 windows altogether.	Each house has 10 windows.
There are 4 streets.	There are 9 houses.
The cars each have 4 wheels.	There are 49 chairs.
There are 7 chairs around each table.	There are 3 pots of flowers outside each house.

© HarperCollinsPublishers 2017

- Ask: *Which pairs of statements contain related information? Put the pairs of linked facts together.* (Pupils should have five pairs of linked facts.) Ask: *Which statements could you ask multiplication questions about? Are there some that you could turn into good division questions?* Discuss pupils' answers.

- On the board write: There are 10 bags of large balls, each bag contains 5 balls. Ask: *What question could you write using this information?* Pupils should discuss in pairs. Share responses.

- Write the question on the board: *How many balls are there in total?* Ask whether multiplication or division is needed to answer this question. Why? Ask pupils to write a number sentence on their mini whiteboards to help them answer the question using the information.

- Ask pupils to choose another pair of linked facts and write a question and a number sentence. Select pupils to come to the front of the class to share the linked facts they chose and how they used this information to write a question and number sentence.

- Pupils should make up some linked facts of their own and write them onto a blank row for a partner to solve.

(i) Problems are not always clearly either multiplication or division. The two processes are very closely linked and therefore pupils may think they were doing one when they are actually doing the other. This is to be expected and should not be treated as an error but welcomed as an opportunity to reinforce the relationship between multiplication and division.

- Pupils are now ready to complete Question 1 in the Practice Book.

Same-day intervention

- Ask pupils to complete **Resource 3.2.8b** Linked facts.

Resource 3.2.8b

Linked facts

There is a lot of sports equipment on the school field.	There are 48 bibs of assorted colours.	
There are 10 bags of large balls, each bag contains 5 balls.	There are 56 small balls shared equally into 8 bags.	
There are 3 tubs of hockey sticks. Each tub contains 12 hockey sticks.	There are 6 empty bib bags.	

© HarperCollinsPublishers 2017

Same-day enrichment

- Write these questions on the board:
  - In a sports lesson, 45 children are asked to get into groups of 5. How many groups of children will there be?
  - In a sports lesson, children are in 9 groups of 5. How many children are there altogether?
- Ask: *What's the same? What's different?*
- Show the following image or draw it on the board. Ask the pupils to look at it.

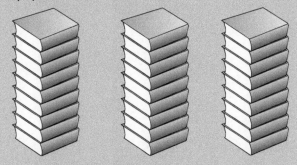

- Can pupils write a multiplication word problem and a division word problem?

Question 2

> 2 Application problems.
>
> (a) Ethan has made 10 origami cranes. Ben, Sofia and Abena have each made as many origami cranes as Ethan. How many origami cranes have they made altogether?
>
> Answer: _____
>
> (b) 9 metres was cut from a ribbon. It is now 5 times as long as the piece that was cut off. How long is the remaining piece?
>
> Answer: _____
>
> How long was the original ribbon?
>
> Answer: _____
>
> (c) May and her parents visited the cinema over the weekend. The ticket price was £10 per person. How much did they pay?
>
> Answer: _____
>
> (d) There were 6 birds in each tree. How many birds were there in 5 trees?
>
> Answer: _____
>
> After 18 birds flew away, how many birds were left?
>
> Answer: _____

What learning will pupils have achieved at the conclusion of Question 2?

- Pupils will have applied their knowledge of known multiplication and division facts to solve problems.

Activities for whole-class instruction

- Write the following problem on the board: *A clown was holding 32 balloons. He shared the balloons equally amongst 8 children. How many balloons did they have each?*
- Give pupils counters. Tell them that each counter represents 1 balloon. Ask them to use the counters to represent the problem.
- Ask: *How many groups of balloons do you have? How many balloons are there in each group?*

 Dividing 32 into 8 equal groups, means that each group has 4.

- Ask pupils to write a number sentence. ($32 \div 8 = 4$)
- Tell pupils that some balloons blew away. The balloons could still be shared equally between the 8 children. Ask: *How many balloons might have blown away?* In mixed ability pairs, ask pupils to use their counters to explore the possibilities. (8, 16 and 24)
- Ask pupils to share their thinking with the rest of the class. What do they do to work it out? How do they know they have all possibilities? Say: *Convince me you are correct.*
- Pupils are now ready to complete Question 2 in the Practice Book.

Same-day intervention

- On the board write: There were 8 bowls of strawberries. Each bowl had 8 strawberries. How many strawberries are there altogether?
- Give pupils cubes to represent the strawberries. Ask them to use the cubes to solve the problem. Ask pupils to write the number sentence.
- On the board write: There are 24 books on a book shelf. There are 4 shelves. How many books are there on each shelf?
- Give pupils cubes to represent the books. Ask them to use the cubes to solve the problem. Ask pupils to write the number sentence.

Same-day enrichment

- Ask pupils to write their own multiplication and division problems for a partner to solve.

Challenge and extension questions

Questions 3 and 4

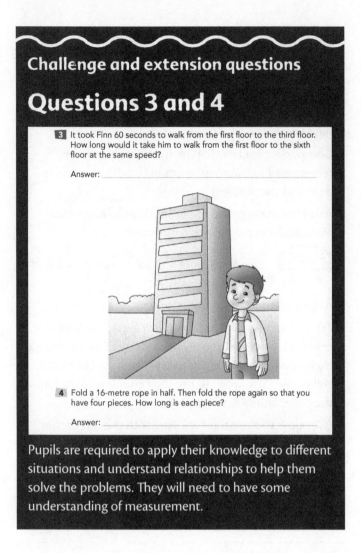

3 It took Finn 60 seconds to walk from the first floor to the third floor. How long would it take him to walk from the first floor to the sixth floor at the same speed?

Answer: _____

4 Fold a 16-metre rope in half. Then fold the rope again so that you have four pieces. How long is each piece?

Answer: _____

Pupils are required to apply their knowledge to different situations and understand relationships to help them solve the problems. They will need to have some understanding of measurement.

Unit 2.9
Use multiplication and addition to express a number

Conceptual context

In this unit, pupils will use multiplication and addition to express a number. They should now have a secure understanding of multiplication and addition facts and should start exploring relationships between the two operations. It is important that pupils are exposed to the structure of the number through images so that they can 'see' the multiplication and addition within the number. There are lots of opportunities for pupils to apply and deepen their existing knowledge within this unit and it relies upon their rapid recall of multiplication facts.

Learning pupils will have achieved at the end of the unit

- Pupils will have explored the structure of numbers in relation to multiplication and addition using arrays (Q1)
- Pupils will have used multiplication facts to help them conceptualise a number as the total of a multiplication and an addition (Q1, Q2)
- Pupils will have identified patterns and relationships when solving problems involving multiplication and addition (Q3)
- When drawing upon known multiplication facts, pupils will have used reasoning to make decisions in order to write calculations (Q4)

Resources

mini whiteboards; interlocking cubes; counters; 3 × 4 cup cake tray; 0–9 digit cards; number cards; **Resource 3.2.9 Multiplication and division**

Vocabulary

multiplication, multiply, times, factor, addition, add, greatest, smallest, equal to

Question 1

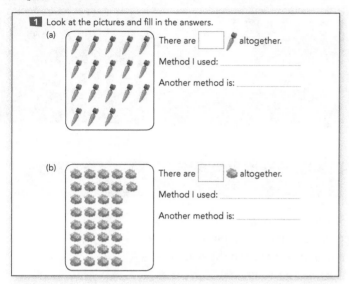

What learning will pupils have achieved at the conclusion of Question 1?

- Pupils will have explored the structure of numbers in relation to multiplication and addition using arrays.
- Pupils will have used multiplication facts to help them conceptualise a number as the total of a multiplication and an addition.

Activities for whole-class instruction

Show pupils a 3 × 4 cup cake tray. Ask: *What multiplication fact does the tray represent?* (3 × 4 = 12 and 4 × 3 = 12)

- Using cubes (or similar), fill 10 of the spaces by placing one cube in each space. Ask pupils how many cubes there are altogether. Ask: *What method did you use to work it out?*

  Look out for … pupils who count the cubes in ones. Pupils should be able to use the tray to help them identify the multiplication fact and then use addition to calculate how many cubes there are altogether.

- Draw the following images on the board. Can pupils work out how many objects there are in each image? Tell them to explain their method to a partner.

- Ask: *What number sentences are represented by these arrays?*
- For the first array, pupils should see 5 × 5 + 2 = 27. They might also see:

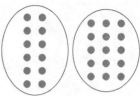

- They might express this as 2 × 6 + 3 × 5. (You should not be concerned with correcting the way pupils express this as long as they can explain that they can distinguish two arrays in the image.)

   Whenever a number can be expressed as the total of the product of a multiplication and another number added together, it can be represented as an array with an additional incomplete row or column. The position of the additional dots in the array (that is, whether they are placed to the right or below the array showing the 'multiplication' element of the number) might influence the number of rows or columns that the array is perceived to have. For example, is this 5 × 3 + 2 or 6 × 2 × + 5?

Of course, it is both. They both describe 17. In your teaching, ensure that pupils have opportunities to interpret arrays representing multiplication and addition in which the added number is not always shown in the same place so they develop useful visual imagery and flexible understanding.

- Pupils are now ready to complete Question 1 in the Practice Book.

Same-day intervention

- Using counters, make the array below.

- Cover the additional counters with a piece of paper. Ask pupils what multiplication the array represents. (2 × 6 = 12 and 6 × 2 = 12 or 4 × 4 = 16) Remove the paper and ask how many additional counters there are. (4) Ask: *How many counters are there altogether?*
- Repeat for the array below:

Same-day enrichment

- Draw these two arrays on the board. Ask: *What's the same? What's different?*

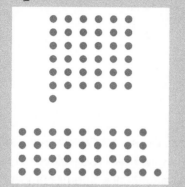

- Ask pupils to draw any other representations for 37.

Question 2

> 2 What is the greatest number you can write in each box?
>
> (a) 6 × ☐ < 45
> (b) 7 × ☐ < 40 − 5
> (c) 27 + 8 > 6 × ☐
> (d) ☐ × 9 < 32
> (e) ☐ × 5 < 4 × 7
> (f) 8 × ☐ < 20 + 27
> (g) 22 = ☐ × 5 + ☐
> (h) 63 = ☐ × 8 + ☐
> (i) 53 = ☐ × 5 + ☐
> (j) 21 = ☐ × 4 + ☐
> (k) 69 = ☐ × 7 + ☐
> (l) 25 = ☐ × 6 + ☐
> (m) 59 = ☐ × 9 + ☐
> (n) 67 = ☐ × 7 + ☐
> (o) 47 = ☐ × 4 + ☐
> (p) 79 = ☐ × 8 + ☐

What learning will pupils have achieved at the conclusion of Question 2?

- Pupils will have used multiplication facts to help them conceptualise a number as the total of a multiplication and an addition.

Activities for whole-class instruction

- Write 5 × ☐ < 28 on the board. Ask: *What do you know about multiples of 5? How many more is 28 than the nearest multiple of 5?* (3)

- On the board, model using a number line as a tool to support pupils:

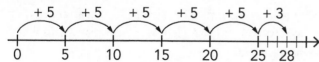

- Pupils will be familiar with multiplication being represented as repeated addition and, although we want

to encourage pupils to recall known multiplication facts for fluency, using a number line is an effective tool to model using both multiplication and addition.

- Write 3 × ☐ > 25 − 6 on the board. Ask: *What could go in the space to make the sentence correct?*

- Ask pupils to discuss in pairs the method that they could use to find the missing value. Without asking for the answer, share pupils' responses. What did pupils do first? (25-6 = 19) Ask them for all the possible answers.

- Write 4 × 6 < 15 + 8 on the board. Ask: *Is the number sentence true or false? How do you know?*

- Tell pupils that they can change one of the numbers to make the sentence correct. What could they change? Ask pupils to work in pairs and record their sentence on their mini whiteboards.

- Select several pupils with different correct answers to come to the front of the class to explain what they did to make sentence correct. Ask: *Why?*

- Write 17 = 2 × 8 + ☐ on the board. Using counters, ask pupils to represent the multiplication fact 2 × 8 as an array. Ask: *How many counters do you need to add to your array to make a total of 17?* (1)

- Write 29 = 3 × 9 + ☐ on the board. Ask pupils to use their counters to help them find the missing value.

- Write 30 = 7 × ☐ + ☐ on the board. Ask pupils to use their counters to help them find the two missing values. Ask: *Is there only one possible answer?*

- Pupils are now ready to complete Question 2 in the Practice Book.

Same-day intervention

- Give pupils a stepping stone to help them solve a problem with both multiplication and division by giving them a problem with only multiplication first. Ask pupils to use counters (or similar) to first represent the multiplication question and then add further counters to the array to complete the multiplication and addition question. Using manipulatives and presenting the question in two steps should help pupils see the structure and understand the relationship.

30 = ☐ × 6

34 = ☐ × 6 + ☐

21 = 3 × ☐

23 = 3 × ☐ + ☐

54 = 6 × ☐

59 = 6 × ☐ + ☐

Same-day enrichment

- Can pupils apply their knowledge to solve the following problems? Ask them to write a number sentence for each problem. Say:

  - Chocolates were sorted into seven boxes with ten chocolates in each box. There were three left over. How many chocolates were there altogether?

  - Katie ran six laps of a running track every day for six days. On the 7th day, she only ran four laps. How many laps did she run altogether in the week?

- Ask pupils to write their own problem for a partner to solve.

Question 3

3 Fill in the boxes.

(a) $19 = 2 \times \boxed{} + \boxed{}$ (b) $32 = 3 \times \boxed{} + \boxed{}$

(c) $19 = 3 \times \boxed{} + \boxed{}$ (d) $32 = 4 \times \boxed{} + \boxed{}$

(e) $19 = 4 \times \boxed{} + \boxed{}$ (f) $32 = 5 \times \boxed{} + \boxed{}$

(g) $19 = 5 \times \boxed{} + \boxed{}$ (h) $32 = 6 \times \boxed{} + \boxed{}$

(i) $19 = 6 \times \boxed{} + \boxed{}$ (j) $32 = 7 \times \boxed{} + \boxed{}$

(k) $19 = 7 \times \boxed{} + \boxed{}$ (l) $32 = 8 \times \boxed{} + \boxed{}$

(m) $19 = 8 \times \boxed{} + \boxed{}$ (n) $32 = 9 \times \boxed{} + \boxed{}$

(o) $19 = 9 \times \boxed{} + \boxed{}$ (p) $32 = 10 \times \boxed{} + \boxed{}$

(q) $19 = 10 \times \boxed{} + \boxed{}$ (r) $32 = 11 \times \boxed{} + \boxed{}$

(s) $19 = 11 \times \boxed{} + \boxed{}$ (t) $32 = 12 \times \boxed{} + \boxed{}$

What learning will pupils have achieved at the conclusion of Question 3?

- Pupils will have identified patterns and relationships when solving problems involving multiplication and addition.

Activities for whole-class instruction

- Give pupils 29 counters. Ask the to use their counters to make the arrays represented by the number sentences below.
- Firstly, $29 = 2 \times \boxed{} + \boxed{}$
- Ask: What did you do with the counters? Model their suggestions on the board.

- Continue with:

  $29 = 3 \times \boxed{} + \boxed{}$

  $29 = 4 \times \boxed{} + \boxed{}$

  $29 = 5 \times \boxed{} + \boxed{}$

  $29 = 6 \times \boxed{} + \boxed{}$

  $29 = 7 \times \boxed{} + \boxed{}$

  $29 = 8 \times \boxed{} + \boxed{}$

  $29 = 9 \times \boxed{} + \boxed{}$

  $29 = 10 \times \boxed{} + \boxed{}$

- Ask: What do you notice? Can any pupils say why something always needs to be added to the product of the multiplication? (29 is a prime number.)

- Ask pupils to do the same with the number 24. Before they start using their counters, can they say which number sentences will end with + 0? How do they know?

- Pupils are now ready to complete Question 3 in the Practice Book.

Same-day intervention

- Ask pupils to complete **Resource 3.2.9** Multiplication and division.

Resource 3.2.9

Multiplication and division

Same-day enrichment

- Ask pupils to solve the following problem:

  Jake was laying slabs on a patio. First, he arranged them in rows of 3 and had 2 left over. He then arranged them in rows of 5 and had 2 left over. How many slabs might he have had?

Question 4

> 4 Use the numbers below to write number sentences with both multiplication and addition.
>
> | 15 | 26 | 6 | 2 | 4 | 3 | 7 | 14 |
>
> Here is an example: 6 × 4 + 2 = 26

What learning will pupils have achieved at the conclusion of Question 4?

- When drawing upon known multiplication facts, pupils will have used reasoning to make decisions in order to write calculations.

Activities for whole-class instruction

- Give pairs of pupils a set of digit cards 0–9.

- Write the following calculations on the board one at a time. Ask pupils to discuss with their partner and hold up the correct card. Ask: *How do you know?*

  3 × 6 + ☐ = 20

  ☐ × 5 ☐ 4 = 29

  8 × 6 + ☐ = 30

  ☐ × 4 + ☐ = 22

  ☐ × ☐ + ☐ = 25

- Pupils are now ready to complete Question 4 in the Practice Book.

Same-day intervention

- Start by giving pupils the number cards 23, 5, 3 and 4. Ask: *Which number do you think is the total? How do you know?*

- Ask pupils what they know about the number 23. Some pupils will know that it is a prime number. What number would they be left with if they subtracted 5? 3? 4? Pupils might be able to identify that 19 is also a prime number. This will help them.

- Give pupils the opportunity to explore possible number sentences with the numbers 5, 4 and 3. Some pupils may adopt a trial and error approach. Ask pupils to discuss what facts they know using these numbers.

- Repeat with number cards 45, 6, 7 and 3. Say: *Three cannot be one of the factors in this number sentence. Why not?* Can pupils use this information to help them?

Same-day enrichment

- Give pupils the answer of 39. How many number sentences with both multiplication and addition can they write?

Challenge and extension question

Question 5

> 5 A box of biscuits contains fewer than 40 biscuits. It is exactly enough to share equally with 6 children. If it is shared equally with 7 children, there is 1 biscuit left over.
>
> There are ☐ biscuits in the box.

This question requires pupils to draw upon known facts and consider relationships between them. Some pupils may find it useful to use counters to represent the biscuits.

Unit 2.10
Division with a remainder

Conceptual context

Pupils have previously explored division with no remainders and have used manipulatives and other images to represent division in different contexts. Pupils should recognise that they can share equally or group equally to help calculate and that sometimes this is determined by the context of the question.

In this unit, a conceptual understanding of a remainder as a leftover will be developed through the use of manipulatives and arrays. Pupils will also explore the relationship between division and multiplication and make use of known facts to help calculate.

Learning pupils will have achieved at the end of the unit

- A conceptual understanding of division with remainders will have been developed through the use of manipulatives (Q1, Q2, Q3, Q4)
- Pupils will have been introduced to remainders and will be able to describe them as leftovers (Q1, Q2, Q3, Q4)
- Pupils will have practised making equal groups and will be able to explain why there is a remainder (Q1, Q3, Q4)
- Knowledge and application of multiplication facts will have been consolidated by relating them flexibly to division (Q1, Q2, Q3, Q4)
- Pupils will have practised representing a remainder in different ways, that is, practically, pictorially and as part of a number sentence (Q1, Q2, Q3, Q4)
- Division as equal sharing will have been explored (Q2, Q4)
- Pupils will have applied their knowledge of the relationship between addition and multiplication, using the inverse as appropriate (Q1, Q3, Q4, Q5)
- Pupils will have interpreted arrays and used them to represent division (Q1, Q2, Q3, Q4)
- Pupils will have practised applying their knowledge of division with remainders to word problems in context (Q5)

Resources

counters; cubes; one pence coins; dice; pencils; pots; number rods; **Resource 3.2.10a** Cauliflowers; **Resource 3.2.10b** Coffee shop problems; **Resource 3.2.10c** Coin arrays; **Resource 3.2.10d** Arrays; **Resource 3.2.10e** Division match up

Vocabulary

equal groups, division, multiplication, dividend, divisor, quotient, remainder, leftover

Question 1

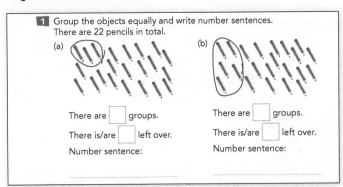

1 Group the objects equally and write number sentences.
There are 22 pencils in total.

(a)

There are ☐ groups.

There is/are ☐ left over.

Number sentence:

_____

(b)

There are ☐ groups.

There is/are ☐ left over.

Number sentence:

_____

What learning will pupils have achieved at the conclusion of Question 1?

- A conceptual understanding of division with remainders will have been developed through the use of manipulatives.
- Pupils will have been introduced to remainders and will be able to describe them as leftovers.
- Pupils will have practised making equal groups and will be able to explain why there is a remainder.
- Knowledge and application of multiplication facts will have been consolidated by relating them flexibly to division.
- Pupils will have practised representing a remainder in different ways, that is, practically, pictorially and as part of a number sentence.
- Pupils will have applied their knowledge of the relationship between addition and multiplication, using the inverse as appropriate.
- Pupils will have interpreted arrays and used them to represent division.

Activities for whole-class instruction

- Before introducing remainders here, the first activity revisits equal groups and the language used to describe division. Pupils should relate division to multiplication and so explain that $20 \div 4 = 5$ because $4 \times 5 = 20$ and that $20 \div 3 = 6$ r 2 because $6 \times 3 = 18$ and 20 is 2 more than 18.
- Give groups of 3–4 pupils a set of 20 counters or cubes. Ask them to organise the counters into groups of 5 so that all counters are used. (4 groups of 5 counters.)
- Ask: *How many counters are there altogether? How many counters in each group? How many equal groups?*

 There are 4 groups of 5 in 20.

- Ask pupils to identify the dividend (20), the divisor (5) and the quotient (4). Ask them to write the matching division sentence on their whiteboards. ($20 \div 5 = 4$)
- Now ask the pupils to arrange the 20 counters into groups of 3.
- Ask: *What is different this time? What does this tell you about 20 and the divisor 3?*
- Pupils should be able to explain that 3 does not go into 20 exactly or that 20 is not a multiple of 3. Record this as $20 \div 3 = 6$ r 2.

 20 divided by 3 is 6 remainder 2 because there are 2 left over.

- Relate this to $6 \times 3 = 18$ and that 20 is 2 more than 18.
- Ask pupils to arrange 20 counters into equal groups of a different number so there is a remainder.

Look out for … pupils who struggle to make connections to multiplication facts. Remember to make the links to multiplication explicit, perhaps counting in threes to 18 or looking at a 3×6 array.

- Show the pupils **Resource 3.2.10a** Cauliflowers.

Resource 3.2.10a

Cauliflowers

- Can pupils use a quick way to find the total number of cauliflowers without having to count them all?
- Pupils should suggest 3 groups of 5 and then add 4, or 4 groups of 4 and then add 3. If pupils suggest repeated addition sentences, encourage them to re-think as multiplicative sentences.
- Ask pupils to explain what the remainder is when 19 is arranged in equal groups of 5.
- Use the representation on page 98 to show 3 groups of 5 and the remainder of 4.

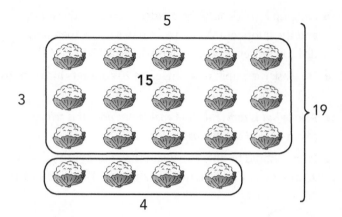

 There are 3 groups of 5 cauliflowers with 4 cauliflowers left over.

- Ask pupils to record this as a division sentence on their mini whiteboards.

- The inverse can be used to check that the answer is correct by finding 3 groups of 5 (15) and then adding on the remainder of 4. (15 + 4 = 19)

- Ask: *What is the remainder when 19 is arranged in equal groups of 4? How do you know?*

 There are 4 groups of 4 cauliflowers with 3 cauliflowers left over.

- Ask pupils to record this new division sentence.

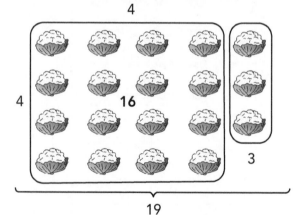

- Pupils are now ready to complete Question 1 in the Practice Book.

Same-day intervention

- Pupils may find it helpful to represent each of the questions practically using counters or cubes so they can physically make equal groups. Moving counters and making their own groups will reinforce the effect of division. They should be encouraged to identify the dividend and the divisor each time.

Same-day enrichment

- **Resource 3.2.10b** Coffee shop problems, provides opportunities for pupils to apply their knowledge of division as equal groups to real-life problems.

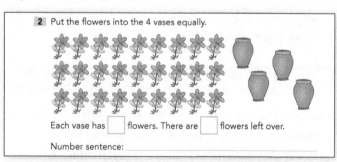

- Pupils should complete each question, making use of multiplication facts to help calculate.

- For problems, like these, that have no illustrations, pupils should talk about how each one would be represented concretely and pictorially.

- Encourage them to talk about the remainder each time and how it relates to the problem, for example does the answer require the remainder to be identified or not?

Question 2

> **2** Put the flowers into the 4 vases equally.
>
> Each vase has ☐ flowers. There are ☐ flowers left over.
>
> Number sentence: _____

What learning will pupils have achieved at the conclusion of Question 2?

- Division as equal sharing will have been explored.

- Pupils will have practised representing a remainder in different ways, that is, practically, pictorially and as part of a number sentence.

- Pupils will have interpreted arrays and used them to represent division.

Activities for whole-class instruction

- Give pupils copies of **Resource 3.2.10c** Coin arrays, and sets of counters or one pence coins.

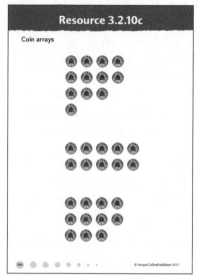

- Ask pupils to explain which groups can be shared equally between four people. They should use counters or coins to model the rearrangement of the arrays.

- Pupils should notice that the first arrangement is divisible by four and, by moving the penny on the fourth row to complete the third row, a 3 by 4 array is completed. This can be compared with the third arrangement that shows there is one penny less.

- Ask: *Where can we see four equal groups in the second arrangement?*

- Agree that 8 pennies can be shared equally between four people but there are two pennies left over.

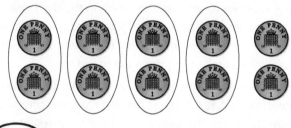

 10 divided by 4 is 2 remainder 2 because 10 shared equally between 4 gives two each and two left over.

- Write the number sentence on the board as $10 \div 4 = 2 \text{ r } 2$.

- Ask: *How can we use the same arrangements of pennies to decide which amounts can be shared equally between three people?*

(**i**) It is important that pupils recognise that the grouping model of division used in Question 1 and the sharing model used here will both result in the same answer. It is useful to discuss contexts for grouping, for example grouping eggs into boxes of six, and contexts for sharing, for example sharing a box of 12 cakes between 6 people. Sharing is also the model used when working with fractions, that is, finding a $\frac{1}{4}$ is the same as dividing into 4 equal parts or groups.

- Pupils are now ready to complete Question 2 in the Practice Book.

Same-day intervention

- Give pupils a set of division calculations, for example: $21 \div 3$, $28 \div 4$, $28 \div 5$.

- Use connecting cubes to represent the dividend each time and pots to represent the divisor.

- Pupils share the cubes equally between the required number of pots and explain when the number can be divided with no leftovers and when a remainder arises. For example: draw 5 empty beakers and show how 28 divided by 5 gives 5 groups of 5 and 3 left over so $28 \div 5 = 5 \text{ r } 3$.

- Ask pupils to connect the cubes in each pot to draw/build an array and show the remainder.

- They could skip-count in fives to 25 and then add the final remainder to total 28.

Same-day enrichment

- Provide pupils with a number of division statements with and without remainders. Challenge pupils to show the matching array and remainder each time. They can sketch, describe or use counters to represent their ideas.

| $22 \div 3 = 7 \text{ r } 1$ | $23 \div 5 = 4 \text{ r } 3$ | $21 \div 6 = 3 \text{ r } 3$ |

- For example:

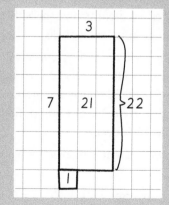

Question 3

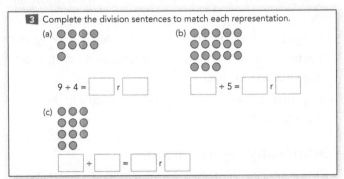

3 Complete the division sentences to match each representation.

(a) $9 \div 4 = \boxed{}$ r $\boxed{}$

(b) $\boxed{} \div 5 = \boxed{}$ r $\boxed{}$

(c) $\boxed{} \div \boxed{} = \boxed{}$ r $\boxed{}$

What learning will pupils have achieved at the conclusion of Question 3?

- Pupils will have been introduced to remainders and will be able to describe them as leftovers.

- Pupils will have practised making equal groups and will be able to show why there is a remainder.

- Pupils will have consolidated multiplication facts by relating them flexibly to division.

- Pupils will have applied their knowledge of the relationship between addition and multiplication, using the inverse as appropriate.

- Pupils will have practised representing a remainder in different ways, that is, practically, pictorially and as part of a number sentence.

Activities for whole-class instruction

- Share the first image of the shells from **Resource 3.2.10d** Arrays with pupils.

Resource 3.2.10d

Arrays
Image 1

- Ask them to describe what they notice, for example that there are eight shells altogether; three shells are in one group together; there are five shells that are not in a group; another group of three shells can be made.

- Ask: *How many shells are in the group shown? How many more groups of three shells can be made from the remaining five shells? How many shells are left over?* (Record this as: There is 3 in one group. There are two groups of 3 and 2 left over.)

- Ask pupils to write a matching division sentence. $(8 \div 3 = 2 \text{ r } 2)$

Explore using the inverse to check $(2 \times 3 = 6)$ and then add the remainder 2 to equal 8.

- Now show pupils the second image from **Resource 3.2.10d** and ask them to describe what they notice this time. Look for pupils who also notice that there are double the number of shells altogether and double the number of shells in each group.

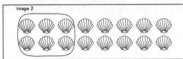

Image 2

- Ask: *How many shells are in the group shown? How many more groups of six shells can be made from the remaining ten shells? How many shells are left over?*

 All say ... *There are two groups of six shells and four shells left over.* (Record this in the same way as before, asking pupils to write a matching division sentence, $16 \div 6 = 2 \text{ r } 4$.)

- Share the following word problem with the class.

  Anita collects 16 shells at the beach. She puts five shells on each of her sandcastles.

  How many sandcastles have five shells on them? How many shells are left over?

- Ask pupils to discuss the problem and to explain how the problem can be represented using the third image on **Resource 3.2.10d**.

Image 3

- Can they record the matching division sentence and explain which number in the problem is the divided, which is the divisor and what is the quotient?

- Use counters to represent the shells in the third image and arrange them into rows of 5 to make an array and leaving the final row incomplete to represent the leftover. Ask: *Why is the last row not a full row of 5?*

- Introduce the fourth representation of the array from **Resource 3.2.10d**. Suggest that the rows and columns of equal length on the array can be used to represent two different division sentences.

Image 4

- Can pupils identify the dividend, divisor and quotient for each division sentence?

- Explore the arrays together and establish that $14 \div 4 = 3 \text{ r } 2$ and $14 \div 3 = 4 \text{ r } 2$ are represented by the columns and rows.

- Invite pupils to create two different word problems, for example about stickers, to match each of the division sentences.

- Pupils are now ready to complete Question 3 in the Practice Book.

Same-day intervention

- Give each pupil a copy of **Resource 3.2.10e** Division match up.

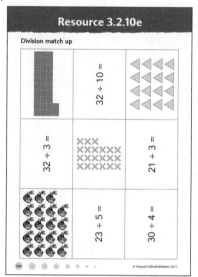

- Pupils should match the division calculations with the different arrays and find the answer each time. They should notice that there are more calculations than arrays and should explain why more than one calculation can be match with an array.

Same-day enrichment

- Ask pupils to write a set of word problems to match the three division sentences in Question 3 of the Practice Book. They should share the word problems with others and challenge them to match the array, sentence and word problem each time.

Question 4

4 Calculate.
(a) $17 \div 3 =$ ☐ r ☐   (b) $23 \div 7 =$ ☐ r ☐
(c) $38 \div 6 =$ ☐ r ☐   (d) $31 \div 4 =$ ☐ r ☐
(e) $64 \div 9 =$ ☐ r ☐   (f) $47 \div 8 =$ ☐ r ☐
(g) $41 \div 7 =$ ☐ r ☐   (h) $55 \div 8 =$ ☐ r ☐
(i) $92 \div 10 =$ ☐ r ☐   (j) $58 \div 9 =$ ☐ r ☐

What learning will pupils have achieved at the conclusion of Question 4?

- A conceptual understanding of division with remainders will have been developed through the use of manipulatives.

- Pupils will have been introduced to remainders and are able to describe them as leftovers.

- Pupils will have practised making equal groups and will be able to explain why there is a remainder.

- Knowledge and application of multiplication facts will have been consolidated by relating them flexibly to division.

- Pupils will have practised representing a remainder in different ways, that is, practically, pictorially and as part of a number sentence.

- Division as equal sharing will have been explored.

- Pupils will have applied their knowledge of the relationship between addition and multiplication, using the inverse as appropriate.

Activities for whole-class instruction

- Quickly revisit the multiplication table for 4 and for 7 and record them on the board.

- Suggest that there will be no remainder or leftover when 28 is divided by 4. Ask: *Do you agree? How do you know?*

- Ask: *When 30 is divided by 4, what will the remainder be? How do you know?* Can pupils explain that since 7×4 is exactly 28, there must be 2 left over if 30 is divided by 4 because 30 is 2 more than 28.

 7 lots of 4 are 28 and 2 more makes 30.

- Ask: *Will the remainder also be 2 when 30 is divided by 7?* Pupils should notice that $7 \times 4 = 28$ and $4 \times 7 = 28$ (or $7 \times 4 = 4 \times 7$) and so the remainder must be the same.

 4 lots of 7 are 28 and 2 more makes 30.

- Challenge pupils to make up a division using multiplication facts that they know (where there is no remainder) and then use this to write another calculation with a remainder of 3.
- Pupils are now ready to complete Question 4 in the Practice Book.

Same-day intervention

- Use number rods to explore equal grouping by giving pupils the dividend each time and the size of the equal groups. (If number rods are not available, use interlocking cubes.)
- They should use the rods that represent the divisor and find how many of these are equivalent or close to the dividend.
- Pupils then identify any remainder by finding the rod that is needed to complete the value of the dividend.
- For example, $19 \div 3 = 6$ r 1 as:

10			9			
3	3	3	3	3	3	1

- Encourage the pupils to check answers by skip-counting from zero. They should keep track of the number of equal groups that they have counted, and then add on the remainder. Is the answer the same as the original dividend?
- For example, 3, 6, 9, 12, 18, and then add 1 is 19.

Same-day enrichment

- Pupils explore different division calculations that result in a given remainder. They should explain any patterns they notice.
- Give pupils interlocking cubes or counters to support decision making. Ask them to focus on the first division sentence initially. They should think about numbers that are exactly divisible by 6 and then reason about a leftover of 2. For example, 12 is divisible by 6 so the number that is two more than 12 will have a remainder of 2 so $14 \div 6 = 2$ r 2 because $2 \times 6 + 2 = 14$.
- For pupils who are struggling, encourage them first to represent numbers using counters that are divisible by 6 and then ask them to show a leftover of 2. They should notice that adding another two counters to their representation will show the remainder.
- Challenge pupils to find at least two possible calculations for each division sentence.

$\square \div 6 = \square$ r 2 $\square \div 7 = \square$ r 2

Question 5

5 Application problems.
(a) Mr Bruce shared a box of water equally between Kit, Marlon, Ava and Amina. There were 9 bottles of water in the box. How many bottles did each child get?

Answer: _____

How many were left over?

Answer: _____

(b) A monkey gave her 7 baby monkeys some peaches. Each baby monkey got 6 peaches. There were 5 peaches left over. How many peaches were there altogether?

Answer: _____

(c) Mahmud has 50 | to make ⬡ without sharing edges. How many ⬡ can he make? How many | will be left over?

Answer: _____

What learning will pupils have achieved at the conclusion of Question 5?

- Pupils will have applied their knowledge of the relationship between addition and multiplication, using the inverse as appropriate.
- Pupils will have practised applying their knowledge of division with remainders to word problems in context.

Activities for whole-class instruction

- Set up a practical problem that requires division and will give rise to a leftover. For example: There are 26 pencils to be shared equally between 6 tables in the classroom or to be put in pots of 6.
- Pupils should explore the two models of division and recognise that the answer is the same each time although the distribution of the pencils will look a little different.
- Ask pupils to work in pairs to make up a word problem to match each of the models. Discuss how the leftover should be described or used each time in the context of the problem. For example:

Mrs Duke has a box of 26 new pencils.

She shares the pencils equally between 6 tables.

How many pencils are given to each table? $\square$

How many pencils are left over? $\square$

Mrs Duke has a box of 26 new pencils and some pots to put them in.

She puts 6 pencils in each pot.

How many pots of 6 can she make? $\square$

How many more pencils does she need to make another pot 6? $\square$

Present another division problem that requires pupils to use the inverse to solve it. Discuss what is the same and what is different this time. For example:

Mrs Duke has a box of new pencils and some pots to put them in.

She uses 4 pots and puts 6 pencils in each.

She has 2 pencils left over in her box.

How many pencils did she have altogether? ☐

- Ask pupils to sketch the problem or represent it in some way using counters, cubes or arrays.

- Pupils are now ready to complete Question 5 in the Practice Book.

Same-day intervention

- Work through the pencil problems together using pencils or cubes. Ask pupils to explain whether the problem requires us to share the pencils equally between groups or to group the pencils into sets of an equal size.

- Make up similar problems requiring sharing or grouping that pupils can represent, for example 28 sweets put in packets of 5, and identify the leftover each time. Check calculations using the inverse.

Same-day enrichment

- Give pupils these division calculations and a multiplication calculation. Challenge them to make up different word problems for each division calculation. Ask pupils to explain whether their problem requires us to share equally between groups or to group things into sets of an equal size.

$32 \div 7$ $45 \div 6$ $51 \div 8$ 4×9

Challenge and extension question

Question 6

6   All the pupils in Class 3G are grouped to take part in a school activity. If they are grouped in sevens, there is one pupil left over. If they are grouped in sixes, there is also one pupil left over. Given that there are fewer than 50 pupils in the class, the number of pupils in the class is ☐.

This question requires pupils to think logically to solve the problem as the division calculation changes as the children are regrouped. Pupils need to recognise that the child left over each time is the remainder after the division calculation.

They should reason about remainders and recognise that a remainder will only arise when the dividend is not a multiple of the divisor. In this case the total number of children is not a multiple of 6 or a multiple of 7.

Give pupils an opportunity to explore the problem for themselves and represent it as required. They may find it useful to write out the multiplication tables for 6 and 7 and note which give a product that is less than 50.

As the leftover each time is the same, pupils will be looking for a number that is both a multiple of 6 and a multiple of 7. (42 in this case) With a leftover of 1, there must be 43 children in the class.

Unit 2.11
Calculation of division with a remainder (1)

Conceptual context

Pupils have been introduced to division with a remainder and have practised describing these as leftovers.

They have used manipulatives and other images to explore equal groups and have used this concept to recognise and explain why another equal group cannot be made with the value left over.

They have begun to explore remainders in context to give them the opportunity to decide when the remainder is of significance, depending on the requirements of the question.

In this unit, division with a remainder will be explored further, looking carefully at the way in which multiplication facts can be used to find multiples of the divisor that are smaller than, but closest to, the dividend. This helps to move pupils from physical sharing towards more mental strategies

Learning pupils will have achieved at the end of the unit

- Pupils will have developed their understanding of remainders and will be able to describe them as leftovers (Q1, Q2, Q4)
- Knowledge and application of multiplication facts will have been consolidated by relating them flexibly to division (Q1,Q2)
- Pupils will have continued to practise representing a remainder in different ways, that is, practically, pictorially and as part of a number sentence (Q1, Q2)
- Pupils will be more confident in identifying the remainder using multiplication facts to help them (Q1, Q3, Q4)
- Pupils will have interpreted arrays and used them to represent division (Q2)
- Pupils will have used knowledge of multiplication facts to find the largest multiple that is smaller than, but closest to, a given number (Q3, Q4)
- Pupils will have practised applying their knowledge of division with remainders to solve word problems in context (Q1, Q2, Q5)

Resources

counters; cubes; pots; pencils; Multiplication squares; **Resource 3.2.11a** Plants and flower beds; **Resource 3.2.11b** Strawberry arrays; **Resource 3.2.11c** Balances 1; **Resource 3.2.11d** Balances 2; **Resource 3.2.11e** Pencil pots; **Resource 3.2.11f** Match up word problems

Vocabulary

equal groups, division, multiplication, dividend, divisor, quotient, remainder, leftover, multiple, factor

Question 1

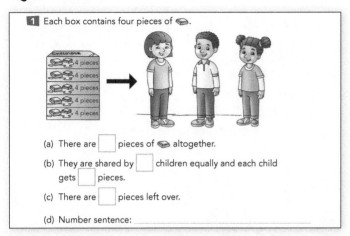

1. Each box contains four pieces of 🍫.

(a) There are ☐ pieces of 🍫 altogether.

(b) They are shared by ☐ children equally and each child gets ☐ pieces.

(c) There are ☐ pieces left over.

(d) Number sentence: _____

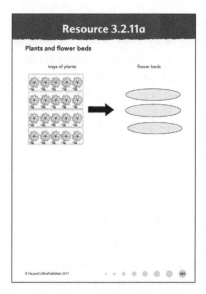

What learning will pupils have achieved at the conclusion of Question 1?

- Pupils will have developed their understanding of remainders and will be able to describe them as leftovers.

- Knowledge and application of multiplication facts will have been consolidated by relating them flexibly to division.

- Pupils will have continued to practise representing a remainder in different ways, that is, practically, pictorially and as part of a number sentence.

- Pupils will be more confident in identifying the remainder using multiplication facts to help them.

- Pupils will have practised applying their knowledge of division with remainders to solve word problems in context.

Activities for whole-class instruction

- Show pupils the picture of the trays of plants and flower beds on **Resource 3.2.11a** Plants and flower beds, and ask them to discuss what they think the problem is about. Can they make up a word problem to match the picture?

- Ask pupils to write the multiplication sentence to match the product of four trays of five plants and then the division sentence needed to find how many plants will be in each of the three equal groups. Discuss and agree that there are two steps to this problem – first a multiplication and then a division. Record the two steps as $4 \times 5 = 20$ and $20 \div 3 = \square$.

- Ask: *How do you know that there will be a leftover when 20 is divided by 3? What will the remainder be?*

 All say … *3 groups of 6 is 18 and 20 is 2 more than 18.*

- Suggest that by adding another tray of flowers and another flower bed to the problem, there will still be a remainder of two plants. Sketch the following representation of the problem or ask pupils to sketch their own, making sure to record each tray as '5 plants' rather than drawing this time.

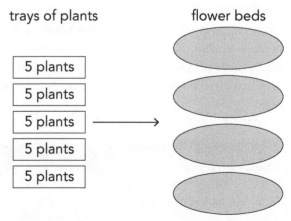

trays of plants   flower beds

5 plants
5 plants
5 plants →
5 plants
5 plants

- Ask pupils to create a new word problem (or adapt their original one) to match the new criteria. Can they write the multiplication sentence and division sentence that they will need to complete the two steps of the problem? (Record the two steps as $5 \times 5 = 25$ and $25 \div 4 = \square$)

- Discuss strategies to divide 25 by 4. Can pupils explain why there will be a remainder before actually calculating or carrying out the division practically? (Agree that 4 × 6 = 24 so each flower bed will have 6 plants in it with one left over and not two like the previous question.)

 4 groups of 6 is 24 and 25 is 1 more.

- Pupils are now ready to complete Question 1 in the Practice Book.

Same-day intervention

- Give pupils a set of interlocking cubes and ask them to make six sticks of four cubes. Arrange the cubes in rows and discuss the multiplication fact they represent. (6 × 4 = 24) Pupils can check by skip-counting in fours.

- Show pupils the six cups that the cubes must be shared equally between. Ask: *How do we know that the cubes will divide exactly and there will be no remainder?* Agree that there are six sticks of four cubes so one stick can go in each cup. (Record the division sentence as 24 ÷ 6 = 4.)

- Put the cubes back in the original rows and remove one of the cups.

- Ask: *How do we know that there will be a left over when we divide 24 between five cups?* (Better to talk about 'multiple of 5' than 'in the 5 times table'.)

- Invite pupils to use their own strategy to complete the problem, either by sharing out cubes or drawing on multiplication facts. Can pupils explain that there are 5 groups of 4 cubes and there are 4 cubes left over? Refer to 5 × 4 is 20 and 24 is 4 more.

- Work together to make up a maths story about the second problem above using the same structure as used in Question 1 in the Practice Book, that is, 'There are 24 cubes altogether. They are shared between 5 cups and each cup has 5 cubes. There are 4 cubes left over.'

Same-day enrichment

- Introduce the enrichment task before working with pupils on the intervention activity.

- Give pupils a set of multiplication and division sentences to represent as two-step problems.

- The first example should have some of the missing numbers added so pupils can see the structure of the calculations.

- For example:

| 5 × 7 = 35 | 7 × 6 = ☐ | 4 × 8 = ☐ | 5 × 9 = ☐ |
| 35 ÷ 4 = ☐ | ☐ ÷ 5 = ☐ | ☐ ÷ 3 = ☐ | ☐ ÷ 7 = ☐ |

- They should take one set of sentences at a time and represent it pictorially before writing a maths story to go with it, as in Question 1 of the Practice Book. For example, for the first example above:

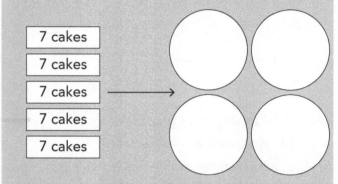

| 7 cakes |
| 7 cakes |
| 7 cakes |
| 7 cakes |
| 7 cakes |

- There are 35 cakes altogether. They are shared equally between 4 plates. Each plate has 8 cakes. There are 3 cakes left over.

Question 2

2 There are 45 🍎.
(a) 6 🍎 are put onto each plate. It is enough to fill ☐ plates.
(b) There are ☐ 🍎 left over.
(c) Number sentence: _____

What learning will pupils have achieved at the conclusion of Question 2?

- Pupils will have developed their understanding of remainders and will be able to describe them as leftovers.

- Knowledge and application of multiplication facts will have been consolidated by relating them flexibly to division.

- Pupils will have continued to practise representing a remainder in different ways, that is, practically, pictorially and as part of a number sentence.

- Pupils will have interpreted arrays and used them to represent division.

- Pupils will have practised applying their knowledge of division with remainders to solve word problems in context.

Activities for whole-class instruction

- This activity works well in a larger space where groups can arrange themselves easily.

- Show the following division story using the number of pupils in the class as the total number, for example 32.

> There are 32 pupils in the class.
> There are ☐ pupils in each group.
> There are ☐ groups of ☐ pupils.
> There are ☐ pupils left over

- First suggest that groups of 3 pupils should be made.

- Can pupils explain what multiplication fact could be used to quickly work out the number of groups that will be made? Ask: *Will there be a remainder? How do you know?*

- Invite pupils to quickly make groups of 3 to check the multiplication fact and the predicted remainder.

- Skip-count in threes to confirm the total number of pupils grouped in threes and then add the remainder to give the total of pupils in the whole class.

- Fill in the missing values in the division story and represent this as an array, for example for 32.

🚶🚶🚶
🚶🚶🚶
🚶🚶🚶
🚶🚶🚶
🚶🚶🚶
🚶🚶🚶
🚶🚶🚶
🚶🚶🚶
🚶🚶🚶
🚶🚶🚶
🚶🚶

 All say ... *There are 10 groups of 3 pupils and 2 left over because 10 groups of 3 is 30 and 32 is 2 more than 30.*

- Next, suggest that groups of 6 pupils should be made. Work through the problem in the same way, identifying the multiplication fact and the remainder.

All say ... *There are ☐ groups of 6 pupils and ☐ left over because ☐ groups of 6 is ☐ and ☐ is ☐ more than ☐.*

- Can pupils explain why there are half the number of groups this time? (Relate 6 to double 3.)

- Explore other group sizes.

- Pupils are now ready to complete Question 2 in the Practice Book.

Same-day intervention

- Ask pupils to complete the following multiplication facts for the 3 times table.

$3 \times 8 = $ ☐   ☐ $\times 3 = 30$ $9 \times 3 = $ ☐

$3 \times 7 = $ ☐ $6 \times 3 = $ ☐

- Share the following division story with the pupils and use the numbers 31, 26, 29 and 20 as the dividend to create four different examples. For example:

> There are <u>31</u> horses altogether.
> There are 3 horses in each field.
> There are ☐ fields of 3 horses.
> There are ☐ horses left over.

- Pupils should identify and explain which of the five multiplication facts they will use to help solve the division story.

Look out for ... pupils who choose a multiplication fact that would leave a remainder that is large enough to make another group, for example, for 26, choose 7×3, which would leave a remainder of 5 so another group of 3 can still be made. Use cubes to represent the division and show that another group of 3 could be made.

- Can pupils make up their own story to match the remaining multiplication fact?

Same-day enrichment

- Give pupils **Resource 3.2.11b** Strawberry arrays.

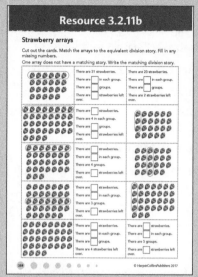

- Pupils should match each array with its division story. Any missing information should be inserted to make all stories correct. One array is left without a matching story so pupils must write their own.

Question 3

> **3** What is the greatest number you can write in each circle?
>
> (a) $7 \times \bigcirc < 41$ (b) $6 \times \bigcirc < 35$ (c) $3 \times \bigcirc < 26$
>
> (d) $9 \times \bigcirc < 38$ (e) $\bigcirc \times 5 < 43$ (f) $\bigcirc \times 2 < 20$
>
> (g) $\bigcirc \times 7 < 32$ (h) $4 \times \bigcirc < 43$ (i) $\bigcirc \times 8 < 65$

What learning will pupils have achieved at the conclusion of Question 3?

- Pupils will be more confident in identifying the remainder using multiplication facts to help them.
- Pupils will have used knowledge of multiplication facts to find the largest multiple that is smaller than, but closest to, a given dividend.

Activities for whole-class instruction

- Begin by rehearsing the 4 times table and related division facts, using a counting stick to keep track of the count or using an interactive times table activity on the whiteboard.
- Show pupils a number line with the tens boundaries marked. Ask pupils to explain which multiples of 4 are between 20 and 30. Record 24 and 28 on the number line.

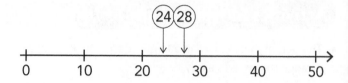

- Ask: *Which of these multiples of 4 is closest to 30? How much larger is 30 than 28? What do we now know about the remainder when 30 is divided by 4.*
- Now ask pupils to write down the three multiples of four that are less than 45 but greater than 35. Ask them to explain which of these is closest to 45. Record the multiples on the number line.
- Ask: *What do we know about the remainder when 45 is divided by 4?*
- Write the statements $9 \times 4 < 45$, $10 \times 4 < 45$, $11 \times 4 < 45$, $12 \times 4 < 45$. Ask the pupils to explain which statement is false and why.
- Return to the number line and challenge pupils to find multiples of other numbers that sit between 20 and 30. Can they complete the three different statements below to make them true?

  $\square \times \square < 24$ $\square \times \square < 27$ $\square \times \square < 30$

- Pupils are now ready to complete Question 3 in the Practice Book.

Same-day intervention

- Give the pupils **Resource 3.2.11c** Balances 1, to further practise identifying multiplication facts that give a product that is larger or smaller than a given number. Pupils should have access to a multiplication square to support their decisions.

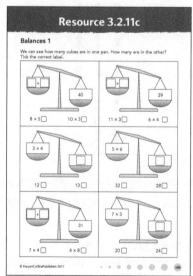

Same-day enrichment

- Provide the pupils with **Resource 3.2.11d** Balances 2.

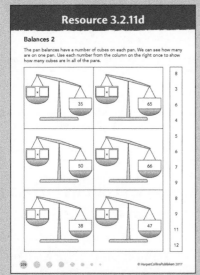

- Introduce the problem so that pupils are clear about the aim of the activity. There are six balancing scales showing inequalities and a set of numbers that should be used to make the incomplete multiplication sentences correct.

Question 4

4 Are these calculations correct? Put a tick (✓) for yes and a cross (✗) for no in each box.

(a) 46 ÷ 8 = 5 r 6 ☐   (b) 36 ÷ 6 = 5 r 6 ☐

(c) 55 ÷ 7 = 7 r 6 ☐   (d) 44 ÷ 5 = 8 r 4 ☐

(e) 80 ÷ 12 = 7 r 4 ☐   (f) 99 ÷ 10 = 9 r 9 ☐

What learning will pupils have achieved at the conclusion of Question 4?

- Pupils will have developed their understanding of remainders and will be able to describe them as leftovers.

- Pupils will be more confident in identifying the remainder using multiplication facts to help them.

- Pupils will have used multiplication facts to find multiples of the divisor that are smaller than, but closest to, the dividend.

Activities for whole-class instruction

- Show pupils the following multiplication sentences and discuss what is the same and what is different about them. (They both have 8 as one of the factors; both

multiplications give a product that is less than the number shown; the 8 is in a different position; neither 43 nor 50 is a multiple of 8 and so on.)

☐ × 8 < 50 and 8 × ☐ < 43

- Suggest that using the 8 times table will be useful for both problems. Revisit commutativity of multiplication.

- As a class, say the 8 times table and stop at points along the way to ask about the related division fact, for example 3 times 8 is 24. How many eights are in 24? What is 24 divided by 8?

- Tell pupils you are going to draw a double-sided number line together. Create one showing multiples of 8, as below, writing bottom row first, then top row, discussing with pupils as you go along.

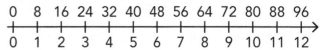

- Locate the multiples of 8 that are less than, but closest to, the numbers 50 and 43 from the original questions. Agree that the largest products for each will be found using 6 × 8 < 50 and 8 × 5 < 43. Ask: *How much larger is 50 than 6 times 8? So the remainder is ...?*

 6 groups of 8 are 48 and 50 is 2 more than 48.

(Record this as 50 ÷ 8 = 6 r 2.)

- Look together at 8 × ☐ < 43 in the same way.

- Show pupils three different division sentences to be checked, each giving rise to a remainder. For example:

43 ÷ 4 = 10 r 3   58 ÷ 7 = 8 r 3   65 ÷ 9 = 8 r 2

- Suggest that we can check each of the calculations using multiplication and not division. Invite pupils to discuss why this is possible and why it may be a useful strategy here.

- Refer back to the previous task and how we found the largest product using our multiplication facts.

- Ask: *What multiple of 4 is smaller than but closest to 43?*

 10 groups of 4 are 40 and 43 is 3 more than 40. So 43 divided by 4 is 10 with 3 left over.

- Agree that the first division sentence is correct.

- Look at 58 ÷ 7 = 8 r 3 in the same way. Agree that the second division sentence is false and should be 58 ÷ 7 = 8 r 2.

- Finally, invite pupils to work with a partner to make a decision about the third calculation. They should be able to explain that the multiple of 9 that is smaller than but closest to 65 is 63, but this is 7 groups of 9 and not 8 groups.

● Pupils are now ready to complete Question 4 in the Practice Book.

Same-day intervention

● Using interlocking cubes, work together to make a double-sided number line for the multiples of 6 up to 36. Alternate the colours of each set of six so they can be easily recognised.

● Begin by giving pupils a division calculation to solve, for example 27 ÷ 6 = ? Can pupils explain that 27 is not exactly divisible by 6 because 27 is not labelled on this number line?

● Ask pupils to locate 27, and to find the multiple of 6 that is smaller than, but closest to, 27.

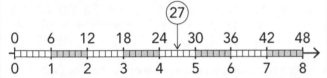

● Agree that 27 sits between the multiples 24 and 30 but that 30 is larger than 27.

● Together, work through the calculation to agree that 27 ÷ 6 = 4 r 3 because 24 is 4 groups of 6 and 27 is 3 more.

● Work through 33 ÷ 9 = ☐

● Finally, give the pupils the complete calculations to check, ensuring that one is incorrect, for example 45 ÷ 6 = 7 r 2 and 34 ÷ 6 = 5 r 4.

Same-day enrichment

● Pupils should use a ruler to draw and complete two double-sided number lines (as used in the whole-class instruction) for the multiplication tables of 7 and 9.

● For each number line, pupils should label four numbers that are between labelled points to use as dividends for division calculations. For example:

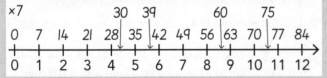

● Pupils write out their two sets of four division sentences and then use the number line to locate the closest multiple each time. For example, for 30 ÷ 7, find that 4 groups of 7 (28) is closest to 30, but 30 is 2 more than 28. Record as 30 ÷ 7 = 4 r 2.

● Pupils can also challenge their friends to spot any mistakes by providing them with a correct and incorrect division sentence to check.

Question 5

> **5** Write a number sentence for each question.
>
> (a) There are 31 days in May and 7 days in a week. How many weeks plus how many days are there in May?
>
> Number sentence: _____
>
> (b) If each child gets 6 sweets, then how many children can share 50 sweets? How many sweets are left over?
>
> Number sentence: _____
>
> (c) A tailor makes coats with 5 buttons. How many coats will 49 buttons go on? How many buttons will be left over?
>
> Number sentence: _____

What learning will pupils have achieved at the conclusion of Question 5?

● Pupils will have practised applying their knowledge of division with remainders to solve word problems in context.

Activities for whole-class instruction

● Give pupils **Resource 3.2.11e** Pencil pots.

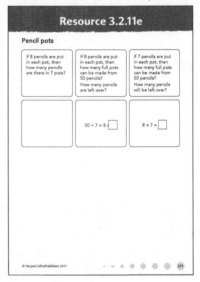

● Ask: *Which word problem matches each of the number sentences? How do you know?*

● Can pupils quickly find the answer to the two calculations? Discuss the strategies they use.

⟨Look out for⟩ … pupils who are only looking at the numbers in the division sentences and not what the context is describing. Model the problems using pencils and cups, making sure that pupils are clear about the number that represents the divisor and quotient each time.

● Invite pupils to find the number sentence that should be used to represent the remaining word problem, explaining the choice they made.

- Invite pupils to make up three more problems to match the following calculations. They can use a context of their choice.

  9 × 6 = 54   61 ÷ 9   61 ÷ 6

- Can pupils represent and explain each of their calculations concretely or pictorially?

- Pupils are now ready to complete Question 5 in the Practice Book.

Same-day intervention

- Return to the problems in Question 5 and ask pupils to represent each one in a different way.

- Suggest that an array would be useful for the first problem about the days in May, making reference to a calendar with rows of seven.

- The second and third problem can be represented using counters or cubes. Pupils may decide to use a pictorial representation. Can they use a double-sided number line to help?

- They should explain why their representations match each problem.

Same-day enrichment

- Give pupils **Resource 3.2.11f** Match up word problems, showing three different pictorial representations.

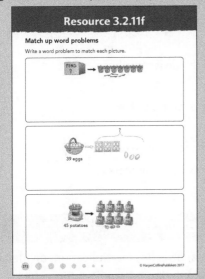

Resource 3.2.11f

Match up word problems
Write a word problem to match each picture.

- Can they write a division word problem to match each of the representations?

- Can they write another word problem that uses the same numbers but does not match the representation?

- They should swap word problems with a partner who should then reason about and explain why a problem does or does not match each representation.

Challenge and extension question

Question 6

6 A class of 26 pupils plans to go boating.

Small boat – 4 people

Big boat – 6 people

(a) If all the pupils take small boats, how many small boats will they need?

Answer: ☐

(b) If all the pupils take big boats, how many big boats will they need?

Answer: ☐

(c) Can you offer some other suggestions for renting the boats?

Answer: _____

In this question, pupils are given information about the dividend and the possible divisors each time as a group of 26 children choose to rent small boats for four people or big boats for six people.

The task could be done practically as pupils group themselves as if in the small or big boats.

Can they explain that some children will be left over each time as 26 is not a multiple of 4 or 6?

Challenge pupils to prove this on a double-sided number line or using another representation of their choice.

Finally, pupils are asked to offer other suggestions about boat rental. Ask pupils to explain why other ideas should be explored here. Can they explain that both previous groupings resulted in a remainder and a solution where all 26 children can go boating is needed?

Pupils should explore combinations of groups of 4 and 6 using multiplication and division facts to help them. The problem can be explored practically or represented pictorially.

All 26 children can go boating in three big boats (3 × 6 = 18) and two small boats (2 × 4 = 8) or in one big boat (1 × 6 = 6) and five small boats (5 × 4 = 20).

Unit 2.12
Calculation of division with a remainder (2)

Conceptual context

Pupils have developed their understanding of division with a remainder and have explored the way that multiplication facts can be used to find multiples of the divisor that are smaller but closest to the dividend. This had developed pupils' efficiency in division by helping them to move from physical sharing of amount towards more mental strategies.

In this unit, pupils continue to solve a range of division problems and further develop their conceptual understanding of the relationship with multiplication. Pupils will explore multiplication with addition and be able to explain how this relates to division with remainders.

Learning pupils will have achieved at the end of the unit

- Pupils will be more confident in identifying the remainder using multiplication facts to help them (Q1, Q2)
- Pupils will have used knowledge of multiplication facts to find the largest multiple that is smaller than, but closest to, a given number (Q1, Q2)
- Pupils will have used manipulatives, images and number sentences to show division and multiplication (Q1, Q2, Q4)
- Pupils will have identified a missing dividend using what they know about the divisor and quotient to help them (Q2, Q4, Q5)
- Pupils will be able to demonstrate why a number sentence such as, $5 \times 4 + 3$ relates to the division $23 \div 4 = 5 \text{ r } 3$ (Q3, Q4)
- Pupils will be able to explain what the largest and smallest remainder will be for a given divisor (Q3)
- Pupils will have continued to practise applying their knowledge of division with remainders to word problems in context (Q3, Q5)

Resources

cubes; number rods; counters; dice; 0–9 digit cards;
Resource 3.2.12a Multiples of 6; **Resource 3.2.12b**
Missing information; **Resource 3.2.12c** Number sentences;
Resource 3.2.12d Spot arrays; **Resource 3.2.12e**
Word problems

Vocabulary

equal groups, division, multiplication, dividend, divisor, quotient, remainder, leftover, multiple, factor

Question 1 and Question 2

1 Calculate mentally.

(a) 37 ÷ 9 = ☐   (b) 47 ÷ 5 = ☐   (c) 43 ÷ 8 = ☐

(d) 19 ÷ 7 = ☐   (e) 69 ÷ 8 = ☐   (f) 62 ÷ 7 = ☐

(g) 26 ÷ 3 = ☐   (h) 27 ÷ 6 = ☐   (i) 9 ÷ 2 = ☐

(j) 14 ÷ 4 = ☐   (k) 95 ÷ 10 = ☐   (l) 63 ÷ 6 = ☐

2 Fill in the boxes.

(a) ☐ ÷ 6 = 4 r 4   (b) ☐ ÷ 7 = 6 r 2   (c) ☐ ÷ 5 = 9 r 1

(d) ☐ ÷ 9 = 3 r 8   (e) ☐ ÷ 3 = 7 r 2   (f) ☐ ÷ 7 = 7 r 4

What learning will pupils have achieved at the conclusion of Question 1 and Question 2?

- Pupils will be more confident in identifying the remainder using multiplication facts to help them.
- Pupils will have used knowledge of multiplication facts to find the largest multiple that is smaller than, but closest to, a given number.
- Pupils will have used manipulatives, images and number sentences to show division and multiplication.
- Pupils will have identified a missing dividend using what they know about the divisor and quotient to help them.

Activities for whole-class instruction

- Begin by quickly rehearsing the multiplication tables for 3, 6 and then for 9. As a class, first say the 6 times table and stop at points along the way to ask about the related division fact, for example 4 times 6 is 24. Ask: *How many sixes are in 24? What is 24 divided by 6?* Discuss the products and focus on the fact that they are all even.
- Relate the multiples of 6 to double the multiples of 3 and *vice versa*. **Use Resource 3.2.12a** Multiples of 6, to show that 12 × 3 is half of 12 × 6 and 12 × 6 is double 12 × 3.

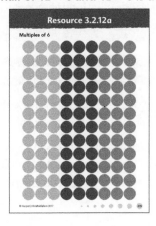

- Rehearse the 9 times table in the same way, again referring to the array. Can pupils see how 12 × 9 is made up of three lots of 12 × 3 arrays? Can they explain why 12 × 9 gives the same product as 12 × 3 add 12 × 6?
- Draw this on the board, but do not share the actual calculation it represents.

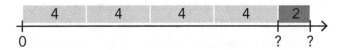

- Can pupils identify the divisor and the remainder? Can they explain how they can find the value of the dividend? Finally, ask them to write down the division sentence the rods represent.
- Draw this on the board and ask pupils to recreate it using rods or interlocking cubes.

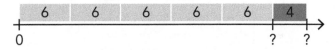

- Ask: *What is the divisor this time? And the remainder?* Using pupils' responses, record this as ☐ ÷ 6 = 5 r 4.
- Can pupils now explain how the dividend can be identified? Take feedback and fill in the missing dividend of 34 in the division sentence.

 All say ... *5 groups of 6 are 30 and 34 is 4 more.*

- Show pupils the following calculations with missing dividends to explore.

  ☐ ÷ 5 = 4 r 2   ☐ ÷ 7 = 3 r 5   ☐ ÷ 2 = 10 r 1

- Ask pupils to identify the divisor (5) and the number of times this must go into the dividend (4).
- Can pupils explain that the multiplication fact 4 × 5 must give the multiple of 5 that is closest to the dividend?

  Look out for ...pupils who struggle to see this connection and use four rods of 5 to demonstrate that the resulting product 20 can be divided by 5 again to give 4 equal groups as shown in the original division sentence.

- Explain that the missing dividend has not been found because the division sentence was not simply 20 ÷ 5 = 4 as we know that there is a remainder of two. Discuss what we know about the dividend, that is, it must be two more than 20 to leave a remainder of 2. Together, build the matching representation using number rods.

 All say ... *4 groups of 5 are 20 and 22 is 2 more.*

- Record this as 22 ÷ 5 = 4 r 2.

- Challenge pupils to use rods to represent the second problem, identifying the missing dividend each time.
- Pupils are now ready to complete Questions 1 and 2 in the Practice Book.

Same-day intervention

- Using **Resource 3.2.12b** Missing information, pupils should fill in the missing information from each division sentence using the images to support their decisions.

Resource 3.2.12b

Missing information

1

$17 \div 3 = \square \, r2$

2

$\square \div 6 = 6 \, r3$

3

$\square \div 10 = 3 \, r4$

4. Draw your own rods to show $27 \div 4 = 6 \, r3$

© HarperCollinsPublishers 2017

Same-day enrichment

- Present the following problems.

  I'm thinking of a number between 40 and 60. When divided by 7, there is a remainder of 3. What number could it be?

  I'm thinking of a number between 26 and 46. When divided by 6, there is a remainder of 5. What number could it be?

- Ask pupils to discuss the first problem and to explain what information is useful and why.
- Ask: *What represents the dividend in the problem? Do we know the whole quotient or only part of it?* (We only know the remainder but not the number of times that 7 goes into the dividend.)
- For the first problem, record the pupils' ideas as $\square \div 7 = \square \, r\,3$.
- Can pupils explain that the number for the first problem must be 3 more than a multiple of 7 that sits between 40 and 60 on a number line?

- Explore strategies to solve the problem, perhaps sketching rods or other images to represent possible solutions.
- Pupils should work together to solve second problem.

Question 3

3 Find the answers.

(a) Maria's plates are big enough to fit 6 apples.

15 apples can be put on $\square$ plates with $\square$ apples left over.

Number sentence: $\square \div \square = \square \, r \, \square$

(b) Some sweets were shared equally between 8 children. After each child got 6 sweets, there were 4 sweets left over.

How many sweets were there altogether?

Number sentence: $\square \times \square + \square = \square$

(c) 24 bananas were given to 5 monkeys equally, and each monkey got $\square$ bananas, with $\square$ bananas left over.

Number sentence: $\square \div \square = \square \, r \, \square$

(d) In $\square \div 6 = 7 \, r \, \square$, the remainder could be $\square$.
The greatest possible remainder is $\square$.
When the remainder is the greatest, the dividend is $\square$.

(e) From the number sentence $5 \times 7 + 5 = 40$, we can write another number sentence:

$40 \div \square = \square \, r \, \square$.

What learning will pupils have achieved at the conclusion of Question 3?

- Pupils will be able to demonstrate why a number sentence such as $5 \times 4 + 3$ relates to the division $23 \div 4 = 5 \, r \, 3$.
- Pupils will be able to explain what the largest and smallest remainder will be for a given divisor.
- Pupils will have continued to practise applying their knowledge of division with remainders to word problems in context.

Activities for whole-class instruction

- Give each group of 3–4 pupils a tray of cubes or counters to explore and represent different problems.
- Ask pupils to make a group of six cubes or counters. Tell pupils that each cube represents one book. Ask: *How many books does your group of cubes represent?*

Chapter 2 Multiplication and division (II)

- Say: *A box of books is arranged in piles of six. Five piles of six books can be made. There are three books left over.*

- Can pupils represent the division story using their cubes? Can they explain how they know that their representation matches the division story? Ask: *How many books were in the box to begin with? How do you know?*

- Write the following number sentences on the board and ask pupils to discuss how each one relates to the book problem.

  $\square \div 6 = 5 \text{ r } 3 \qquad 5 \times 6 + 3 = \square$

- Agree that both number sentences represent this maths story and that the second example helps us to quickly determine the dividend. Look together at the use of multiplication and addition in the same sentence. Refer back to previous activities where the remainder was also added to the closest multiple of the divisor to find the missing dividend, but that the number sentence had not been recorded in this way before.

- Explore a second problem in a similar way, using counters or cubes to represent it.

- Ask pupils to make five groups of cubes with one cube in each group. This time explain that each cube represents a marble. Agree that the five marbles have been shared between five groups.

- Say: *A container of marbles is shared equally between five children. Each child gets seven marbles. There are some marbles left over in the container.*

- Ask pupils to represent the division story using their cubes. Can they explain how they know that their representation matches the division story? Record the division sentence $\square \div 5 = 7 \text{ r } \square$.

- Ask: *Why is it difficult to say how many marbles were in the container to begin with? What do we know?*

- Work together to agree that there must have been more than 35 marbles because we know some are left in the container after 5 groups of 7 have been made.

- Ask pupils to explain why there cannot be 40 marbles in the container. (Each child would have received 8 marbles and not 7.)

- Explore possible solutions to prove that there could be 36, 37, 38 or 39 marbles in the container giving rise to a remainder 1, 2, 3 or 4 respectively. The smallest remainder is one and the largest is four.

- Return to the division sentence to show that:

  $36 \div 5 = 7 \text{ r } 1 \qquad 37 \div 5 = 7 \text{ r } 2$
  $38 \div 5 = 7 \text{ r } 3 \qquad 39 \div 5 = 7 \text{ r } 4$

- Pupils are now ready to complete Question 3 in the Practice Book.

Same-day intervention

- Choose two of the word problems from Question 3 that pupils found more difficult or were unable to complete.

- In the same way as during the whole-class instruction, invite pupils to represent each problem using counters or cubes. In each case, break the problem down into smaller parts so that pupils are secure with the way that division has been implied each time through the different language used and what values represent the dividend, the divisor and the quotient. Look closely at any remainder and how it is represented or used to answer each problem.

Same-day enrichment

- Provide pupils with page one of **Resource 3.2.12c** Number sentences. (Answers below.)

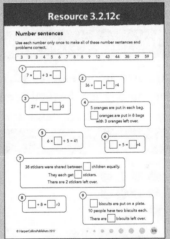

- Pupils can work individually or with a partner to make decisions about which numbers must be placed in the empty boxes to make all division sentences or problems correct. Each number can be used only once.

$7 \times 5 + 3 = 38$	$36 \div 8 = 4 \text{ r } 4$	5 oranges are put in each bag. 43 oranges are put in 8 bags with 3 oranges left over.
$27 \div 3 = 3 \text{ r } 3$	$6 \times 6 + 5 = 41$	$44 \div 5 = 8 \text{ r } 4$
38 stickers were shared between 3 children equally. They each get 12 stickers. There are 2 stickers left over.	$59 \div 8 = 7 \text{ r } 3$	29 biscuits are put on a plate. Ten people have two biscuits each. There are 9 biscuits left over.

Question 4

> **4** Use the number sentences of multiplication and addition to write number sentences of division with remainders.
>
> (a) $4 \times 3 + 1 = 13$
>
> $13 \div \square = 4 \text{ r } \square$
>
> $13 \div \square = 3 \text{ r } \square$
>
> (b) $4 \times 6 + 2 = 26$
>
> $\square \div \square = \square \text{ r } \square$
>
> $\square \div \square = \square \text{ r } \square$

What learning will pupils have achieved at the conclusion of Question 4?

- Pupils will be able to demonstrate why a number sentence such as, $5 \times 4 + 3$ relates to the division $23 \div 4 = 5 \text{ r } 3$.
- Pupils will be able to use manipulatives, images and number sentences to show division and multiplication.
- Pupils will have identified a missing dividend using what they know about the divisor and quotient to help them.

Activities for whole-class instruction

- Show pupils the array from **Resource 3.2.12d** Spot arrays. Can they quickly find the total number of spots in each group?

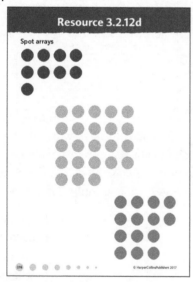

Resource 3.2.12d

Spot arrays

- Ask: *How did you find the total number?* Discuss the strategies used, focusing on multiplication and addition, for example the first array can be described as two rows of four or four columns of two and then add one.
- Draw around the equal rows or columns in each array so the multiplication facts are clear. (Save this image as it will be used again in the next lesson.)

 ...pupils who are not starting with the product of the equal rows/columns and adding the incomplete row or column at the end. Some pupils might, for example, simply add the number in each column from left to right – they will arrive at the correct total but without understanding of using the properties of arrays for efficiency.

- Ask: *What two number sentences can we write to match each array? Why will we need to include both multiplication and addition?*
- Record the number sentences for each as:

  $2 \times 4 + 1 = 9$ $4 \times 5 + 3 = 23$ $4 \times 3 + 2 = 14$

  $4 \times 2 + 1 = 9$ $5 \times 4 + 3 = 23$ $3 \times 4 + 2 = 14$

- Suggest that each of the number sentences above can be written as a division sentence. Look together at the sentences for the first array. Ask: *Which number will become the dividend? How do you know?* Record this as $9 \div \square = \square \text{ r } \square$ Ask: *What is the remainder? How do you know?* Write the remainder of one into the number sentence as $9 \div \square = \square \text{ r } 1$.
- Can the pupils explain what the divisor could be? Is there more than one possibility? Agree that two division sentences can actually be written about this array using 4 and then 2 as the divisor. Complete the calculations as $9 \div 4 = 2 \text{ r } 1$ and $9 \div 2 = 4 \text{ r } 1$.
- Invite pupils to work together to find the matching division sentences for the second and third array. Record the sentences on the board under each array.
- Pupils are now ready to complete Question 4 in the Practice Book.

Same-day intervention

- Give pupils enough cubes to make six sticks of 3 and three sticks of 6. Then give pupils two more cubes. Can another stick of six or three cubes be made?
- Use the sticks to make the following arrangements.

- Ask: *What multiplication and division sentences do the cubes represent?* Make a list together on the board.

- Focus on the division sentence 20 ÷ 3 = 6 r 2. Ask: *How does this relate to the first arrangement of cubes? Where can you see the numbers that are in the sentence in the arrangement of cubes?* Work together to draw the cubes and label the 20 (the dividend), the 3 (the divisor), the 6 (the quotient) and the 2 (the remainder). Agree that the number sentence tells us that: 20 can be grouped into 6 lots of 3 with 2 left over.

- Pupils should draw, label and describe the second arrangement as 20 ÷ 6 = 3 r 2 in a similar way. Now join the two leftover cubes to make a stick of two, rather than two single 'ones'. Place it under the 'rows' of sixes or alongside the 'columns' of threes so it becomes another (incomplete) stick. Can pupils still see: 20 ÷ 3 = 6 r 2 and 20 ÷ 6 = 3 r 2? Agree that they can.

- Pupils should make other arrangements of cubes and write number sentences that describe them.

Same-day enrichment

- Give pupils the following division sentences to complete. Can they find and write the related multiplication and addition fact each time?

- 27 ÷ 6 = 4 r ☐   27 ÷ 4 = ☐ r 3
  51 ÷ 9 = 5 r ☐   51 ÷ 5 = ☐ r ☐

- Pupils should then choose one or two of the examples and represent them concretely or pictorially as arrays.

Question 5

> **5** Application problems.
>
> (a) In a flower shop, a bouquet should have 6 flowers. How many bouquets can be made with 25 flowers? How many flowers will be left over?
>
> Answer: _____
>
> (b) 58 oranges are to be shared among 7 people equally. How many oranges will each person get? How many oranges will be left over?
>
> Answer: _____
>
> If each person should get 9 oranges, how many more oranges are needed?
>
> Answer: _____

What learning will pupils have achieved at the conclusion of Question 5?

- Pupils will have continued to practise applying their knowledge of division with remainders to word problems in context.

Activities for whole-class instruction

- Begin by returning to the arrays that were used as part of the previous lesson.

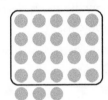

- Share the following division story with the pupils. Can they work in pairs to use one of the arrays to complete the missing information and make the story true?

  ☐ *bottles are arranged in boxes of* ☐*. There are enough bottles to fill* ☐ *boxes. There are* ☐ *bottles left over.*

- Ask: *What division sentence can you write to match the story?* (For example, 23 ÷ 5 = 4 r 3.)

- Can pupils also write a number sentence as ☐ × ☐ + ☐ = ☐ ?

- Tell pupils three different word problems can be made about this maths story.

  Bottles are arranged in boxes of 4. There are 5 of these boxes and three bottles left over.

  How many bottles are there altogether?

  23 bottles are arranged in groups of equal size.

  5 groups can be made and there 3 bottles are left over.

  How many bottles are in each group?

  23 bottles are put into boxes of 4.

  5 boxes can be made.

  How many bottles are left over?

- Ask pupils to discuss which piece of information is missing each time and how the question shows this.

- Relate this to the question that has been asked. Agree that the first question asks about the dividend in the related division sentence, which can be found using the number sentence 5 × 4 + 3.

- Look at the other problems in the same way to show that the group size is required in the second example and the remainder in the third.

- Invite pupils to work together to create their own division story for the third array using a context of their choice. Can they make up some different word problems to match the story? Remind them to try to change the information that is missing each time.

- Introduce the new problem:

  - *23 cherries are shared equally between 5 people. Each person gets 4 cherries. How many cherries are left over?*

  - *Each person would like to have 5 cherries. How many more cherries are needed?*

- Look together at how the arrays clearly show the answers to each of these problems.

- Agree that two more cherries are needed.

- Pupils are now ready to complete Question 5 in the Practice Book.

Same-day intervention

- Give pupils **Resource 3.2.12e** Word problems, and explain that they should use the representation each time to complete the missing information in each of the related word problems.

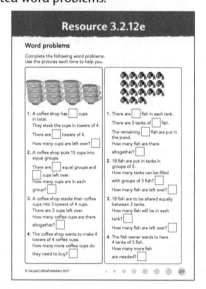

- Pupils can use cubes or counters to represent the problem each time if required.

Same-day enrichment

- Ask a pupil in the group to roll a 1–9 (or similar) dice three times (or roll three dice) to generate three numbers. The three numbers should be used to complete the number sentence ☐ × ☐ + ☐ = ☐.

- Tell pupils that the number sentence will be used to create two related division sentences, so they will need to think carefully about which number they use for the remainder.

- Pupils should quickly calculate the answer to the number sentence, for example roll 5, 3 and 7 and create the sentence $7 × 5 + 3 = 38$ or $5 × 7 + 3 = 38$, and then write the matching division sentences.

- Ask pairs of pupils to make up a division problem where the dividend has to be found.

- Share the problems together and check that this is true.

- Can they now make up a problem where the remainder needs to be found? Share the problems again.

- Finally ask them to think about another question that could be asked that requires another group of 5 (or 7) to be made using what is known about the remainder, for example, *Sam wants 8 marbles in each of his five bags. How many more marbles does he need?*

- Pupils work in pairs to roll the dice and generate a new number sentence to explore.

Challenge and extension question

Question 6

> **6** In the list of numbers 1, 3, 5, 1, 3, 5, 1, 3, 5, …,
>
> the 26th number is ☐.
>
> The sum of these 26 numbers is ☐.

This question requires pupils to interpret the start of a list of numbers with repeating numbers and relate this to what they know about division and remainders.

The abstract style of the question is rather different from those seen previously so it should be broken down into steps to ensure that pupils make sense of the way the list builds up as more sets of three numbers are added.

Unit 2.13
Calculation of division with a remainder (3)

Conceptual context

Pupils have solved a range of division problems and have further developed their conceptual understanding of the relationship with multiplication. Pupils have explored multiplication with addition, for example $4 \times 5 + 2 = 22$, and should be able to explain how this relates to division with remainders, for example $22 \div 5 = 4$ r 2.

In this unit, pupils apply their knowledge of 'a divisor' and further explore the smallest and largest possible remainder to find that the smallest possible remainder is one but that the largest possible remainder must be one less than the divisor. Their language to describe multiplication is also developed as pupils explain that, for example, 24 is 3 times as many as 8, and that 26 is 2 more than 3 times as many as 8.

Learning pupils will have achieved at the end of the unit

- Robust conceptual knowledge will have been developed when pupils use multiple representations using manipulatives and images to show division and multiplication (Q1, Q2, Q3, Q4)
- Pupils will have used knowledge of multiplication facts to find the largest multiple that is smaller but closest to a given number (Q1, Q2, Q3)
- Pupils will have identified a missing dividend using what they know about the divisor and quotient to help them (Q1, Q2, Q3, Q5)
- Through explaining what the largest and smallest remainder will be for a given divisor, pupils will have demonstrated depth of understanding (Q2, Q3, Q5)
- Pupils will have recognised that they may need to round up or down after a division or use the reminder, depending on the what is being asked (Q4)
- The language of multiplication will have been developed to include 'times as many as', 'times the number of' (Q4)
- Pupils will have continued to practise applying their knowledge of division with remainders to a range of problems with and without a context (Q3, Q4)

Resources

cubes; number rods; counters; dice; 0–9 digit cards; paper clip and pencil for a spinner; **Resource 3.2.13a** Number grid and table; **Resources 3.2.13b** Remainders; **Resources 3.2.13c** Grids

Vocabulary

equal groups, division, multiplication, dividend, divisor, quotient, remainder, leftover, multiple, factor, times as many as

Question 1

> **1** Calculate mentally.
>
> (a) $5 \times 4 =$ ☐　　　　　(b) $48 \div 6 =$ ☐
>
> (c) $28 \div 4 =$ ☐　　　　　(d) $6 \times 9 =$ ☐
>
> (e) $16 \div 4 =$ ☐　　　　　(f) $58 \div 8 =$ ☐
>
> (g) $4 \times 8 =$ ☐　　　　　(h) $56 - 8 =$ ☐
>
> (i) $17 \div$ ☐ $= 8 \, r \, 1$　　　(j) ☐ $\div 9 = 4 \, r \, 2$
>
> (k) ☐ $\div 6 = 8 \, r \, 2$　　　(l) ☐ $\div 8 = 9 \, r \, 7$
>
> (m) ☐ $\div 7 = 4 \, r \, 5$　　　(n) ☐ $\div 10 = 7 \, r \, 1$

What learning will pupils have achieved at the conclusion of Question 1?

- Robust conceptual knowledge will have been developed when pupils use multiple representations using manipulatives and images to show division and multiplication.

- Pupils will have used knowledge of multiplication facts to find the largest multiple that is smaller but closest to a given number.

- Pupils will have identified a missing dividend using what they know about the divisor and quotient to help them.

Activities for whole-class instruction

- Show pupils the grid of numbers and table on **Resource 3.2.13a** Number grid and table. Pupils can have their own copy of the table or simply draw a quick version of their own.

Resource 3.2.13a

Number grid and table

41	25	36	30
12	32	7	63
48	40	24	45
100	18	80	27

Multiples of 3	Multiples of 4	Multiples of 5
Multiples of 6	Multiples of 8	Multiples of 9

378 ● ● ● ● ● ● ● ● ● 　　© HarperCollinsPublishers 2017

- Give pupils five minutes to find and record all the different multiples from the number grid on their table.

- Select a multiple from each section on the table. Ask pupils to write the related multiplication and division sentences, for example select 27 from the first section and record as $9 \times 3 = 27$ and $27 \div 3 = 9$.

- Can pupils explain why some numbers appear in more than one section?

- Ask: *What is 27 – 9? Why does this give the same results as 2 × 9?*

- Invite pupils to use the question mark on the grid to represent another multiple of their choice that can be used in their table.

- One number from the grid has not been used (31). Can pupils explain why?

- Explore using 31 as the dividend using the following division sentences:

  $31 \div 5 =$ ☐ 　　 $31 \div$ ☐ $= 5 \, r \, 1$

- Can pupils find the multiple of 5 that is smaller than, but closest to, 31 for the first example?

- Continue to explore 31 using other divisors from the table. Pupils can use number rods or interlocking cubes to create a number line to show the closest multiple of the divisor and the remainder each time. Rearrange the rods in rows to show a different representation.

- Draw the image below.

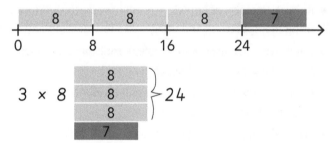

- Can pupils suggest other numbers from their tables that will leave a remainder when divided by 3, 4, 5, 6, 8 or 9?

  Look out for … pupils who think that, for example, a multiple of 3 will give a remainder when divided by 3. Pupils should also know that if a number is not a multiple of 3 it will give rise to a reminder when divided by 3. The arrays and number rods that pupils have worked with are the best models and images for instilling this understanding. Alternatively, use a double-sided number line or put flowers in groups of 3 into vases or similar.

- Draw the image below

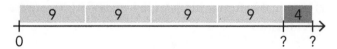

- Tell pupils that the number rods in your drawing represent a division using one of the numbers from the grid **Resource 3.2.13a** as the dividend.

- Ask: *What information do we know so far? How can we show this in a division sentence?* Record this as ☐ ÷ 9 = 4 r 2.

- Ask: *Which number has been used as the dividend? How do you know?* Revisit finding 4 × 9 + 4 to find the dividend.

- Pupils are now ready to complete Question 1 in the Practice Book.

Same-day intervention

- Explore some of the calculations discussed in the whole-class enrichment session in more depth.

- Pupils should use number rods or interlocking cubes for the divisor and represent the nearest multiple each time and then the remainder.

- Encourage them to describe the divisions as, for example, '7 groups of 5 are 35 and 36 is 1 more'.

- Ensure that some examples are set up so the dividend has to be found.

Same-day enrichment

- Pupils should draw the table below.

remainder of 1 31 ÷ 3 = 10 r 1	remainder of 2	remainder of 3 or 4

- Tell pupils to use the grid from the whole-class instruction activity and their table of multiples (both on **Resource 3.2.13a**) to create division sentences to complete the table.

- Encourage them to reason about possible divisions and what they know about multiples rather than simply carrying out each calculation.

- Can pupils explain that a dividend that leaves a remainder of:

    - 1, must be 1 more than a multiple of the divisor?

    - 2, must be 2 more than a multiple of the divisor?

    - 3, must be 3 more than a multiple of the divisor?

    - and so on?

Question 2 and Question 3

2 Find the remainder from each set of numbers and then work out the dividend.

(a) ⑤⑥⑦⑧
   ☐ ÷ 6 = 6 r ☐

(b) ④⑤⑥⑦
   ☐ ÷ 5 = 3 r ☐

(c) ③④⑤⑥
   ☐ ÷ 4 = 7 r ☐

(d) ①②③④
   ☐ ÷ 2 = 10 r ☐

3 Complete each calculation and find the answers.

(a) ◯ ÷ 4 = 4 r ☐
   The greatest possible number in the ◯ is ☐.
   The smallest possible number is ☐.

(b) ◯ ÷ 9 = 3 r ☐
   The greatest possible number in the ◯ is ☐.
   The smallest possible number is ☐.

What learning will pupils have achieved at the conclusion of Question 2?

- Robust conceptual knowledge will have been developed when pupils use multiple representations using manipulatives and images to show division and multiplication.

- Pupils will have used knowledge of multiplication facts to find the largest multiple that is smaller but closest to a given number.

- Pupils will have identified a missing dividend using what they know about the divisor and quotient to help them.

- Through explaining what the largest and smallest remainder will be for a given divisor, pupils will have demonstrated depth of understanding.

Activities for whole-class instruction

- Ask pupils to arrange 18 cubes into equal groups in different ways so there is a remainder each time. Collect different examples and record the related division facts, for example 18 ÷ 4 = 4 r 2 and 18 ÷ 5 = 3 r 3.

- Ask: *Why can we not arrange 18 cubes in groups of 3 here?* Agree that 18 is a multiple of 3 so there will be no remainder.

- Look together at the remainder and the divisor in each of the recorded division sentences. Say: *I have noticed something about the size of the remainder compared to the size of the divisor.* Ask pupils to discuss what you may have noticed.

- Explore the idea, using the pupils' grouping of cubes and the division sentence so confirming that the remainder is always smaller than the divisor – if it is the same or more, then another group can be made.

- Show pupils the following arrays for the multiplication facts 6×4 and 5×7. Can pupils write the matching division sentences? ($24 \div 4 = 6$ and $35 \div 7 = 5$)

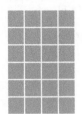 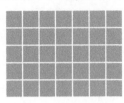

- Now start a new row by adding the first square. Ask: *How does this change the division sentence?* Agree that the dividend is now 25 and the remainder is 1 so $25 \div 4 = 6$ r 1 because $6 \times 4 + 1 = 25$.

- Add a second and then a third square, confirming the division sentences as $26 \div 4 = 6$ r 2 and $27 \div 4 = 6$ r 3.

- Add the fourth square that now completes the row. Write the division sentence $28 \div 4 = 6$ r 4. Ask pupils to discuss whether they agree with the sentence. Establish that $28 \div 4 = 7$ because another whole group of four has been made.

- Agree that the smallest remainder when dividing by four is 1 and the largest is 3.

- Now look at each dividend. The smallest dividend (25) leaves a remainder of 1, whereas the largest dividend (27) leaves a remainder of 3.

- Invite pupils to add one square at a time to make up the bottom row of the second array. Can they explain how the dividend and divisor changes each time? Agree as a class that the smallest remainder is again 1 but this time the largest remainder is 6.

- Look at the second array again. This time add one square to the top row. Ask: *What division sentence does the array represent now?* Agree $36 \div 5 = 7$ r 1 because $7 \times 5 + 1 = 36$

- Add more squares to the top row, amending the division sentence to show what is represented each time.

- Write ☐ $\div 4 = 6$ r ☐ on the board. Ask: *What can we say about the remainder if we know that the divisor is 4?* Can pupils tell you that the smallest remainder is 1 and the largest is 3?

- Ask: *So, what is the smallest possible dividend? And the largest possible dividend?*

- Pupils are now ready to complete Questions 2 and 3 in the Practice Book.

Same-day intervention

- Give pupils **Resource 3.2.13b** Remainders.

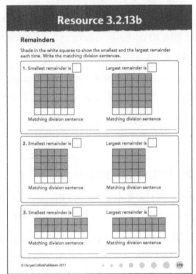

- Pupils should shade in extra squares to represent the division that gives the largest and smallest remainder.

- Pupils can then work together using interlocking cubes or squared paper to explore the smallest and largest remainders to match other arrays of their choice.

Same-day enrichment

- Share the following problem with the group.

  Ben plants 19 cabbage plants in the garden.

  He plants an equal number of cabbage plants in each row.

  He also plants an extra row that has fewer cabbage plants in than the others.

  In how many different ways can Ben plant his cabbage plants to give:

  a) the smallest number of cabbage plants in the extra row

  b) the greatest number of cabbage plants in the extra row

- Pupils should discuss the problem in pairs or small groups and decide what they need to do.

- Suggest that it may be useful to sketch or to represent the problem using cubes or counters.

- Ask: *What do we know about the smallest possible remainder each time? How can we use this to help think about a starting point for the problem?*

- Ask: *If you start with a remainder of 1, Ben must have planted a single cabbage plant in one of his rows and the rest must all be in equal rows. How might those rows be set out? (2 rows of 9, 9 rows of 2, 3 rows of 6 and so on.)*

- Can pupils write the division and multiplication sentences to match their solutions? (For example, 19 ÷ 3 = 6 r 1 and 6 × 3 + 1 = 19)

- Next, explore how the cabbage plants might be set out if the remainder is 2, 3, 4 and so on.

- Do pupils recognise that the size of the divisor will need to increase in order to make the remainder bigger?

- Pupils may suggest one row of 18 as this is the longest possible row that will still leave a reminder, but should notice that this will leave a remainder of one. Explore and reject other examples in the same way.

- The largest divisor that will leave the largest remainder for the dividend 19 is 10.

- Ask pupils to record the division and multiplication sentences. (19 ÷ 10 = 1 r 9 and 1 × 10 + 9 = 19)

Question 4

> 4 Application problems.
>
> (a) Jo's family has raised 3 ducks and 4 times as many chicks as ducks. How many chicks has the family raised?
>
> Answer: _____
>
> (b) The family has also raised 23 rabbits. A hutch can house 4 rabbits. How many hutches does the family need to house all the rabbits?
>
> Answer: _____
>
> (c) Jo has £50 to buy some hamsters for her family. Each hamster costs £9. How many hamsters can she buy?
>
> Answer: _____
>
> (d) The family has 5 black goldfish and 13 red goldfish.
>
> The number of red goldfish is ☐ more than ☐ times the number of black goldfish.

What learning will pupils have achieved at the conclusion of Question 4?

- Robust conceptual knowledge will have been developed when pupils use multiple representations using manipulatives and images to show division and multiplication.

- Pupils will have recognised that they may need to round up or down after a division or use the remainder, depending on the what is being asked.

- The language of multiplication will have been developed to include 'times as many as', 'times the number of'.

- Pupils will be more confident demonstrating and explaining why a number sentence such as, 5 × 4 + 3 relates to the division 23 ÷ 4 = 5 r 3.

Activities for whole-class instruction

- Group pupils at the front of the class and label them as shown.

Group A   Group B   Group C

Ask: *How many times larger is group B than group A? How do you know?* Establish that the number of pupils in group B is 2 times the number or twice as many as in group A because 2 × 2 = 4.

 4 is 2 times as many as 2.

- Also agree that there are 2 groups of 2 in 4 so 4 ÷ 2 = 2.

- Can pupils compare group B and C in the same way? What do they notice?

 8 is 2 times as many as 4.

- Record the related number sentences as 2 × 4 = 8 and 8 ÷ 2 = 4.

- Suggest that the number of pupils in group C is 4 times as many as in group A. Ask pupils to explain whether they agree.

- Invite a third pupil to join group A. Ask: *Are there still 4 times the number of pupils in group C as in group A?* Explore this practically, by adding a line of three to group C each time. Establish that C is not large enough to show 3 times the number as group A:

	Group A	Group C	
i)	X X X	X X X	group C is 1 times the number of pupils in group A so 3 × 1 = 3
ii)	X X X	X X X X X X	group C is 2 times the number of pupils in group A so 3 × 2 = 6
iii)	X X X	X X X X X X	group C is 2 more than 2 times the number of pupils in group A so 3 × 2 + 2 = 8

- Combine group B and C to become group D.

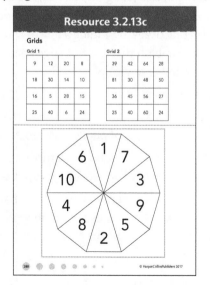

- Can pupils complete the sentence, '12 is ☐ times as many as ☐?'

 All say ... *12 is 4 times as many as 3.*

- Record the number sentences as $3 \times 4 = 12$ and $12 \div 4 = 3$.
- Show everyone a stick of five cubes. Say that you want to give each pupil in group D a cube. Ask: *How many sticks of cubes will I need to use?* (3)
- Write the division sentence $12 \div 5 = 2$ r 2 on the board. Ask: *Why is the answer to the division is 2 r 2 but the answer to the question asked is three?*
- Can pupils explain that while you use two whole sticks of five you must begin a third stick so that all pupils have a cube?
- Say that the cubes can be used to buy pencils. Each pencil costs five cubes. Invite pupils to gather their cubes to purchase pencils. Ask: *How many pencils can we buy with 12 cubes?* Establish that although the division sentence is again, $12 \div 5 = 2$ r 2, this time the answer needed does not include the remainder.
- Talk about other division problems that require us to make a decision about how to use the remainder.
- Pupils are now ready to complete Question 4 in the Practice Book.

Same-day intervention

- Give pupils grid 1 from **Resource 3.2.13c** Grids.

Resource 3.2.13c

Grids

Grid 1
9	12	20	8
18	30	14	10
16	5	28	15
25	40	6	24

Grid 2
39	42	64	28
81	30	48	50
36	45	56	27
25	40	60	24

(Spinner with numbers 1, 7, 3, 9, 5, 2, 8, 4, 10, 6)

© HarperCollinsPublishers 2017

- Ask them to choose a number from the grid and represent it using number rods (or interlocking cubes) of the same length, for example pick 15 and show this as 5 rods of 3.
- Pupils can check other's representations and say why they agree or disagree.
- Introduce the following game. Explain that the aim is to place as many of their team's counters on the grid. Each number can be used more than once.
- Pupils work in pairs. Each pair takes turns to spin a 1–10 spinner (using a pencil and paper clip and the spinner provided on **Resource 3.2.13c**) and take the matching number rod, for example spin 3 and take a 3 rod. They choose a number from the grid that is a multiple of their number rod figure, for example 12. Pupils use the rods to show how many times larger the number on the grid is than the rod, for example 12 is 4 times as many as 3. They place a counter on 12 if they are correct and write, for example $3 \times 4 = 12$ and $12 \div 4 = 3$.

Same-day enrichment

- Introduce the game while other pupils independently complete the first intervention activity. Give pupils grid 2 from **Resource 3.2.13c** to play a game with remainders. Pupils may find it useful to work in pairs.
- The game starts in the same way as for the intervention activity but, having spun a number, for example 3, pupils choose a value from the grid that will leave a remainder when divided by the number on the spinner, for example 20. Pupils can use number rods or their own pictorial representations to show that '20 is 2 more than 6 times as many as 3'. They record this as $3 \times 6 + 2 = 20$ and as $20 \div 3 = 6$ r 2. If they are correct, they place a counter on the value on the grid.

Challenge and extension questions

Question 5

> **5** Fill in the boxes.
>
> (a) 35 ÷ ☐ = ☐ r 5   (b) 57 ÷ ☐ = ☐ r 3
>
> (c) 35 ÷ ☐ = ☐ r 3   (d) 57 ÷ ☐ = ☐ r 1

For Question 5, pupils must draw on their knowledge of multiples and use this to make decisions about a divisor that will leave the given remainder. In each of the examples, pupils are also given the dividend, for example 35 ÷ ■ = ■ r 5.

They should explain that the multiple of the divisor that is smaller than but closest to the dividend must, in the example above, be 5 less than the dividend, so is 30. They can then choose factors of 30 to make the number sentence correct, for example 35 ÷ 6 = 5 r 5. Can they explain why 35 ÷ 5 = 6 r 5 or 35 ÷ 3 = 10 r 5 are not possible solutions?

Question 6

> **6** Six children are sharing some stickers. If 5 more stickers are added, then each child can have 5 stickers.
>
> There are ☐ stickers.

In Question 6, pupils may find it useful to represent the problem practically or pictorially. They will need to determine that the number of stickers is 5 less than 6 × 5. You could model this as 6 groups of 5 cubes showing that each child has five stickers and then subtract 5 by taking away one sticker from five of the children .

Chapter 2 test (Practice Book 3A, pages 63–68)

Test question number	Relevant unit	Relevant questions within unit
1	Unit 2.1	Q3
	Unit 2.2	Q2
	Unit 2.3	Q4
	Unit 2.9	Q2, Q3
	Unit 2.10	Q4
	Unit 2.12	Q1
	Unit 2.13	Q1
2	Unit 2.2	Q3
	Unit 2.5	Q3

Test question number	Relevant unit	Relevant questions within unit
3	Unit 2.6	Q3
	Unit 2.7	Q1, Q2
	Unit 2.8	Q1, Q2
	Unit 2.10	Q5
	Unit 2.12	Q2, Q3
	Unit 2.13	Q2, Q3
4	Unit 2.6	Q3
	Unit 2.7	Q1, Q2
	Unit 2.8	Q1, Q2
	Unit 2.10	Q5
	Unit 2.11	Q1, Q2, Q5
	Unit 2.12	Q5
	Unit 2.13	Q4
5	Unit 2.10	Q5, Q6
	Unit 2.11	Q3, Q4
	Unit 2.12	Q3, Q4
	Unit 2.13	Q2, Q3, Q4

Chapter 3
Knowing numbers up to 1000

Chapter overview

Area of mathematics	National Curriculum statutory requirements for Key Stage 2	Shanghai Maths Project reference
Number – number and place value	Year 3 Programme of study: Pupils should be taught to:	
	■ count from 0 in multiples of 4, 8, 50 and 100; find 10 or 100 more or less than a given number	Year 3, Units 3.5, 3.6
	■ recognise the place value of each digit in a three-digit number (hundreds, tens, ones)	Year 3, Units 3.1, 3.2, 3.5, 3.6
	■ compare and order numbers up to 1000	Year 3, Units 3.3, 3.4
	■ identify, represent and estimate numbers using different representations	Year 3, Units 3.1, 3.2, 3.3, 3.5, 3.6
	■ read and write numbers up to 1000 in numerals and in words	Year 3, Units 3.1, 3.2, 3.3, 3.5, 3.6
	■ solve number problems and practical problems involving these ideas.	Year 3, Units 3.3, 3.4, 3.5, 3.6

Unit 3.1
Knowing numbers up to 1000 (1)

Conceptual context

This chapter develops pupils' ability to read, write and partition numbers up to 1000, extending their understanding of the positional aspect of place value. This understanding is vital for success in mathematics.

Various visual representations are used to assist understanding. This unit uses the following

- Diagrams showing hundreds, tens and ones.

  Squares represent 100, lines represent tens and dots represent ones.

   Diagram shows 243

  These 2-D diagrams correspond to the elements of concrete base 10 apparatus that are used in the next unit.

- Place value stick abacus

  By placing counters on the sticks, three-digit numbers can be represented with the correct numbers of counters in each position. Blank place value stick abacuses can also be reproduced on paper and the 'counters' drawn as circles on each stick.

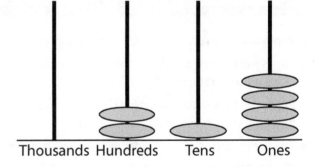

- Place value grid

Thousands	Hundreds	Tens	Ones
1	0	0	0

The use of place value grids enables pupils to build up numbers in any order and then add zeros as placeholders where required.

- Place value arrow cards

These are sets of hundreds, tens and ones cards that have an 'arrow' on the right-hand side. Pupils can select a hundreds card, a tens card and a ones card. By lining up the arrows on the cards they can make a three-digit number and by separating them they can show the number in expanded form.

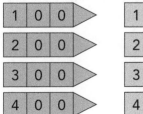 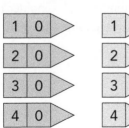

Pupils need to appreciate the relationship between the representations and the numerals in the three-digit number. Using a variety of different visual representations helps pupils to master the position concept of place value.

While the unit focuses on three-digit numbers, there is mention of the number 1000, thus introducing the concept of four-digit numbers. The thousands place has been shown in the place value grid representation. If appropriate, the thousands place can be shown to pupils for the place value stick abacus and the place value arrow card representations so that pupils appreciate that there is a potential column there.

 Zero as a placeholder

Hundreds	Tens	Ones
4	0	7

This number is 407. If the zero was not there, the number would appear as 47, not the value it is intended to represent at all. So the zero in 407 is important; it is a placeholder, keeping the 4 in the correct hundreds column, and not in the tens column.

Hundreds	Tens	Ones
5	3	0

This number is 530. This number has no 'ones' so it is important to show that the 3 does not represent ones, but tens so the zero is placed there for this purpose – it holds everything in the correct place.

Learning pupils will have achieved at the end of the unit

- Pupils will have revised addition, subtraction, multiplication and division within 100 (Q1)
- Pupils will have explored how to write three-digit numbers in numerals and in words from visual representations (Q2)
- Pupils will have used place value charts to write three-digit numbers and occasional four-digit numbers in numerals and in words (Q2)
- The use of zero as a placeholder will have been understood (Q2, Q3, Q4)
- Knowledge of place value in three-digit numbers will have been developed (Q2, Q3, Q4)
- Pupils will have partitioned three-digit numbers into hundreds, tens and ones and four-digit numbers into thousands, hundred, tens and ones (Q3, Q4)
- Pupils will have begun to communicate about place value in abstract terms with understanding (Q3)

Resources

1–50 class number cards; place value grids; 0–9 dice; place value stick abacus; place value arrow cards; mini whiteboards

Vocabulary

ones, tens, hundreds, placeholder, place value, place value arrow cards, place value stick abacus

Question 1

> **1** Calculate mentally.
>
> (a) 7 × 7 = ☐   (b) 80 ÷ 10 = ☐   (c) 45 − 18 = ☐
>
> (d) 50 ÷ 9 = ☐   (e) 74 + 6 = ☐   (f) 5 × 8 = ☐
>
> (g) 44 + 13 = ☐   (h) 54 ÷ 5 = ☐   (i) 0 ÷ 3 = ☐
>
> (j) 10 × 10 = ☐   (k) 70 − 7 = ☐   (l) 63 ÷ 9 = ☐

What learning will pupils have achieved at the conclusion of Question 1?

- Pupils will have revised addition, subtraction, multiplication and division within 100.

Activities for whole-class instruction

- Practise the 7 times table with examples of multiplication and division.
- Ask:

  How can I use what I know about my 10 times table to help me work out answers to question about the 9 times table?

  How can I use what I know about my 10 times table to help me work out answers to question about the 5 times table?

  How can I use what I know about my 5 times table to help me work out answers to question about the 6 times table?

  How can I use what I know about my 2 times table to help me work out answers to question about the 8 times table?

- Use a set of 1–50 number cards. Choose a pupil to select a card and ask the class to calculate what must be added to make 70. Ask individual pupils to describe their method.
- Pupils complete Question 1, working mentally to revise the four operations.

Same-day intervention

- In pairs, make up eight different number statements that equal 30. There should be two addition, two subtraction, two multiplication and two division statements.
- Repeat with the number 24 or 50.

Same-day enrichment

- Ask pupils to complete the same-day intervention task.
- Ask: *Is it possible to carry out this task with any two-digit number?* Establish that you can always make up addition and subtraction statements. There is a finite number of possible addition statements and an unlimited number of subtraction statements possible. However, not all numbers allow the formation of two multiplication or two division statements using whole numbers.

Question 2

> **2** Look at each diagram and write the number. The first one has been done for you.
>
> (a) (i)   In numerals: 243
>   In words: *two hundred and forty-three*
>
> (ii)   In numerals: _____
>   In words: _____
>
> (iii)   In numerals: _____
>   In words: _____
>
> (iv)   In numerals: _____
>   In words: _____
>
> (v)   In numerals: _____
>   In words: _____
>
> (b) (i)
>
> | Hundreds | Tens | Ones |
> |----------|------|------|
> | 6 | 0 | 5 |
>
> In numerals: _____
> In words: _____
>
> (ii)
>
> | Hundreds | Tens | Ones |
> |----------|------|------|
> | 8 | 2 | 4 |
>
> In numerals: _____
> In words: _____

What learning will pupils have achieved at the conclusion of Question 2?

- Pupils will have explored how to write three-digit numbers in numerals and in words from visual representations.
- Pupils will have used place value charts to write three-digit numbers in numerals and in words.
- The use of zero as a placeholder will have been practised.

Activities for whole-class instruction

- Write 100 on the board. Ask: *How many digits does this number have?* Establish that it has three digits and it is 100.
- Draw a place value chart and choose a pupil to write 100 on it.

Hundreds	Tens	Ones
1	0	0

- Ask: *How many tens are there in 100?* Agree that there are 10. (10 × 10 = 100) Count together in tens to 100.
- Now ask if anyone can count in hundreds. Listen to what they say. They may count, '1 hundred, 2 hundred, 3 hundred, … 9 hundred, 10 hundred'. Ask: *What is the special name for 10 hundred?* Explain that is a 'thousand' and that 10 × 100 = 1000.

 10 times a 100 is a 1000.

- Display a place value grid showing a thousand.

Thousands	Hundreds	Tens	Ones
1	0	0	0

- Discuss the concept that this is a big number and that it would take quite a long time to count 1000 items. Ask if they can think of something that could be used to do this and how it could be shown. Listen to their ideas. They may suggest counters, beads, sweets, leaves, beans or perhaps drawing crosses, circles and so on.
- Choose something suitable to count together to make 1000. Split pupils into 10 groups and say that you would like each group to count 100 items. Ask them how they can work as a group to make sure that they exactly 100 items. Elicit that each 100 could be displayed in a pattern, for example as 4 sets of 25.

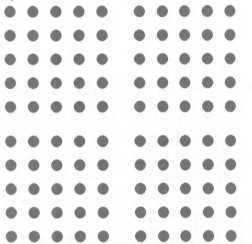

- Display this 1000 that pupils have made somewhere in the classroom. This visual display will help pupils appreciate the magnitude of 1000.
- Show a place value grid and ask pupils to explain how to show 200, 300 and so on. Confirm that for 300 there is a 3 in the hundreds column and zeros in the tens and ones columns. The zeros are placeholders so that the hundreds numeral is in the correct position.

Hundreds	Tens	Ones

- When there are numbers in the tens and ones columns, the whole number is read by saying 'and' after the hundred so that, for example 467 is read as 'four hundred and sixty-seven'; 540 is read as 'five hundred and forty'. Ask pupils to practise reading a variety of numbers.
- Give pupils mini whiteboards and ask them to draw a place value chart. Choose one pupil to come and look at a card with a three-digit number on it, for example 306. They should say the number out loud and pupils should write it on their mini whiteboards.

Hundreds	Tens	Ones
3	0	6

- Choose another pupil and repeat with a new number. Continue until pupils demonstrate confidence reading three-digit numbers, including numbers with a zero in the tens or ones position.
- Pupils complete Question 2, working independently.

Same-day intervention

- Introduce pupils to the place value stick abacus. Show them the number 214 on the abacus and ask them to read the number. Repeat with more numbers.
- Now reverse the process by asking them to show a number on the abacus.

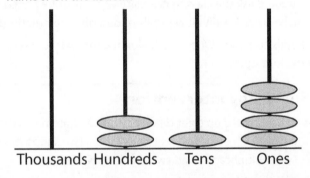

Thousands Hundreds Tens Ones

Same-day enrichment

- In pairs, take turns to roll a 0–9 dice and put digits into a place value chart. Once placed the digit cannot be moved. Pupils take turns to write their digit into a column on their place value grid, aiming to finish with the highest three-digit number they can. The pupil with the highest number after three rolls of the dice each scores 10. The first pupil to reach a 100 wins the game.

Question 3

> **3** Complete the sentences.
>
> (a) 856 is made up of ☐ hundreds, ☐ tens and ☐ ones.
>
> (b) Counting from the right in a 4-digit number, the first place is
>
> the _____ place, the tens place is the _____ place
>
> and the thousands place is the _____ place.
>
> (c) 707 is written in words as _____ .
>
> _____
>
> 7 in the ones place means ☐ ones.
>
> 7 in the hundreds place means ☐ hundreds.
>
> The difference between them is ☐ .
>
> (d) 4 hundreds and 3 ones make ☐ .
>
> It is written in words as _____ .
>
> _____

What learning will pupils have achieved at the conclusion of Question 3?

- Pupils will have partitioned three-digit numbers into hundreds, tens and ones.
- Knowledge of place value in three-digit numbers will have been developed.
- Pupils will have begun to communicate about place value in abstract terms with understanding.
- The use of zero as a placeholder will have been practised.

Activities for whole-class instruction

- Use a place value stick abacus to show three-digit numbers, for example 481. Ask pupils to say which number is in:
  - the hundreds position: 4 – 400
  - the tens position: 8 – 80
  - the ones position: 1
- Now write the number in the form – 400 + 80 + 1
- Repeat with more numbers.
- Show pupils a diagram of 326. Establish that it is 300 + 20 + 6. This is a way of writing numbers to see the value of each digit.
- Explain that the two ways of writing the number are equivalent, so 300 + 20 + 6 = 326.

> (All say...) *Counting from the right in a three-digit number, the first place is the ones place, the second place is the tens place and the third place is the hundreds place.*

- Pupils complete Question 3 in the Practice Book.

Same-day intervention

- Ask pupils to consider these pairs of numbers, reading the numbers and then writing them in words.
  - 102 and 120
  - 340 and 304
  - 506 and 605
  - 480 and 840
  - 199 and 200
- An additional challenge would be to work out the difference between the pairs of numbers.

Same-day enrichment

- Using the digits 1, 2 and 3, challenge pupils to write as many three-digit numbers as they can. Ask them to explain whether they have found all the possible numbers and how they know. (123, 132, 213, 231, 312, 321)

Question 4

> **4** Fill in the boxes.
>
> (a) 462 = ☐ + ☐ + ☐
>
> (b) 1050 = ☐ + ☐ + ☐ + ☐
>
> (c) 788 = ☐ + ☐ + ☐
>
> (d) 300 + 90 + 0 = ☐
>
> (e) 800 + 8 = ☐

What learning will pupils have achieved at the conclusion of Question 3?

- The use of zero as a placeholder will have been practised.
- Pupils will have partitioned three-digit numbers into hundreds, tens and ones.

Activities for whole-class instruction

- Introduce pupils to place value arrow cards.

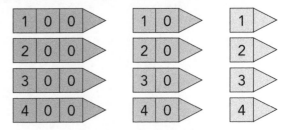

- Give pupils time to explore the cards freely.

- Explain that by lining up the arrows on the cards they can make a three-digit number and by separating them they can show the parts of the number.
- Ask them to find the number 400, the number 20, and the number 3. Put them together to make 423.
- Repeat with other numbers, including numbers with zero in the tens or the ones position.
- Pupils complete Question 4 in the Practice Book.

Same-day intervention

- Give pupils the following arrow cards, 700, 800, 50, 60, 3 and 4. Ask pupils to make at least 10 different three-digit numbers.
- A pupil was asked to write the number 534 and wrote 500304. Ask pupils to explain the error to a partner and then listen to one or two explanations.

Same-day enrichment

- Ask pupils to use these clues to find each three-digit number:

  - This three-digit number is an odd number. The sum of the digits is 2. (101)

  - The numerals in this three-digit number are 6, 3 and 1. It is the smallest possible odd number. (163)

  - The three numerals in this three-digit number are all the same. The sum of the digits is 9. (333)

  - This three-digit number has the same numeral in the ones place and in the hundreds place. The numeral in the tens place is 2 less than the numeral in the ones place. 7 is in the hundreds place. (757)

  - This three-digit number is even. The hundreds numeral is 2 more than the tens numeral, which is 2 more than the ones numeral. The numeral in the tens position is 6. (864)

- Challenge pupils in pairs to write clues for another three-digit number and try it on another pair.

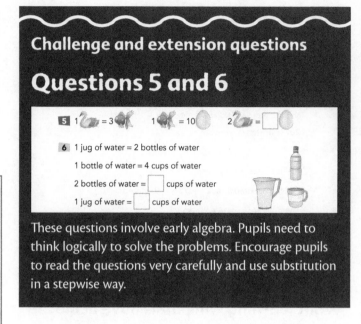

Challenge and extension questions

Questions 5 and 6

These questions involve early algebra. Pupils need to think logically to solve the problems. Encourage pupils to read the questions very carefully and use substitution in a stepwise way.

Unit 3.2
Knowing numbers up to 1000 (2)

Conceptual context

This unit continues to build pupils' knowledge of numbers to 1000. Base 10 blocks are again used as a concrete representation to aid understanding of place value. The pieces in base 10 blocks look exactly like their value.

Place value grids and place value arrow cards are also used and will sometimes include thousands as well as hundreds, tens and ones.

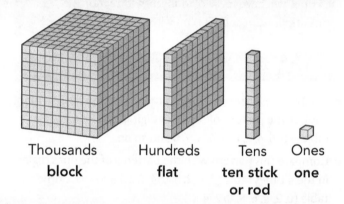

Thousands	Hundreds	Tens	Ones
block	**flat**	**ten stick or rod**	**one**

Learning pupils will have achieved at the end of the unit

- Pupils will have revised addition, subtraction, multiplication and division within 100 (Q1)
- Pupils' number world will have been expanded to include thousands (Q2, Q3)
- Pupils will have explored different ways to represent three-digit numbers (Q2, Q5)
- Use of zero as a placeholder will have been practised (Q2, Q3, Q4, Q5)
- Reading three-digit numbers will have been consolidated (Q2, Q3, Q4, Q5)
- Writing three-digit numbers in numerals and words will have been practised (Q3, Q4, Q5)
- Pupils will have begun to communicate about place value in abstract terms with understanding (Q4)
- Pupils will have practised drawing diagrams for three-digit numbers (Q5)

Resources

0–9 digit cards (individual and class); 0–100 number cards; base 10 blocks; mini whiteboards; place value arrow cards; **Resource 3.3.2** Food masses; food items (showing their mass), place value grids, place value stick abacus

Vocabulary

ones, tens, hundreds, thousand, placeholder, base 10 blocks (one, rod, stick, flat, big block), place value, place value arrow cards, place value stick abacus

Question 1

> **1** Calculate mentally.
>
> (a) $4 \times 7 \times 2 =$ ☐
>
> (b) $51 - 30 + 5 =$ ☐
>
> (c) $45 + 91 + 29 =$ ☐
>
> (d) $60 \div 6 + 29 =$ ☐
>
> (e) $4 \times 3 - 0 =$ ☐
>
> (f) $0 \times 4 + 62 =$ ☐
>
> (g) $51 + 37 - 59 =$ ☐
>
> (h) $1 \times 8 - 7 =$ ☐
>
> (i) $3 \times 0 + 48 =$ ☐
>
> (j) $16 + 75 - 41 =$ ☐

What learning will pupils have achieved at the conclusion of Question 1?

- Pupils will have revised addition, subtraction, multiplication and division within 100.

Activities for whole-class instruction

- Practise the 8 times table with multiplication and division. Look at the pattern of the ones digit in the 8 times table – 8, 6, 4, 2, 0, 8, 6, 4, 2, 0 and so on.

- Compare this pattern with the patterns of the ones digits in the 4 times table (4, 8, 2, 6, 0, 4, 8 …) and the 2 times table (0, 2, 4, 6, 8, 0, 2, 4 …)

- Use the 50–100 number cards. Pick one and ask pupils to calculate what must be subtracted to make 20. Ask individual pupils to describe their method.

- Pupils complete Question 1 in the Practice Book.

Same-day intervention

- In pairs, ask pupils to shuffle a set of 1–9 digit cards and to pick four cards. They should then use them to make two two-digit numbers and then write addition and subtraction sentences using these numbers. For example, picking 2, 5, 7 and 1, pupils could make 25 and 71.

  $25 + 71 = 96$

  $71 - 25 = 46$

- Rearrange the four digits to make new numbers, for example 57 and 12, and repeat.

- Use multiplication and division facts to write three more number sentences for each of the following:

  $8 \times 5 = 40$ ($5 \times 8 = 40$; $40 \div 8 = 5$; $40 \div 5 = 8$)

  $6 \times 9 = 54$ ($9 \times 6 = 54$; $54 \div 9 = 6$; $54 \div 6 = 9$)

  $7 \times 6 = 42$ ($6 \times 7 = 42$; $42 \div 7 = 6$; $42 \div 6 = 7$)

Same-day enrichment

- Ask pupils, in pairs, to shuffle a set of 1–9 digit cards and to pick four cards. They should then use them to make two two-digit numbers and then write addition and subtraction sentences using these numbers. For example, picking 2, 5, 7 and 1, pupils could make 25 and 71.

  $25 + 71 = 96$

  $71 - 25 = 46$

- Using the same four numbers, challenge pupils to make two two-digit numbers that have:

  - the largest total ($71 + 52$ or $72 + 51$)

  - the smallest total ($17 + 25$ or $15 + 27$)

- Ask pupils to explain their reasoning.

- Challenge pupils to write at least ten multiplication and division facts that contain the number 24.

Question 2

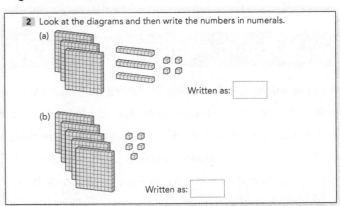

> **2** Look at the diagrams and then write the numbers in numerals.
>
> (a)
>
> Written as: ☐
>
> (b)
>
> Written as: ☐

What learning will pupils have achieved at the conclusion of Question 2?

- Pupils' number world will have been expanded to include thousands.

- Pupils will have explored different ways to represent three-digit and four-digit numbers.

- Use of zero as a placeholder will have been practised.

Activities for whole-class instruction

- Give pupils base 10 blocks and review together how 10 ones make 1 ten stick (or rod). The 100 'flats' and 1000 'big blocks' are new to pupils. Explore how 10 sticks make 1 flat, 10 flats make 1 big block and practise the names of the different pieces.

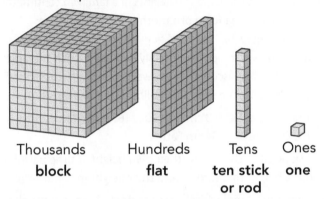

Thousands **block** Hundreds **flat** Tens **ten stick** Ones **one or rod**

- Make the number 324 using base 10 blocks. Choose pupils to state the number of hundreds (300), the number of tens (20) and the number of ones (4).

- As pupils describe each part and write the matching number statement:

  $300 + 20 + 4 = 324$ or $324 = 300 + 20 + 4$

- Repeat this process with five more three-digit numbers, for example 271, 165, 436, 509 and 350. Ensure that numbers using zero as a placeholder are included.

- Carry out the process in reverse by showing pupils a three-digit number, for example 185, and working together, make it using base 10 blocks. Ask pupils to articulate the pieces they have chosen.

 To make 185, you need 1 flat, 8 sticks and 5 ones.

- Repeat until pupils show confidence in handling base 10 blocks to represent three-digit numbers.

- Pupils complete Question 2 in the Practice Book.

Same-day intervention

- Give pupils base 10 blocks and allow them to check the pattern of the blocks again.

- 10 ones make 1 rod (or ten stick), 10 rods make 1 flat, 10 flats make 1 big block.

- So, 10×1 is 10, 10×10 is 100, 10×100 is 1000.

- Challenge pupils to make the number 239 using base 10 blocks. Ask: *What is 1 more than 239 and how is that shown using base 10 blocks?*

- Continue with other pairs of numbers, for example:

  - 109 and 1 more than 109
  - 300 and 1 less than 300
  - 430 and 1 less than 430

Same-day enrichment

- Say: *The digits in a three-digit number add up to ten. Two of the digits are odd.*

- Challenge pupils to find as many possible numbers as possible that fit these criteria. (118, 181, 811; 336. 363, 633)

- Ask pupils to make some of the numbers with base 10 blocks.

Question 3

> 3 Read the numbers and then write them in words or in numerals.
>
> (a) 635 in words: _____
>
> (b) 302 in words: _____
>
> (c) Nine hundred and thirty-six in numerals: ☐
>
> (d) 1000 in words: _____
>
> (e) Four hundred in numerals: ☐

What learning will pupils have achieved at the conclusion of Question 3?

- Pupils' number world will have been expanded to include thousands.

- Reading three-digit numbers will have been consolidated.

- Writing three-digit numbers in numerals and words will have been practised.

Activities for whole-class instruction

- Choose four pupils to sit in chairs marked Th, H, T and O in order. Give the H pupil a set of 1–9 digit cards and T and O pupils a set of 0–9 digit cards each. Ask them, without looking, to choose a card to show to the rest of the class. Ask the class to read the three-digit number together. Repeat with a new selection of numbers and choose a different selection of pupils to read the number.

- Ask different questions about the numbers, for example:
  - *How many hundreds are there?*
  - *Is the number even?*
  - *How many tens are there?*
- Ask the H T and O pupils to show their '9' cards. Ask the class: *What will happen if I add one to this number?* Support the class to consider how each digit in turn will change and a '1' will appear in the Thousands column.

 10 ones are 10. 10 tens are 100. 10 hundreds are 1000.

- Use base 10 blocks to show a three-digit number. Give pupils mini whiteboards and ask them to write the numbers in numbers and words. If necessary, display words with tricky spellings, for example eighty.
- Pupils complete Question 3 in the Practice Book.

Same-day intervention

- Challenge pupils to use place value arrow cards to make a three-digit number and to write the number on a mini whiteboard. Repeat with more numbers, including numbers with zero in the tens or ones position.

Same-day enrichment

- Read out these numbers: 364, 418, 904, 540, 491, 643. Ask pupils to identify the value of the '4' in each of them, writing the value on mini whiteboards.

Question 4

> 4 Fill in the boxes.
>
> (a) Four hundred and eight is written in numerals as ☐.
>
>   It is a ☐-digit number.
>
>   It consists of ☐ hundreds and ☐ ones.
>
> (b) The number consisting of 6 tens and 4 hundreds is ☐.
>
> (c) 10 one hundreds is ☐. ☐ one hundreds is 500.
>
> (d) The numbers that come before and after 300 are ☐ and ☐ respectively.
>
> (e) There are ☐ tens in 470.

What learning will pupils have achieved at the conclusion of Question 4?

- Pupils will have begun to communicate about place value in abstract terms with understanding.

Activities for whole-class instruction

- Write these measurements on the board. Tell pupils that they are height measurements in centimetres. Ask them to read the numbers.

  126 cm   143 cm   168 cm   105 cm   184 cm

- Explain that these are the heights of a family in centimetres. The family is made up of a mother, father and three children aged 10, 7 and 3. Ask pupils to talk to a partner to work out which height probably belongs to each member of the family. Discuss their answers and how they decided. Establish that the most likely allocation of heights is:

  126 cm Age 7   143 cm Age 10   168 cm Mother
  105 cm Age 3   184 cm Father

- Ask pupils if they know their own height. If time permits, use a height measure to find their height in centimetres.
- Give pupils mini whiteboards and ask them to write the number 300 and then to write the numbers that come before and after it.
- Repeat with these numbers: 210, 999, 700 and 409.
- Pupils should work in small groups with base 10 blocks and answer two questions written on the board.
- What is the tens digit in the number 126? (The tens digit is 2.)
- How many tens are there in 126? (There are 12 tens in 126.)
- Elicit that the answers to these questions are different. Repeat with other numbers.
- Pupils complete Question 4 in the Practice Book.

Same-day intervention

- Give pupils **Resource 3.3.2** Food masses, which examines the mass in grams of items of food and ask them to complete it.
- If possible, bring some real foods with labels into the classroom for pupils to look at.

> **Resource 3.3.2**
>
> **Food masses**
>
> Here are some items of food with the mass in grams shown on them.
>
> Find and write the mass of each item in the table.
>
> | Biscuits | Beans | Honey | Tuna | Coffee |
> |---|---|---|---|---|
> | | | | | |
>
> Don't forget to write the 'g' for grams!
> Look at the mass and answer these questions.
>
> 1. Biscuits: What is the tens digit? _____
> 2. Beans: How much more than 400 g does the tin weigh? _____
> 3. Honey: What does the zero in the ones column mean? _____
> 4. Tuna: What is the hundreds digit? _____
> 5. Coffee: How many lots of 10 g are there in the jar of coffee? _____
> 6. What is the difference in mass between the coffee and the biscuits? _____
> 7. How much do two tins of beans weigh? _____
> Make up and answer three questions of your own. Try them out on a friend.
> 8. _____
> 9. _____
> 10. _____
> Look at some real items of food to find out what they weigh.
>
> © HarperCollins Publishers 2017

Same-day enrichment

- The sum of the digits in the numbers 214 is 7 (2 + 1 + 4 = 7).
- Challenge pupils to find five more three-digit numbers with a digit sum of 7. They should draw a place value stick abacus for each of the three-digit numbers.

Question 5

5 Draw diagrams to represent numbers and then write them in numerals. The first one has been done for you.

One hundred and eighty-seven

Draw: Written as: | 187 |

(a) Six hundred and six

Draw: Written as: | |

(b) Two hundred and eighty

Draw: Written as: | |

What learning will pupils have achieved at the conclusion of Question 5?

- Pupils will have practised drawing base 10 block diagrams for three-digit numbers.
- Use of zero as a placeholder will have been practised.

Activities for whole-class instruction

- Revise with pupils the use of base 10 block diagrams to represent three-digit numbers. Squares represent 100, lines represent tens and dots represent ones.

- Practise drawing a few three-digit numbers together, including numbers with zero in the tens or ones column.
- Divide the class into three groups. Give the first group place value arrow cards, the second group place value stick abacuses and the third group place value grids. Write a three-digit number on the board and ask each group to show the number using their representation and then to draw the number as a diagram on mini whiteboards.

- Pupils complete Question 5 in the Practice Book.

Same-day intervention

- Use base 10 blocks to make a selection of three-digit numbers and then draw the matching diagram. Challenge pupils to make 20 three-digit numbers, including some that include zeros.

Same-day enrichment

- Challenge pupils to find as many ways as they can of representing the number 134.
- If pupils run out of ideas, suggest possible ways such as counters, base 10 blocks, place value stick abacus, place value arrow cards, base 10 block diagrams or a number line.

Challenge and extension question

Question 6

6 The numbers 4, 0 and 2 can be used to make ☐ 3-digit numbers. Write these numbers and put them in order starting from the greatest, using > to link them.

4 **0** **2**

Answer: _____

Here, pupils are challenged to write and order all possible three-digit numbers using the digits 4, 0 and 2. The number of possibilities is reduced as a result of the inclusion of 0 because three-digit numbers cannot begin with zero. Thus the arrangements, 042 and 024, are not permissible.

Unit 3.3
Number lines (to 1000) (1)

Conceptual context

In Book 2 pupils explored number lines to 100. In this unit pupils compare and order numbers up to 1000, using a 0–1000 number line or part of it. Comparing and ordering require numerical reasoning and evaluation – these are higher-order skills.

A large scale 0–1000 number line mounted on the wall is a very useful resource at this stage of pupils' learning. Number lines should have an arrow at the right end to indicate that numbers continue. At a later stage in learning, an arrow will also be shown at the left end to acknowledge numbers less than 0.

As numbers become larger, using number lines that start at zero becomes unwieldy so pupils can simply use the relevant part of a number line. The use of a number line is a visual way for pupils to consolidate understanding of the relative size of numbers. Numbers increase to the right and pupils quickly learn to compare one number with another by knowing which one is further to the right on the number line.

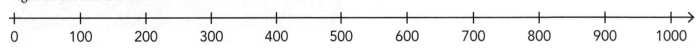

| 0 | 100 | 200 | 300 | 400 | 500 | 600 | 700 | 800 | 900 | 1000 |

To order three-digit numbers, pupils first need to look at the size of the hundreds digit.

Learning pupils will have achieved at the end of the unit

- Pupils will have applied number knowledge to infer the value represented by unlabelled marks on number lines where only hundreds or tens are labelled (Q1)
- Pupils will have identified the tens numbers and hundreds numbers on either side of a three-digit number (Q1, Q3)
- Pupils will have reinforced their knowledge of counting to 1000, particularly crossing tens and hundreds boundaries (Q2, Q4)
- Number sequences where the numbers increase and decrease in ones, tens and hundreds, including crossing tens and hundreds boundaries as appropriate, will have been explored (Q3)
- Overall fluency in number sense will have been developed, enabling pupils to determine where any three-digit number is located and the pattern of surrounding numbers (Q2, Q3, Q4, Q5)
- Pupils will have developed their understanding that to order three-digit numbers, they should begin by looking at the hundreds digit (Q4)
- Pupils will have solved word problems involving comparison and ordering of numbers up to 1000 (Q4, Q5)

Resources

0–1000 class number lines; blank number lines (individual and class); ten 1-metre sticks; small blank cards; **Resource 3.3.3a** Sequences; **Resource 3.3.3b** Sequencing errors; **Resource 3.3.3c** Number questions

Vocabulary

number line, ones, tens, hundreds, thousand, three-digit number, number pattern, number sequence

Question 1

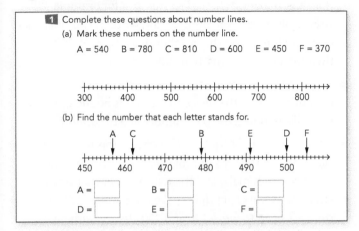

1 Complete these questions about number lines.
(a) Mark these numbers on the number line.
A = 540   B = 780   C = 810   D = 600   E = 450   F = 370

(b) Find the number that each letter stands for.

A = ☐   B = ☐   C = ☐
D = ☐   E = ☐   F = ☐

What learning will pupils have achieved at the conclusion of Question 1?

- Pupils will have applied number knowledge to infer the value represented by unlabelled marks on number lines where only hundreds or tens are labelled.
- Pupils will have identified the tens numbers on either side of a three-digit number.

Activities for whole-class instruction

- Show pupils part of a blank number line, from 500 to 550, with 500 marked as the first tens number.

500

- Discuss with pupils how number lines are constructed, beginning at zero and getting bigger to the right. Explain that we do not always need to see the whole of the number line but can focus on a particular section. Here we are going to look at the section from 500 to 550.

- Work together and write in the rest of the tens numbers, 510, 520, 530, 540 and 550.

- Spend time locating different numbers and discussing how to do this for example:

531 – I found 530 and counted on 1.

525 – I know 525 is half way between 520 and 530. I found 520 and looked for the halfway point.

539 – I know 539 is one less than 540. I found 540 and counted back 1.

- For numbers ending in 2 or 3, find the previous 10 and count on.

- For numbers ending in 4, they may choose to find the previous 10 and count on four or to find the appropriate number ending in 5 and count back one. Listen and share pupils' methods.

- Pupils may count on from 5, for numbers ending in 6, 7 and 8.

- Ask them to locate 504, 536, 543, 518 and 527.

- Give pupil pairs a blank part number line with the first 10 labelled 780 and ask them:

  – to label the rest of the tens (790, 800, 810, 820 and 830) below the number line.

  – to locate the position of the following numbers using a small arrow pointing down the line from above: 815, 801, 806, 793, 817, 784, 799.

- Pupils complete Question 1 in the Practice Book.

Same-day intervention

- To develop pupils' understanding of the magnitude of 1000 and how numbers grow, make a large number line using 10 one-metre sticks

      ... up to 1000

- Prepare small cards as shown above for 0, 100, 200, 300, 400, 500, 600, 700, 800, 900 and 1000.

- Working together as a group, lay the metre sticks out in a straight line and place the small cards below the intersections of the metre sticks.

- Choose a pupil to walk from zero along the number line and stop at 200. Tell another pupil to walk from zero along the number line to 600. Ask: *Which number is bigger and by how much?* Ask the pupil at 200 to walk slowly along the number line, counting the number of hundreds to reach 600. Establish that it is 400. (200 + 400 = 600)

- Ask individual pupils to walk along the line to intermediate numbers to practise finding numbers on the line, for example 735.

- Take the number line apart and shuffle the cards, then challenge pupils to reassemble it.

Same-day enrichment

- Give pupil pairs a blank part number line (50 units long) and ask them to number it from 600 to 650.
- Mark the number 625 as 'A' on the number line.
- Work out and mark the following on the number line.
  - B = A + 5 (630)
  - C = A – 10 (610)
  - D = A + 21 (646)
  - E = A – 19 (606)

Question 2

> 2 Fill in the boxes with numbers based on the given information.
>
> (a) Write the numbers that come before and after each number.
>
>   (i) [ ] , 278, [ ]
>
>   (ii) [ ] , 999, [ ]
>
>   (iii) [ ] , 406, [ ]
>
> (b) Write the tens numbers that come before and after each number.
>
>   (i) [ ] , 390, [ ]
>
>   (ii) [ ] , 455, [ ]
>
>   (iii) [ ] , 789, [ ]
>
> (c) Write the hundreds numbers that come before and after each number.
>
>   (i) [ ] , 657, [ ]
>
>   (ii) [ ] , 405, [ ]
>
>   (iii) [ ] , 790, [ ]

What learning will pupils have achieved at the conclusion of Question 2?

- Pupils will have reinforced their knowledge of counting to 1000, particularly crossing tens and hundreds boundaries.
- Pupils will have identified the tens numbers and the hundreds numbers on either side of a three-digit number.
- Overall fluency in number sense will have been developed, enabling pupils to determine where any three-digit number is located and the pattern of surrounding numbers.

Activities for whole-class instruction

- Give pupils mini whiteboards. Sit three pupils in chairs at the front of the class with their mini whiteboard and pen. They are the Hundreds (H), Tens (T) and Ones (O) positions for three-digit numbers. Ask them to write a single digit on the mini whiteboard and then reveal it to the class, making a three-digit number, for example 389. Ask pupils to read the number and write the number that comes before it and the number that comes after it on their mini whiteboards, here 388 and 390.

- Repeat with new numbers. Check that pupils can confidently cross the tens and hundreds boundaries. Use the wall-mounted 0–1000 number line to support.

- Choose new pupils for the HTO chairs. This time ask the class to write the tens number that comes before the number pupils make and the tens number that comes after. For example, if pupils make the number 723, the class need to write 720 and 730. Repeat with new numbers, checking that pupils can confidently cross the hundreds boundaries.

- Change the HTO pupils again. This time ask pupils to write the hundreds number that comes before and the hundreds number that comes after.

- Pupils complete Question 2 in the Practice Book.

Same-day intervention

- Tell pupils that this is an example of counting that has gone wrong. Ask them, in pairs, to write the set of five numbers correctly and explain what has happened.

  - … three hundred and ninety-seven, three hundred and ninety-eight, three hundred and ninety-nine, three hundred and ninety-ten, three hundred and ninety-eleven …

- Show pupils a three-digit number, for example 642. Give them mini white boards and ask them to write the answers to questions such as:

  - the numbers either side of 642
  - the whole tens either side of 642
  - the whole hundreds either side of 642
  - the number that is 9 more than 642
  - the number that is 3 fewer than 642

- Repeat with another number, for example 799 that will cross tens and hundreds boundaries.

Same-day enrichment

- Write the whole tens between the following pairs of numbers:
  - 217 and 271 (220, 230, 240, 250, 260, 270)
  - 349 and 394 (350, 360, 370, 380, 390)
  - 738 and 783 (740, 750, 760, 770, 780)
  - 564 and 646 (570, 580, 590, 600, 610, 620, 630, 640)
- Write the whole hundreds between the following pairs of numbers:
  - 175 and 462 (200, 300, 400)
  - 349 and 994 (400, 500, 600, 700, 800, 900)
  - 294 and 763 (300, 400, 500, 600, 700)
  - 499 and 800 (500, 600, 700)

Question 3

3 Count and complete the number patterns.
  (a) 567, 568, 569, ☐ , ☐
  (b) 350, 370, 390, ☐ , ☐ , 450
  (c) 743, 742, 741, ☐ , ☐
  (d) 250, ☐ , ☐ , 550, 650, 750

What learning will pupils have achieved at the conclusion of Question 3?

- Pupils' number world will have been expanded to include thousands.
- Number sequences where the numbers increase and decrease in ones and tens, including crossing tens and hundreds boundaries as appropriate will have been explored.
- Pupils will have explored number sequences where the numbers increase and decrease in hundreds.
- Overall fluency in number sense will have been developed, enabling pupils to determine where any three-digit number is located and the pattern of surrounding numbers.

Activities for whole-class instruction

- In this question, pupils learn to identify number patterns, in which numbers increase or decrease in ones, tens and hundreds. Show pupils the following two number patterns
  - A 567, 568, 569, 570 …
  - B 567, 566, 565, 564 …

- Ask pupils to talk to a partner about what is happening in each number pattern. Establish that in pattern A the numbers increase by 1 and in pattern B the numbers decrease by 1.
- Here are two more number patterns for pupils to analyse.
  - C 360, 350, 340, 330
  - D 360, 370, 380, 390
- Ask what is happening to the ones digit. (It is zero in every number in both C and D.)
- Ask what is happening to the tens digit. (The tens digit is decreasing by one in C and increasing by one in D.)
- In pattern C the numbers decrease by 10 and in pattern D they increase by 10.
  - E 200, 300, 400, 500, 600
  - F 800, 750, 700, 650, 600
- In E the numbers increase by 100 and in F each number decreases by 50.
- Ask pupils to continue the number patterns.
- Try other number sequence until pupils can confidently identify the patterns.
- Ask pupils to make up a series of five three-digit numbers that increase **or** decrease by 10 **or** 5. Their partner should add the next 3 numbers.
- Pupils complete Question 3 in the Practice Book, completing the number patterns.

Same-day intervention

- Give pupils **Resource 3.3.3a** Sequences, and ask them to fill in the missing numbers in the sequences and add the next two numbers.

Resource 3.3.3a

Sequences

For each sequence fill in the missing number, add the next two numbers in the sequence and describe the pattern of the numbers. The first one has been done for you.

1. 458, 459, 460 , 461, 462, 463, 464 , 465

The sequence is going up in ones.

2. 346, ☐ , 348, ☐ , 350, 351, ☐ ☐

3. ☐ , 999, 998, ☐ , 996, 995, ☐ ☐

4. 90, 95, ☐ ☐ , 110, 115, ☐ ☐

5. ☐ , 900, 800, ☐ , 600, ☐ , ☐

6. 670, 680, 690, ☐ ☐ ☐ ☐

© HarperCollinsPublishers 2017

● Answers

1. 458, 459, **460**, 461, **462**, 463, **464, 465** The sequence is going up in ones.

2. 347, **348**, 349, **350**, 351, 352, **353, 354** The sequence is going up in ones.

3. **1000**, 999, 998, **997**, 996, 995, **994, 993** The sequence is going down in ones.

4. 90, 95, **100, 105,** 110, 115, **120, 125** The sequence is going up in fives.

5. **1000**, 900, 800, **700**, 600, **500, 400** The sequence is going down in hundreds.

6. 670, 680, 690, **700, 710, 720, 730** The sequence is going up in tens.

5. 150, 250, 350, 450, ↑, 650, 750
   150, 250, 350, 450, 550, 650, 750   550 is missing.

6. 615, 610, 605, 600, 590, 585
   615, 610, 605, 600, 595, 590, 585   595 is missing.

Question 4

> 4 Choose suitable numbers from the list below to answer these questions.
>
> | 439 | 501 | 92 | 888 | 654 | 499 | 328 | 1000 |
>
> (a) The numbers greater than 400 but less than 500 are:
>
> _____
>
> (b) The number that is 111 less than 999 is:
>
> _____
>
> (c) The numbers that come before and after 500 are:
>
> _____
>
> (d) Write the above numbers in order, starting from the greatest.

Same-day enrichment

● Give pupils **Resource 3.3.3b** Sequencing errors, and ask them to locate and explain the errors.

Resource 3.3.3b

Sequencing errors

There is a mistake in each of these sequences. Sometimes a number is missing and sometimes a number is incorrect.
Use an arrow to show where the missing number should be or circle the incorrect number. Then write the sequence out correctly and explain the error. The first one has been done for you.

1. 160, ↑ 180, 190, 200, 210
   160, 170, 180, 190, 200, 210   170 is missing.

2. 232, 234, 235, 236, 237, 238

_____

3. 380, 390, 400, 401, 420, 430

_____

4. 820, 800, 780, 760, 740, 730

_____

5. 150, 250, 350, 450, 650, 750

_____

6. 615, 610, 605, 600, 590, 585

_____

© HarperCollinsPublishers 2017

What learning will pupils have achieved at the conclusion of Question 4?

● Pupils will have reinforced their knowledge of counting to 1000, particularly crossing tens and hundreds boundaries.

● Overall fluency in number sense will have been developed, enabling pupils to determine where any three-digit number is located and the pattern of surrounding numbers.

● Pupils will have developed their understanding that to order three-digit numbers, they should begin by looking at the hundreds digit.

● Pupils will have solved word problems involving comparison and ordering of numbers up to 1000.

Activities for whole-class instruction

● Ask pupils to order this set of two-digit numbers starting with the biggest number. They explain to a talk partner how they did it.

| 73 | 46 | 81 | 78 | 29 |

● Listen to one pair's explanation. Establish that they need to look at the tens digit first and the largest number has the greatest number of tens, so 81 is the biggest number.

● There are two numbers with 7 tens, 73 and 78. 78 is greater than 73 because 8 > 3, so we now have:

81 > 78 > 73

● Answers

1. 160, ↑180, 190, 200, 210
   160, 170, 180, 190, 200, 210   **170 is missing.**

2. 232, ↑, 234, 235, 236, 237, 238
   232, 233, 234 235, 236, 237, 238   **233 is missing.**

3. 380, 390, 400, 401, 420, 430
   380, 390, 400, 410, 420, 430   **401 is incorrect, it should be 410.**

4. 820, 800, 780, 760, 740, 730
   820, 800, 780, 760, 740, 720   **The numbers are decreasing by 20 each time.**

- Continue by looking at the remaining numbers to establish the following order:

81 > 78 > 73 > 46 > 29

- Now ask pupils how they think they would order these three-digit numbers, again beginning with the largest number:

365   143   282   149   259

- Elicit that they need to begin by comparing the hundreds digits, looking for the largest one. This is 365.

 All say ... *To order three-digit numbers, first look at the hundreds digit and choose the number with the biggest digit.*

- There are two numbers in the 200s; 282 and 239. We can order them by looking at the tens digit, 80 > 50, so the order becomes 365 > 282 > 259.

- The final two numbers both have the same hundreds digit and the same tens digit, so they need to be sorted by the ones digit. The final order is thus:

365 > 282 > 259 > 149 > 143

- Sometimes numbers need to be sorted in increasing size. Ask pupils to explain to a partner how they would sort three-digit numbers, beginning with the smallest.

- Write another set of numbers on the board and ask them to put these numbers in order, this time beginning with the smallest number, for example:

435   987   172   333   123
(Order is 123 > 172 > 333 > 435 > 987)

- Pupils complete Question 4 in the Practice Book, choosing suitable numbers.

Same-day intervention

- Show pupils a three-digit number, for example 631. Here are some statements about this number.

631 is an odd number.

631 comes after 630 and before 632.

The whole tens either side of 631 are 630 and 640.

631 in expanded form is 600 + 30 + 1

The whole hundreds either side of 631 are 60 and 700.

The hundreds digit is 6.

There are 63 tens in 631.

- Give pupils a new three-digit number, for example 496 and challenge them to write five statements about it.

- Repeat with another number.

Question 5

5 Write the numbers.
(a) Write the numbers greater than 498 but less than 505.

_____

(b) Write the hundreds numbers less than 900.

_____

(c) Write all the 3-digit numbers that are less than 200 and have the same digit in the ones place and in the tens place.

What learning will pupils have achieved at the conclusion of Question 5?

- Overall fluency in number sense will have been developed, enabling pupils to determine where any three-digit number is located and the pattern of surrounding numbers.

- Pupils will have solved word problems involving comparison and ordering of numbers up to 1000.

Activities for whole-class instruction

- Remind pupils about the importance of reading word problems very carefully. Write the answer and then read the question again to check that you have answered what was being asked!

- Pupils complete Question 5 in the Practice Book.

Same-day intervention

- Give pupils **Resource 3.3.3c** Number questions

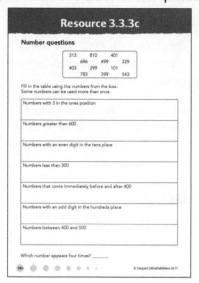

- Answers

Numbers with 3 in the ones position	213, 403, 543, 783
Numbers greater than 600	696, 783, 810
Numbers with an even digit in the tens place	329, 543, 696, 783
Numbers less than 300	101, 213, 299
Numbers that come immediately before and after 400	399, 401
Numbers with an odd digit in the hundreds place	101, 329, 543, 783
Numbers between 400 and 500	401, 403, 499

- Which number appears four times? (783)

Same-day enrichment

- Write all the numbers between 900 and 1000 that have 2 as a digit. (902, 912, 920, 921, 922, 923, 924, 925, 926, 927, 928, 929, 932, 942, 952, 962, 972, 982, 992)

- Write all the numbers between 10 and 1000 that have all digits the same. (11, 22, 33, 44, 55, 66, 77, 88, 99, 111, 222, 333, 444, 555, 666, 777, 888, 999)

Challenge and extension questions

Question 6

> 6 In some 3-digit numbers, the sum of the three digits is 15 and the digit in the hundreds place is twice the digit in the ones place.
>
> These 3-digit numbers are: _____.

Here, pupils need to read the problem carefully and break it down into stages. To solve the problem, they need to focus on finding possibilities that fulfil the requirement 'the hundreds digit is twice the ones digit'. The pairs are 2 – 1, 4 – 2, 6 – 3, 8 – 4. The second part of the problem states 'the sum of the three digits is 15' which means that 2 – 1 is not a possible solution because 15 minus 3 does not give a single digit solution.

Pupils may appreciate using a place value grid to place the digits.

Question 7

> 7 When you write numbers from 200 to 300, you need to write the digits:
>
> 1 ☐ times, 2 ☐ times and 0 ☐ times.

This question requires pupils to count and reason systematically. Suggest that pupils work in pairs so that they can monitor each other.

Unit 3.4
Number lines (to 1000) (2)

Conceptual context

In this unit pupils deepen their abstract understanding of the composition of three-digit numbers, recognising the magnitude of the values in each of the places (columns). This allows them to compare and order three-digit numbers, using greater than and less than symbols appropriately.

To order numbers starting with the largest, pupils need to determine the biggest number. To do this, they should understand that the number with the most hundreds is the largest, and if there is more than one number with the same hundreds digit, then the one with the most tens will be the largest. If two numbers have the same hundreds and tens digits, then the size of the ones digit will enable them to be placed in order; therefore 436364 357 353.

This deeper understanding of place value means that for most pupils, there is no longer confusion over pairs of number such as 365 and 356. Continue to encourage pupils to visualise pictorial representations such as diagrams and number lines. Ensure, too, that concrete and pictorial representations are still readily available for any pupils who still need them. Pupils vary a great deal in the time taken to be able answer questions simply using abstract numbers.

Learning pupils will have achieved at the end of the unit

- Pupils will have applied number knowledge to infer the value represented by unlabelled marks on number lines where only tens are labelled (Q1)
- Pupils will have identified the tens number and hundreds number either side of a three-digit number (Q1)
- Pupils will have practised using letters to represent numbers (Q1)
- Overall fluency in number sense will have been improved, enabling pupils to determine where any three-digit number is located and the pattern of surrounding numbers (Q1, Q3, Q4, Q5)
- Pupils will have developed their understanding that to order three-digit numbers, they should begin by looking at the magnitude of the hundreds digit, then the magnitude of the tens digit and finally the magnitude of the ones digit (Q1, Q3, Q4, Q5)
- Number sequences where the numbers increase and decrease in ones, twos, fives, tens, twenties and hundreds, including crossing tens and hundreds boundaries will have been explored further (Q2)
- Pupils will have explained the pattern of number sequences using full sentences (Q2)
- Pupils will have solved problems involving comparison and ordering of three-digit numbers (Q3, Q5)
- Pupils will have practised writing three-digit numbers as x00 + y0 + z (expanded) (Q4)
- Pupils will have compared pairs of three-digit numbers using >, < and = symbols (Q4, Q5)
- Pupils will have used reasoning to solve word problems involving comparison and ordering of three-digit numbers (Q5)

Resources

0–1000 number line; blank number lines; mini whiteboards; 1–9 digit cards; 1–6 dice; place value arrow cards; **Resource 3.3.4a** Number sequences; **Resource 3.3.4b** Number patterns; **Resource 3.3.4c** Partitioned numbers; **Resource 3.3.4d** Finding 3-digit numbers

Vocabulary

number line, ones, tens, hundreds, thousand, three-digit number, number pattern, number sequence, rule

Question 1

> **1** Complete these questions about number lines.
>
> ```
> |++++++++|++++++++|++++++++|++++++++|++++++++|++++++++|→
> 560     570      580      590      600      610
> ```
>
> (a) Mark the following numbers on the number line.
>
> A = 587   B = 565   C = 599
>
> D = 571   E = 602   F = 618
>
> (b) Fill in the boxes.
>
> (i) A + ☐ = 590   (ii) B + ☐ = 570   (iii) C − ☐ = 590
>
> (iv) D − ☐ = 570   (v) E + ☐ = 610   (vi) F − ☐ = 610
>
> (c) Fill in the boxes.
>
> (i) A + ☐ = 600   (ii) B − ☐ = 500   (iii) C + ☐ = 600
>
> (iv) D − ☐ = 500   (v) E + ☐ = 700   (vi) F − ☐ = 600
>
> (d) Put the six numbers A, B, C, D, E and F in order, starting with the smallest.
>
> ☐ < ☐ < ☐ < ☐ < ☐ < ☐

What learning will pupils have achieved at the conclusion of Question 1?

- Pupils will have applied number knowledge to infer the value represented by unlabelled marks on number lines where only tens are labelled.
- Pupils will have identified the tens number and the hundreds number either side of a three-digit number.
- Pupils will have practised using letters to represent numbers.
- Overall fluency in number sense will have been improved, enabling pupils to determine where any three-digit number is located and make sense of the way other numbers and digits in numbers change.

Activities for whole-class instruction

- Begin by asking pupils to revise making single numbers up to the next 'tens number'. Show 0–9 digit cards and ask what has to be added to 6 to make 10. Repeat with all other numbers, checking that pupils have rapid recall.
- Now practise making up to the next 10, by showing cards from 10–99 number cards and asking pupils to answer in a full sentence, for example show 62.

(All say …) *62 plus 8 makes 70.*

- Repeat with other numbers, including some numbers over 90 to review that the next tens number, in this case, is 100, a three-digit number. For example, 96 plus 4 makes 100.

- Show pupils a section of a blank number line from 670 to 720, with only 670 and 720 marked. Work together to add the missing tens numbers, 680, 690, 700 and 710. Ask individual pupils to indicate individual numbers and to describe how they do this. For example:

694 – The number is six hundred and **ninety**-four, so I find 690 and then I count on 4 to give 694.

699 – I know that the number is one less that 700, so I find 700 and count one back.

699 – The number is six hundred and **ninety**-nine. The nineties come just before the next hundreds numbers so I find 700 and then I count back 1 to give 699.

- Where pupils suggest alternative methods, accept them all and discuss which might be less prone to error.
- Continue placing further numbers, using further number line sections if necessary, until pupils show confidence.
- Look back at some of the chosen numbers, and ask pupils to make them up to the next tens boundary, again articulating their thinking.

682 + 8 makes 690

694 + 6 makes 700

- Finally ask pupils to make the number up to 700 by addition or subtraction, explaining their thinking.
- Pupils should complete Question 1 in the Practice Book.

Same-day intervention

- Here are some pairs of numbers:

449 and 476   598 and 704   867 and 941

- List the tens numbers that lie between the pair of numbers.
- List the hundreds numbers either side of the pair of numbers.

(Answers: 450, 460, 470, 400 & 500; 600, 610, 620, 630, 640, 650, 660, 670, 680, 690, 700, 600, 800; 870, 880, 890, 900, 910, 920, 930 940, 800 & 1000)

Same-day enrichment

● Ask pupils to complete this table:

	10 fewer	10 more	100 fewer	100 more
156				
278				
592				
803				
695				

Answers

	10 fewer	10 more	100 fewer	100 more
156	146	166	56	256
278	268	288	178	378
592	582	602	492	692
803	793	813	703	903
695	685	705	595	795

Question 2

> 2 Count and complete the number patterns.
>
> (a) ☐ , ☐ 290, 285, ☐ , ☐
>
> (b) 486, 488, ☐ , ☐
>
> (c) 123, 223, 323, ☐ , ☐ , ☐

What learning will pupils have achieved at the conclusion of Question 2?

● Number sequences where the numbers increase and decrease in ones, twos, fives, tens, twenties and hundreds, including crossing tens and hundreds boundaries will have been explored further.

● Pupils will have explained the pattern of number sequences using full sentences.

Activities for whole-class instruction

● Show pupils the following two number sequences.

| 350 | 370 | 390 | 410 |
| 350 | 330 | 310 | 290 |

● Ask them what is the same about the two sequences and what is different. Establish that they are the same because the first term in each sequence is 350. However, in the first

sequence the numbers increase by 20, so the rule is 'add 20'. In the second sequence, the numbers decrease by 20, so the rule is 'subtract 20'.

● Challenge pupils to write some sequences on mini whiteboards, for example:

● A sequence in which three-digit numbers go up in tens and the tens digits are all 5.

● A sequence in which three-digit numbers go up in hundreds.

● A sequence in which three-digit numbers decrease in twos.

● Share and discuss some of the sequences. Ask them how to find the next term and explain that this is known as the rule.

 All say ...

To find the rule for a number sequence, find the difference between terms that are next to each other.

● Pupils should complete Question 2 in the Practice Book. While they are working, ask individual pupils to explain the rule for the sequence.

Same-day intervention

● Give pupils **Resource 3.3.4a** Number sequences.

Resource 3.3.4a

Number sequences

Start each number sequence with the number 290 and write the next five numbers in the pattern.

1. Increasing in ones

   290, _____

2. Increasing in tens

   290, _____

3. Decreasing in fives

   290, _____

4. Increasing in twenties

   290, _____

5. Increasing in hundreds

   290, _____

6. Decreasing in ones

   290, _____

© HarperCollinsPublishers 2017 · · · · · · · 285

Answers

- Increasing in ones   (290, 291, 292, 293, 294, 295)
- Increasing in tens   (290, 300, 310, 320, 330, 340)
- Decreasing in fives   (290, 285, 280, 275, 270, 265)
- Increasing in twenties (290, 310, 330, 350, 370, 390)
- Increasing in hundreds (290, 390, 490, 590, 690, 790)
- Decreasing in ones   (290, 289, 288, 287, 286, 285)

Same-day enrichment

- Give pupils **Resource 3.3.4b** Number patterns

Resource 3.3.4b

Number patterns

Look at the number patterns, write the next three numbers and explain the rule. The first rule has been done for you.

1. 340, 345, 350, 355, 360, ☐, ☐.

   The rule is add 5.

2. 840, 830, 820, 810, ☐, ☐, ☐.

   The rule is _____

3. 508, 506, 504, 502, ☐, ☐, ☐.

   The rule is _____

4. 795, 815, 835, 855, ☐, ☐, ☐.

   The rule is _____

5. 951, 851, 751, 651, ☐, ☐, ☐.

   The rule is _____

6. 682, 684, 686, 688, ☐, ☐, ☐.

   The rule is _____

© HarperCollinsPublishers 2017

Answers

1. 340, 345, 350, 355, 360, 365, 370. The rule is add 5.
2. 840, 830, 820, 810, 800, 790, 780. The rule is subtract 10.
3. 508, 506, 504, 502, 500, 498, 496. The rule is subtract 2.
4. 795, 815, 835, 855, 875, 895, 915. The rule is add 20.
5. 951, 851, 751, 651, 551, 451, 351. The rule is subtract 100.
6. 682, 684, 686, 688, 690, 692, 694. The rule is add 2.

Question 3

> 3 Write the numbers in order.
>
> (a) Start with the greatest: 175 715 517 157 751 117
>
> (b) Start with the smallest: 869 886 689 668 969 898
>
> (c) From the greatest to the smallest: all the 3-digit numbers with 7 in the ones place and 6 in the tens place.

What learning will pupils have achieved at the conclusion of Question 3?

- Overall fluency in number sense will have been further developed, enabling pupils to determine where any three-digit number is located and the pattern of surrounding numbers.

- Pupils will have developed their understanding that to order three-digit numbers, they should begin by looking at the magnitude of the hundreds digit, then the magnitude of the tens digit and finally the magnitude of the ones digit.

- Pupils will have solved word problems involving comparison and ordering of three-digit numbers.

Activities for whole-class instruction

- Write the numbers 265 and 256 on the board and ask: *Which is the bigger number and how do they know?* Establish that 265 > 256. They may explain by saying that both numbers have the same hundreds value but 265 has 6 tens while 256 has only 5 tens so it is smaller. They may suggest showing the relative positions of the numbers on a number line. Try further pairs of numbers, for example 613 and 631, 453 and 435.

- Ask pupils to suggest three-digit numbers that have 2 in the hundreds place and 5 in the ones place.

- Determine that the numbers that fit these criteria are 205, 215, 225, 235, 245, 255, 265, 275, 285 and 295.

- Write the numbers randomly over the board as pupils suggest them and then ask them how they would order them, beginning with the least.

- Establish that the order is
205 < 215 < 225 < 235 < 245 < 255 < 265 < 275 < 285 < 295
and that because all the numbers have 2 in the hundreds place they need to look for the smallest value in the tens place and then the smallest remaining value.

- Ask pupils to explain how to order three-digit numbers. Establish that they should begin by looking at the magnitude of the hundreds digit – the value of the number that it represents, then the value of the tens digit and finally the value of the ones digit.

- Challenge them to write as many three-digit numbers as they can using only the digits 6 and 7.

- Work together to elicit that the following numbers are possible 666, 667, 676, 677, 766, 767, 776, 777 and then work together to put them in order starting with the greatest using > symbol.

- Look out for … pupils who are still confused about the place value of numbers. Problems may arise when pupils order numbers such as 326 and 362. To remedy this, give pupils more practice with concrete representations such as base 10 blocks focusing on 3 digit numbers as a number of hundreds, a number of tens and a number of ones.

Ask: *By looking at how many hundreds are in each number can you see which number is bigger?* Pupils should be able to explain that if both have the same number of hundreds they must look at how many tens are in each number. The one with more tens than the other is the greater number, for example 264 is greater than 245.

- Pupils should complete Question 3 in the Practice Book.

Same-day intervention

- With pupils working in pairs, ask each pupil to write three different three-digit numbers each on small cards. Shuffle all six cards and then order them, beginning with the largest.

- Ask each pupil to roll a 1–6 dice three times. (Roll again if any numbers are the same).

- Use the three digits to write six different three-digit numbers and order them, beginning with the smallest.

Same-day enrichment

- In pairs, pupils need a table with three columns and five rows.

- Pupils take turns to throw the dice and write the numbers thrown into the boxes, filling the rows of the table from left to right until all boxes are filled, creating 10 three-digit numbers.

- Each pair swaps their table with another pair and races to put the numbers in order on a separate piece of paper. The winning pair scores a point.

- Pairs then swap again with other pairs so everyone has a new table to order and pairs who have each other's tables race against each other to score a point. Continue as time permits to find a winning pair.

Question 4

4 Write >, < or = in each ◯.

(a) 428 ◯ 482   (b) 789 ◯ 787   (c) 543 ◯ 453

(d) 603 ◯ 630   (e) 1000 ◯ 999   (f) 135 ◯ 125

(g) 155 ◯ 205   (h) 299 ◯ 301

(i) 438 ◯ 400 + 30 + 8

What learning will pupils have achieved at the conclusion of Question 4?

- Overall fluency in number sense will have been developed, enabling pupils to determine where any three-digit number is located and the pattern of surrounding numbers.

- Pupils will have practised writing three-digit numbers as x00 + y0 + z (expanded).

- Pupils will have compared pairs of three-digit numbers using >, < and = symbols.

Activities for whole-class instruction

- Revisit >, < and = symbols by asking individual pupils to write a number sentence using three-digit numbers on the board using each symbol.

- Ask pupils to read the numbers, 345 and 354.

- Challenge them to write them in the form, 345 = 300 + 40 + 5 and 354 = 300 + 50 + 4, and to explain which number is bigger.

- Use > and < symbols to write a number sentence comparing them. Establish that 345 < 354 and 354 > 345.

- Give pupils **Resource 3.3.4c** Partitioned numbers.

Resource 3.3.4c

Partitioned numbers

Use place value arrow cards to make each number.
Write the number and the partitioned number in a number sentence.

1. 435 = 400 + 30 + 5

2. 609 = _____

3. 216 = _____

4. 398 = _____

5. 723 = _____

6. 453 = _____

7. 192 = _____

8. 540 = _____

9. 389 = _____

10 Order the numbers, from greatest to least.

© HarperCollinsPublishers 2017

Answers

2. 192 = 100 + 90 + 2   3. 216 = 200 + 10 + 6

4. 389 = 300 + 80 + 9   5. 398 = 300 + 90 + 8

6. 435 = 400 + 30 + 5   7. 453 = 400 + 50 + 3

8. 540 = 500 + 40   9. 609 = 600 + 9

10. 723 = 700 + 20 + 3

- Pupils complete Question 4 in the Practice Book, using >, < and = symbols.

Same-day intervention

- Show pupils pairs of three-digit numbers, for example 413 and 431 and ask pupils to write any number that lies between those two numbers on the number line. Repeat with further pairs of numbers. Ask pupils to write the three numbers in order, for example 413 < 420 < 431.

- Look at the pairs of numbers in Question 4 and write a number that comes between the two numbers. Explain why this is not possible for two of the pairs of numbers.

Same-day enrichment

- Show pupils the digit cards 3, 5, 1 and 0.

  1. Write down all of the three-digit numbers greater than 500.

- Now put them in order starting with the least. (501 < 503 < 510 < 513 < 530 < 531)

  2. Write down all the possible even numbers.

- Now put them in order starting with the smallest. (130 < 150 < 310 < 350 < 510 < 530)

Question 5

> **5** What is the greatest number you can write in each box?
> (a) ☐65 < 655 (b) 7☐8 > 778 (c) 453 < 4☐3
> (d) 321 > ☐21 (e) 642 > ☐43 (f) 795 < 79☐

What learning will pupils have achieved at the conclusion of Question 3?

- Pupils will have compared pairs of three-digit numbers using >, < and = symbols.

- Pupils will have used reasoning to solve word problems involving comparison and ordering of three-digit numbers.

Activities for whole-class instruction

- Show pupils part of a number line from 500 to 600.

- Using the number line for support, list numbers that would complete this number sentence correctly, 545 < 5☐5.

- Agree that the possible answers are 5,6,7,8 and 9 to give 555, 565, 575, 585, 595.

- Ask: *What answer would give the biggest possible number?* Agree 9, to give 595.

- Ask: *What answer would give the smallest possible number?* This time it is 5, to give 555.

- Ask pupils to find the biggest and smallest possible numbers that satisfy 574 > 57☐. (The numbers that satisfy this number statement are 0, 1, 2 and 3, so the biggest number is 3 and the smallest 0.)

- Ask pupils to find the biggest and smallest possible numbers that satisfy 5☐☐ > 569. (The biggest number is 599 and the smallest number is 570.)

- If pupils are finding the reasoning required to answer these questions challenging, then continue working together on similar problems, perhaps using a different number line section to vary the numbers and for support.

- Pupils should complete Question 5 in the Practice Book. Check that they understand that they are looking for the largest number that makes the number statement true.

Same-day intervention

- Give pupils **Resource 3.3.4d** Making 3-digit numbers.

> **Resource 3.3.4d**
>
> **Making 3-digit numbers**
>
> Use place value arrow cards to make each of the following numbers. Write each number and the partitioned number in a number sentence.
>
> 435 = 400 + 30 + 5
>
> 609 = _____
>
> 216 = _____
>
> 723 = _____
>
> 192 = _____
>
> 540 = _____
>
> 389 = _____
>
> Order the numbers, from largest to smallest.
>
> _____
>
> © HarperCollinsPublishers 2017

Answers

1. FALSE 986; 2. FALSE 236; 3. TRUE; 4. FALSE 268; 5. TRUE; 6. TRUE; 7. FALSE 987; 8. TRUE

Same-day enrichment

● Give pupils Resource 3.3.4e Finding 3-digit numbers

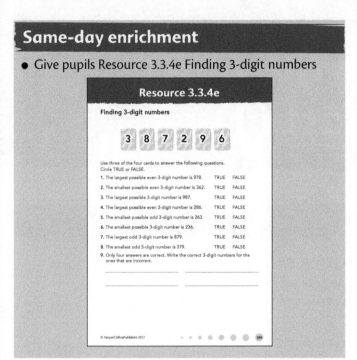

Challenge and extension questions

Question 6

> 6 Fill in the boxes with the same number so the subtraction is correct.
>
> $$1 \quad 9 \quad 8$$
> $$-$$

The first challenge question is a number puzzle in which pupils are asked to solve a subtraction calculation where all four digits in the subtrahend and difference are the same. Consider reminding pupils that addition and subtraction are inverse operations.

Question 7

> 7 Look at this calculation.
>
> $70 \div \Box = \bigcirc$ r 6
>
> Write a suitable 1-digit number in each of the $\Box$ and $\bigcirc$ so that the number sentence is correct.
> Write out the number sentence(s).

Question 7 challenges pupils to find a solution to $70 \div c = O$ r 6. The c cannot be less than 7 because the remainder is 6. Therefore the only single digit possibilities are 7, 8 or 9. Trial and error reveals that the number statement is $70 \div 8 = 8$ r6.

Unit 3.5
Fun with the place value chart (1)

Conceptual context

This is the first of two units exploring the place value chart for hundreds, tens and ones. Pupils are already familiar with place value charts for tens and units. Here they extend their knowledge of place value charts to include a third column so they can construct and 'read' place value charts up to 1000.

Adding a single additional dot to a diagram on the place value chart can add 1, 10 or 100 depending on whether the dot is placed in the ones column (+ 1), in the tens column (+ 10) or in the hundreds column (+ 100). Exploring the effect of adding or moving dots deepens pupils' conceptual understanding of place value.

(i) If there are no dots in a column then the column is empty and a zero is required as a placeholder (literally, to hold the place – to prevent the column collapsing and disappearing). The three-digit numbers shown below are 306 and 360 – without their respective zeros they would both become 36.

Hundreds	Tens	Ones
●●●		●●●●●●

Hundreds	Tens	Ones
●●●	●●●●●●	

Learning pupils will have achieved at the end of the unit

- Pupils will have practised reading three-digit place value diagrams and consolidated their understanding that an empty place will require a zero as a placeholder in the number (Q1)
- Writing three-digit numbers in numerals and words will have been reinforced (Q1, Q2, Q3)
- Pupils will have practised drawing three-digit place value diagrams, consolidating their conceptual understanding of place value (Q2, Q3)
- Pupils will have investigated the effect on a number of adding a dot to each of the place value columns (Q3)

Resources

blank hundreds, tens and ones place value charts; counters; chalk and playground space; mini whiteboards; 0–9 dice; 0–9 digit cards; **Resource 3.3.5** Hundreds, tens and ones

Vocabulary

place value chart, placeholder, ones, tens, hundreds, three-digit number

Question 1

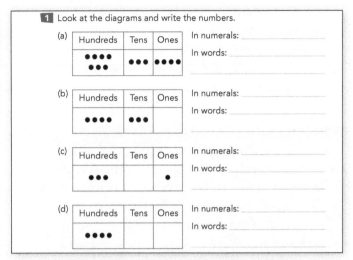

What learning will pupils have achieved at the conclusion of Question 1?

- Pupils will have practised reading three-digit place value diagrams and consolidated their understanding that an empty place will require a zero as a placeholder.
- Writing three-digit numbers in numerals and words will have been reinforced.

Activities for whole-class instruction

- Display a place value diagram as shown:

Hundreds	Tens	Ones
●●●●	●●	●●●●●

- Ask pupils how many hundreds there are and how they know/how many tens/how many ones. Now ask them to read the number, establishing that it is 425.
- Display a place value diagram as shown:

Hundreds	Tens	Ones
●●		●●●●●●

- Ask pupils what number this represents and establish that it is 206. Ask: *Is this what the place value chart for this number should look like?*

Hundreds	Tens	Ones
2		6

- Can pupils see that, unless a zero is written in the tens column, the number will be 26?

  By adding a zero to 'hold the place', the number 206 is correctly represented.

- Repeat with other diagrams until pupils demonstrate confidence reading three-digit numbers, including those that have zeros as placeholders.
- Pupils complete Question 1 in the Practice Book.

Same-day intervention

- Draw a large chalk place value chart (each box about 1 m² and ask half the pupils to walk around the three places until a chosen pupil from the other group claps their hands. At the clap, pupils freeze in the place where they are currently standing. Each pupil represents a dot in the place value chart diagram. Ask how many hundreds/tens/ones there are in the number and what the three-digit number is. Change roles and repeat.
- Using mini whiteboards, ask pupils to write down the number in numerals and words.

Same-day enrichment

- Give pairs of pupils five counters and a blank hundreds, tens and ones place value chart. Take turns to place the counters to make different three-digit numbers. Draw the place value diagrams and write the numbers in numerals and words. Try to find at least six different numbers.
- For example:

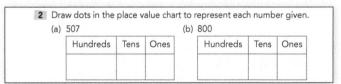

221 – Two hundred and twenty-one

Question 2

2 Draw dots in the place value chart to represent each number given.
(a) 507

Hundreds	Tens	Ones

(b) 800

Hundreds	Tens	Ones

What learning will pupils have achieved at the conclusion of Question 2?

- Writing three-digit numbers in numerals and words will have been reinforced.
- Pupils will have practised drawing three-digit place value diagrams, consolidating their conceptual understanding of place value.

Activities for whole-class instruction

- Display a place value chart. Choose three pupils to be responsible for each of the hundreds, tens and ones places. Give them a set of 0–9 digit cards. In turn ask pupils to shuffle the cards and choose one, unseen, first hundreds, then tens and ones.

- As each number is chosen, invite another pupil to place the correct number of counters on the diagram to make a three-digit number.

- (A zero in the hundreds column would give a two-digit number. To avoid this, remove the 0 from the set of cards given to the pupil managing the hundreds column).

- Read the three-digit number produced, comparing the dots on the place value chart and the number.

- Repeat to make more numbers.

- Pupils complete Question 2 in the Practice Book.

Same-day intervention

- Give pupils mini whiteboards. Read this story and ask pupils to write the three-digit numbers as numerals.

- Draw place value diagrams using dots for each number.

  Sam lived at number two hundred and fifty-three New Road. The silver numbers sparkled in the sunshine as he closed the front door. He walked four hundred and fifty metres down the road to the bus stop and waited for his bus. The first bus to come along was number one hundred and not the bus Sam wanted, but his bus, number three hundred was just behind. Sam jumped on board and took his seat. He made his way to the centre of town and got off outside number four hundred and five High Street. His friend, Ana, was waiting for him and they walked off down the road, chatting happily.

- Pupils could be asked to order the numbers.
  Answers

253	450	100	300	405
100	253	300	405	450

- Give pupil pairs a copy of this table.

100	200	300	400	500	600	700	800	900
10	20	30	40	50	60	70	80	90
1	2	3	4	5	6	7	8	9

- Invite pupils to place a counter randomly in each row to make a three-digit number. Write the three-digit number produced in numerals and words. Draw a place value diagram for this number. Repeat with more numbers.

- Pupils could use a 0–9-sided dice or spinner to determine the numbers in each row.

Same-day enrichment

- Give pupils a place value chart and counters. Take turns to roll a 1–6 dice. Each pupil should place that number of counters on the chart. The pupil with the biggest number after three dice rolls is the winner. After two dice rolls, encourage pupils to predict who will win.

- After two dice rolls, a possible scenario is shown below.
  Pupil 1

Hundreds	Tens	Ones
● ● ● ● ●		● ●

  Pupil 2

Hundreds	Tens	Ones
	● ● ● ● ●	● ●

- When a situation like this arises, pupils might be able to predict that, if the final roll is a 6, Pupil 2 wins. Rolling any other number means that Pupil 1 wins.

- Play for a set time, or until one pupil has won a specific number of times.

Question 3

3 Meera has drawn the dots representing 264 in the place value chart below. Joe adds one more dot. What could the new number be?

Hundreds	Tens	Ones
● ●	● ● ● / ● ● ●	● ● ● ●

Hundreds	Tens	Ones

In numerals: _____
In words: _____

Hundreds	Tens	Ones

In numerals: _____
In words: _____

Hundreds	Tens	Ones

In numerals: _____
In words: _____

Hundreds	Tens	Ones

In numerals: _____
In words: _____

What learning will pupils have achieved at the conclusion of Question 3?

- Writing three-digit numbers in numerals and words will have been reinforced.
- Pupils will have continued to practise drawing three-digit place value diagrams.
- Pupils will have investigated the effect on a number of adding a dot to each of the place value columns.

Activities for whole-class instruction

- Give pupils whiteboards and show them a diagram of 300 (check that this is the best number). Ask: *If you had one more dot to place in the diagram, how many different numbers could you make?* Establish that it could be placed in the ones place to give 301, the tens place to give 310 or the hundreds place to give 400.
- Repeat with a different starting number, for example 287 to give 288, 297 and 387.

All say ... *When the dot is placed in the ones place, the number increases by 1. When the dot is placed in the tens place, the number increases by 10. When the dot is placed in the hundreds place, the number increases by 100.*

- Pupils complete Question 3 in the Practice Book.

Same-day intervention

- Give pupils **Resource 3.3.5** Hundreds, tens and ones.

Answers: 1. tens; 2. ones; 3. tens; 4. hundreds; 5. tens; 6. ones; 7. own answers

- Show pupils any place value chart diagram. Tell pupils to imagine that you are adding another dot and write the numeral after the (invisible) dot is added. Ask: *Where did I place the dot?*

Same-day enrichment

- Challenge pupils to work out the new numbers that can be made if you add two more dots to the number 400. Encourage them to be systematic in their approach. Ask them to draw the place value chart diagrams and work out how much bigger than the original each new number is.
- Agree that the new numbers are +200, +110, +101, +20, +11, +2. So the dots must have been drawn ... (where?)
- Ask whether another starting number, for example 256 would have the same pattern of new numbers. Query whether this would always be the case.

Challenge and extension questions

Questions 4 and 5

4 Hai folded a piece of paper in half three times. He then drew a flower in the centre and cut it out.

Hai cut out ☐ flowers in the paper.

5 After drinking half a cup of milk, Asha filled up the cup with water. She then drank half of the contents of the cup again. After that, she filled up the cup with water a second time. She then drank the whole of the contents of the cup.

In total, Asha drank ☐ cup(s) of milk and ☐ cup(s) of water.

Both challenge and extension questions are unrelated to the mathematics of the unit. The first involves visualising paper folding and cutting flowers. Pupils may be able to predict that the number of flowers doubles with each fold and can carry out the exercise to check.

The second question is a word problem involving fractions of quantities. Encourage pupils to draw diagrams to help their reasoning.

Unit 3.6
Fun with the place value chart (2)

Conceptual context

This is the final unit in the chapter and the second of two units exploring the place value chart for hundreds, tens and ones. The problems in this unit challenge pupils to operate flexibly with number symbols and the value of the quantity they represent.

The unit also revisits partitioning and combining three-digit numbers. For example: 632 = 600 + 30 + 2

Learning pupils will have achieved at the end of the unit

- Pupils will have consolidated linking different representations of three-digit numbers (Q1, Q2)
- Pupils will have secured their understanding of the relationship between quantity value and place value by adding or removing dots from different columns of the place value chart (Q1, Q2)
- Using a logical approach, pupils will have explored problems to find possible solutions (Q1, Q2)
- Writing three-digit numbers in numerals and words will have been practised (Q1, Q2)
- Pupils will have practised partitioning and combining three-digit numbers (Q3)

Resources

place value chart; counters; place value arrow cards; **Resource 3.3.6a** 3-digit numbers; **Resource 3.3.6b** The target is 500!; **Resource 3.3.6c** Start with 400

Vocabulary

place value chart, placeholder, ones, tens, hundreds, three-digit number, partition

Question 1

1 Draw three dots in each place value chart to represent six different 3-digit numbers.

(a)

Hundreds	Tens	Ones

In numerals: _____

In words: _____

_____

(b)

Hundreds	Tens	Ones

In numerals: _____

In words: _____

(c)

Hundreds	Tens	Ones

In numerals: _____

In words: _____

(d)

Hundreds	Tens	Ones

In numerals: _____

In words: _____

(e)

Hundreds	Tens	Ones

In numerals: _____

In words: _____

(f)

Hundreds	Tens	Ones

In numerals: _____

In words: _____

What learning will pupils have achieved at the conclusion of Question 1?

- Pupils will have consolidated linking different representations of three-digit numbers.

- Pupils will have secured their understanding of the relationship between quantity value and place value by adding or removing dots from different columns of the place value chart.

- Using a logical approach, pupils will have explored problems to find possible solutions.

- Writing three-digit numbers in numerals and words will have been practised.

Activities for whole-class instruction

- Show pupils this set of numbers: 451, 460, 550. Explain that these numbers were made by adding a single dot to the same number in different columns. Ask pupils to say what the original number was and explain how they know. Establish that the starting number was 450; 451 is 450 + 1, 460 is 450 + 10 and 550 is 450 + 100.

- Try another set of numbers, this time not in order, for example 617, 707 and 608. Agree that the original number was 607.

- Ask pupils, in pairs, to write a set of numbers and challenge their peers to find the original number and draw a place value diagram for it.

- Pupils should complete Question 1 in the Practice Book. Using a place value chart and counters will allow pupils to try out possible numbers.

Same-day intervention

- Give pupils **Resource 3.3.6a** 3-digit numbers.

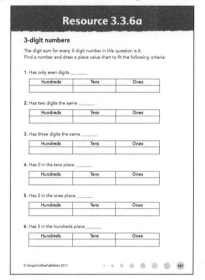

- Pupils could try the same questions with a new digit sum of 9. It will be impossible to find a three-digit number with only even digits when the digit sum is odd. Ask pupils if they can explain why.

Same-day enrichment

- Give pupils **Resource 3.3.6a** 3-digit numbers.

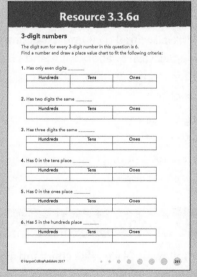

- Pupils could try the same questions with a new digit sum, 7, 8 or 9. It will be impossible to find a three-digit number with only even digits when the digit sum is odd. Ask pupils if they can explain why.

Question 2

2 Draw dots in the first place value chart to represent 153.
Then, in each place value chart below, move one dot into another column and write the number.

Hundreds	Tens	Ones

(a)

Hundreds	Tens	Ones

In numerals: _____

In words: _____

(b)

Hundreds	Tens	Ones

In numerals: _____

In words: _____

(c)

Hundreds	Tens	Ones

In numerals: _____

In words: _____

(d)

Hundreds	Tens	Ones

In numerals: _____

In words: _____

(e)

Hundreds	Tens	Ones

In numerals: _____

In words: _____

(f)

Hundreds	Tens	Ones

In numerals: _____

In words: _____

What learning will pupils have achieved at the conclusion of Question 2?

- Pupils will have consolidated linking different representations of three-digit numbers.
- Pupils will have secured their understanding of the relationship between quantity value and place value by adding or removing dots from different columns of the place value chart.
- Using a logical approach, pupils will have explored problems to find all possible solutions.
- Writing three-digit numbers in numerals and words will have been practised.

Activities for whole-class instruction

- Display the following place value chart and agree that the number represented is 333.

Hundreds	Tens	Ones
● ● ●	● ● ●	● ● ●

- Ask pupils what numbers they could make by moving one dot into another column. Agree that the following numbers could be made: 423, 432, 342, 324, 243, 234.

- Ask whether there are numbers where it is not possible to add another dot? Agree that another dot cannot be out into any column that has already has nine dots.
- Pupils should complete Question 2 in the Practice Book.

Same-day intervention

- Mark one blank 6-sided dice H, H, T, T, O, O and another 1, 1, 2, 2, 3, 3.
- Play this game in pairs. Each pupil has a blank place value diagram and some counters. Take turns to roll both dice and add the counters to their diagram, for example for H2 add two counters to the hundreds column. After each dice roll the pupil should tell their partner the number they have made. The winner is the pupil with the largest number after they have each had four turns. If a dice roll cannot be executed because it would make more than 9 in a column, for example the current number is 190 and the dice roll is T3, then they are declared 'bust' and the other pupil is the winner.

Same-day enrichment

- Give pupils **Resource 3.3.6b** The target is 500!

Answers: 1. 501; 2. 501; 3. 511; 4. 497; 5. 513; 6. 490

Question 3

3 Write in the correct numbers.

(a) 643 = ☐ + ☐ + ☐

(b) 300 + 30 + 8 = ☐

(c) 302 = ☐ + ☐ + ☐

(d) 900 + 0 + 9 = ☐

What learning will pupils have achieved at the conclusion of Question 3?

- Pupils will have practised partitioning and combining three-digit numbers.

Activities for whole-class instruction

- Write the number 283 on the board and ask pupils to discuss with a partner how the number is made up.

 All say... $283 = 200 + 80 + 3$

- Repeat with further numbers, including numbers containing one or more zeros to ensure pupils show confidence partitioning three-digit numbers.
- Pupils should complete Question 3 in the Practice Book.

Same-day intervention

- Ask each pupil to draw a place value diagram for a three-digit number of their choice. Share the diagrams with the group.
- Ask pupils to write the numbers and the expanded form from the place value diagrams.
- For example:

Hundreds	Tens	Ones
●●●●	●●●●●●	●●

$462 = 400 + 60 + 2$

Same-day enrichment

- Give pupils **Resource 3.3.6c** Start with 400.

Resource 3.3.6c

Start with 400

Start with 400 and then follow the instructions. Calculate the numbers and partition them.
The first one has been done for you.

1. Add the number of letters in the alphabet. $400 + 26 = 426$
   $426 = 400 + 20 + 6$

2. Add your age. $400 +$ $=$

3. Subtract your age.

4. Add the number of pupils in your class.

5. Subtract the number of letters in your first and last names.

6. Make up a question of your own.

© HarperCollins Publishers 2017

Answers will vary.

Challenge and extension questions

Questions 4 and 5

4 A group of children form a line for a running race. Lily is in the twelfth place counting from the start. Ellis is in the twelfth place counting from the end. Lily is exactly in front of Ellis.

There are ☐ children in total.

5 A book has 40 pages. If a bookmark is put in every three pages starting from the first page, it will have ☐ bookmarks in total.

The first challenge and extension question requires pupils to solve a problem involving ordinal numbers, while the second question is a word problem that requires them to count in threes. Prompt pupils to read both question very carefully.

Chapter 3 test (Practice Book 3A, pages 86–90)

Test question number	Relevant unit	Relevant questions within unit
1	3.1	1
	3.2	1
2	3.2	5
3	3.2	2
	3.5	1
4	Not specific to unit	
5	3.1	3
	3.2	4
	3.3	1, 2, 4, 5
	3.4	1
6	3.4	4
7	3.1	3
	3.2	4
	3.3	1
	3.5	3
	3.6	1, 2
8	Not specific to unit	
9	Not specific to unit	
10	Not specific to unit	
11	Not specific to unit	

Chapter 4
Statistics (II)

Chapter overview

Area of mathematics	National Curriculum statutory requirements for Key Stage 2	Shanghai Maths Project reference
Statistics	Year 3 Programme of study: Pupils should be taught to:	
	■ interpret and present data using bar charts, pictograms and tables	Year 3, Units 4.1, 4.2, 4.3
	■ solve one-step and two-step questions [for example, 'How many more?' and 'How many fewer?'] using information presented in scaled bar charts and pictograms and tables.	Year 3, Units 4.1, 4.2, 4.3

Unit 4.1
From statistical tables to bar charts

Conceptual context

Statistical tables were introduced in Chapter 3 of Book 2. This unit revisits and revises that learning, looking at tables, pictograms and block diagrams.

Bar charts are the new learning in this chapter. Although pupils have not studied them in maths, they are used across the curriculum and so might be familiar from other contexts, for example science or geography. It is important to discuss the various features of bar charts and gradually build up pupils' ability to construct them correctly.

Bar charts should, at this stage in learning, have a suitable title that fully explains what the chart shows, labels along one axis for individual bars and a suitable numerical scale along the other. A focus on checking these features will help good practice become embedded as pupils begin to draw their own bar charts.

 Discrete data (things that are counted e.g. children, books) should be presented in bars of equal width that do **not** touch and are evenly spaced.

 Continuous data (things that are measured e.g. height, weight) should be presented in bars of equal width that touch.

Learning pupils will have achieved at the end of the unit

- Tables, pictograms and block diagrams will have been revisited, deepening pupils' understanding (Q1)
- Pupils will have developed their knowledge of the composition of bar charts (Q1, Q2, Q3)
- Pupils will have practised reading bar charts with a scale of 1 (Q1, Q2, Q3)
- The transfer of data from a bar chart to a frequency table and vice versa will have been practised (Q1, Q2)
- Pupils will have interrogated the data in bar charts to answer questions, including finding the total number taking part in a survey (Q3)

Resources

squared paper; **Resource 3.4.1a** Toast!; **Resource 3.4.1b** Favourite subjects; **Resource 3.4.1c** Blank bar chart

Vocabulary

statistical chart, pictogram, block diagram, tally chart, bar chart, frequency table, cell, vertical axis, horizontal axis, interval, scale

Question 1

> **1** 60 children are divided into 3 groups for 3 different tasks.
> Group A has 20 children, Group B has 30 children and Group C has 10 children. Present this information by completing the following:
>
> (a) A statistical table
>
> | | Group A | Group B | Group C |
> |---|---|---|---|
> | Number of children | | | |
>
> (b) A pictogram
>
> | Group A | |
> |---|---|
> | Group B | |
> | Group C | |
>
> Each ◯ stands for 10 children.
>
> (c) A block diagram
>
> Each cell stands for 5 children.
>
> 35 30 25 20 15 10 5 0
> Group A Group B Group C

What learning will pupils have achieved at the conclusion of Question 1?

- Tables, pictograms and block diagrams will have been revisited, deepening pupils' understanding.
- Pupils will have been introduced to the composition of bar charts.
- Pupils will have practised reading bar charts with a scale of 1.
- The transfer of data from a bar chart to a frequency table and vice versa will have been practised.

Activities for whole-class instruction

- Share **Resource 3.4.1a** Toast!

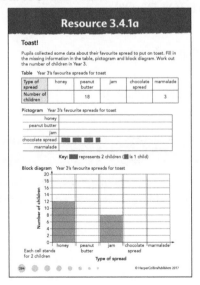

- Give **Resource 3.4.1a** to pupil pairs as a revision aid to refresh the statistical tables explored in Book 2. Ask them to explain how a table is derived from a tally chart. Invite individual pupils to show specific numbers in tally marks. Discuss the key in the pictogram and check that pupils understand that the rectangle represents two children and half a rectangle represents one child.

- In the block diagrams in Book 2 each cell represented one, while in this block diagram each cell stands for two children. This is a good introduction to scales in bar charts. Count together in twos and discuss how to present odd numbers of children by filling half a cell.

- Work together on each type of spread in turn to complete the tables.

- Ask pupils to complete Question 1 in the Practice Book, working independently.

Same-day intervention

- Look at **Resource 3.4.1a** Toast! and ask pupils the following questions about the data.

  - *Which spread is most popular?* (peanut butter)
  - *Which spread is least popular?* (marmalade)
  - *Which spread had seven votes?* (chocolate spread)
  - *How many more children chose honey than jam?* (four)
  - *Make up two questions of your own using the data.*

Same-day enrichment

- Look at the table in Question 1. Ask pupils to discuss and explain:

  - what the pictogram looks like if ◯ stands for five children
  - what the block diagram looks like if the scale has intervals of 10
  - whether it is possible for ◯ to stand for two children in the pictogram and for the scale to have intervals of two in the block diagram.

Question 2

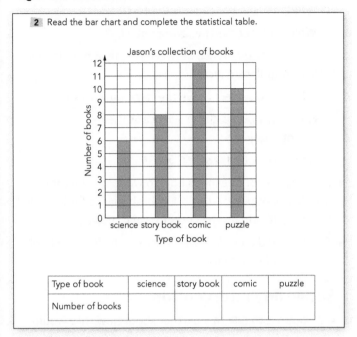

2 Read the bar chart and complete the statistical table.

Jason's collection of books

Type of book	science	story book	comic	puzzle
Number of books				

What learning will pupils have achieved at the conclusion of Question 2?

- Pupils will have developed their knowledge of the composition of bar charts.
- Pupils will have practised reading bar charts with a scale interval of 1.
- The transfer of data from a bar chart to a table and vice versa will have been practised.

Activities for whole-class instruction

Give pupils **Resource 3.4.1b** Favourite subjects.

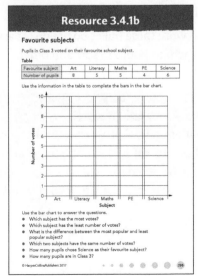

- Give pupils **Resource 3.4.1b** and work together to examine and systematically complete the bar chart. Ask

pupils to point to the title of the bar chart. Confirm that it is 'Year 3's favourite subjects' and that this appears at the top of the bar chart. Discuss how the title describes what the bar chart is about and that the labels on the horizontal axis tell us what is being counted or what is being measured. On this bar chart, it is pupils' votes about their favourite subjects that are being counted, so the labels are the subjects.

- Ask pupils to point to the vertical axis and look at the title 'Number of votes'. This is where it will show how many people voted for each subject. Look carefully at the position of the numbers and establish that the numbers are on the lines (not in the cells) to show the height of the bars. Establish that in this bar chart the numbers go up in ones.

- Invite an individual pupil to tell you how many pupils chose art as their favourite subject. Agree that it is eight. Discuss the position and height of this bar. Ask pupils to colour the bar. Continue with the remaining subjects until the bar chart is completed. Observe that the bars are the same width and evenly spaced.

- Discuss how the table provided the numbers for the bar chart and elicit that the process works in reverse - a table showing how many of something there are (in this case, how many voted for each subject) can be completed from a bar chart.

- Pupils should complete Question 2 in the Practice Book.

Same-day intervention

- Use **Resource 3.4.1c** Blank bar chart to prepare another bar chart with a subject that is relevant to your setting and ask pupils to point out different features of the bar chart, for example title, scale, horizontal axis and so on.

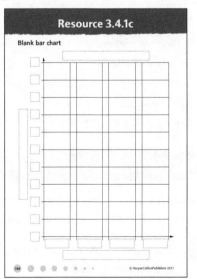

Same-day enrichment

- Using the same bar chart prepared for intervention, ask pupils to write five questions on the data. Challenge them to use the following words

most	least	more	difference	total

Question 3

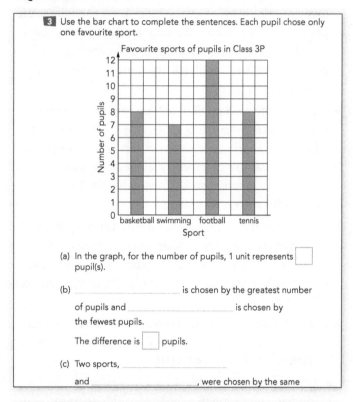

3 Use the bar chart to complete the sentences. Each pupil chose only one favourite sport.

Favourite sports of pupils in Class 3P

(a) In the graph, for the number of pupils, 1 unit represents ☐ pupil(s).

(b) _____ is chosen by the greatest number

of pupils and _____ is chosen by

the fewest pupils.

The difference is ☐ pupils.

(c) Two sports, _____

and _____, were chosen by the same

What learning will pupils have achieved at the conclusion of Question 3?

- Pupils will have practised reading bar charts with a scale of 1.
- Pupils will have developed their knowledge of the composition of bar charts
- Pupils will have interrogated the data in bar charts to answer questions and pupils will have found the total number taking part in a survey.

Activities for whole-class instruction

- Look at **Resource 3.4.1b** Favourite subjects and verify that the bars are all the same width and equally spaced.

All say... The bars in a bar chart are the same width and equally spaced.

- Answer the questions on the data together.

- Pupils should answer the questions:
  - Which subject has the most votes?
  - Which subject has the fewest votes?
  - What is the difference between the most popular and least popular subject?
  - Which two subjects have the same number of votes?
  - How many pupils chose Science as their favorite subject?
  - How many pupils are in Class 3?

- Ask a volunteer to explain how to work out the total number of pupils in Class 3. Establish that to find the total you need to add up the votes for each subject.

- Pupils should complete Question 3 in the Practice Book, working independently.

Same-day intervention

- Ask pupils to talk to a partner about how they would approach carrying out a survey to find out their class's favourite computer game (perhaps limited to a choice of four). Elicit that they would need to:
  - draw a tally chart with the four choices
  - ask pupils their choice and enter it on the tally chart
  - total the tallies to give a frequency for each game
  - construct a bar chart with space on the horizontal axis for four equally-spaced bars of the same width and space on the vertical axis for the highest value
  - give the chart a title (Favourite Computer Games in our Class), label the bars with the names of the games and add the vertical scale and title (Number of votes)
  - carefully colour the bars.

- You could write these bullet points on separate cards and give them to pupils to sort and order.

Same-day enrichment

- Discuss with pupils whether they think this would be the result of a survey on favourite subjects in their class. Challenge them to conduct a survey on their class and to use a blank bar chart to record their findings.

Challenge and extension question

Question 4

4 (a) How many of the following items of stationery do you have?
    Find out and complete the table.

Type	Quantity
pencil	
rubber	
felt-tipped pen	
exercise book	

(b) Now use another suitable statistical tool you have learned
    to present the data.

In Question 4 pupils are asked to count how many
items of stationery they have. Support if necessary by
suggesting that they use tally marks to count and total
the tally marks to give the frequency of each item. Pupils
are then free to choose their own statistical tool to
present the data.

Unit 4.2
Bar charts (1)

Conceptual context

The unit develops knowledge and understanding of bar charts recording discrete data. Discrete data is data that can be counted. When the data is discrete the bars should not touch. Bars should be the same width and equally spaced. Scaled bar charts are introduced, with each interval representing two items.

Learning pupils will have achieved at the end of the unit

- The composition of a bar chart will have been developed further and consolidated (Q1, Q2)
- Pupils will have practised completing and reading bar charts with a scale interval of 2 (Q1, Q2, Q3)
- Pupils will have interrogated data in a bar chart to answer two-stage questions (Q1, Q2, Q3)
- The transfer of data from a bar chart to a frequency table and vice versa will have been further consolidated (Q2, Q3)

Resources

mini whiteboards; **Resource 3.4.1c** Blank bar chart; **Resource 3.4.2** Colour charts

Vocabulary

bar, bar chart, frequency table, cell, vertical axis, horizontal axis, interval, scale

Question 1

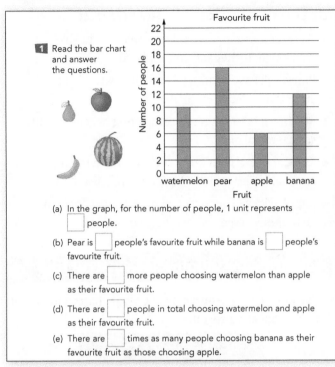

1 Read the bar chart and answer the questions.

Favourite fruit

Number of people (vertical axis: 0, 2, 4, 6, 8, 10, 12, 14, 16, 18, 20, 22)

Fruit: watermelon, pear, apple, banana

(a) In the graph, for the number of people, 1 unit represents ☐ people.

(b) Pear is ☐ people's favourite fruit while banana is ☐ people's favourite fruit.

(c) There are ☐ more people choosing watermelon than apple as their favourite fruit.

(d) There are ☐ people in total choosing watermelon and apple as their favourite fruit.

(e) There are ☐ times as many people choosing banana as their favourite fruit as those choosing apple.

What learning will pupils have achieved at the conclusion of Question 1?

- The composition of a bar chart will have been developed further and consolidated.
- Pupils will have practised reading bar charts with a scale interval of 2.
- Pupils will have interrogated data in a bar chart to answer one stage and two stage questions.

Activities for whole-class instruction

- Give each pupil a copy of **Resource 3.4.1c** Blank bar chart

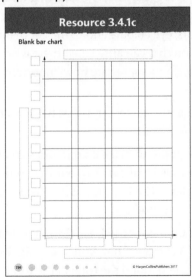

Resource 3.4.1c

Blank bar chart

© HarperCollinsPublishers 2017

- Ask pupils to talk to a partner about what kind of information needs to be written in the blank spaces. Elicit from them that the title is written in the box at the top. The spaces on the horizontal axis are for the things being counted. We call these the labels. There are numbers on the vertical axis to show the number of things. These questions will check and consolidate pupils' knowledge and understanding of the vocabulary of bar charts.

- Show pupils the following frequency table.

Type of punctuation mark	full stop	comma	question mark	exclamation mark
Number	16	20	2	7

- Explain that a Year 3 pupil was investigating punctuation marks. He counted the number of full stops, commas, question marks and exclamation marks on one page of his reading book. Ask pupils to suggest a suitable title for a bar chart of this data and to show you where this is written.

- Establish that the different types of punctuation marks should be written in the boxes for individual bars on the horizontal bar. These are the labels.

- The next step is to label the vertical axis. First ask pupils to look at the numbers and to find the largest one, 20. Ask if there is space to number the vertical scale from 0 to 20 in ones. Establish that there is not, so invite them to suggest a solution. Agree that they can use each square to represent two things and this will allow 20 things to fit on the graph. Practise counting in twos and then complete the scale on the vertical axis.

- Now look at the numbers of each punctuation mark and ask them to show you with their finger where the bar for full stops, commas and question marks would reach.

- There are seven exclamation marks. Choose a pupil to explain how to draw a bar to show seven, halfway between six and eight. You could ask pupils to point out the bar heights for other odd numbers to check understanding.

- Invite pupils to label and complete the bars on the bar chart carefully.

- Ask pupils how you can tell which is the most/least used punctuation mark just by looking at the bars. Establish that the most/least used mark is the tallest/shortest bar. Practise reading the bars and comparing them by asking and answering questions, such as: *What is the difference between the number of exclamation and question marks?* or *What is the total number of full stops, question marks and exclamation marks?*

● Pupils complete Question 1 in the Practice Book.

Same-day intervention

● Ask pupils to draw and complete a frequency table using the data in the bar chart on Favourite fruit, in the Practice Book. Check that pupils are able to label the table correctly. Confirm that the top row of the table shows the type of fruits and the second row the numbers of each fruit.

● Challenge pupils to work out the number of people who took part in the fruits survey and to explain the process.

Same-day enrichment

● Discuss with pupils whether the bar chart of punctuation marks showed the kind of result they expected. Ask whether they think their creative writing would show a similar pattern. Challenge them to investigate the frequency of these four punctuation marks in a page of their reading book. Use **Resource 3.4.1c** Blank bar chart to record the data.

Question 2

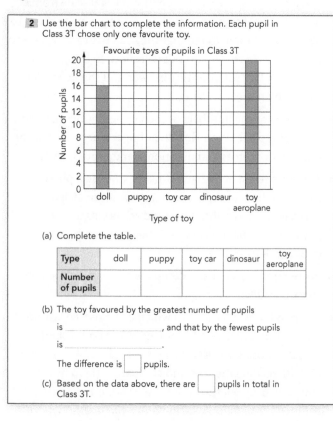

2 Use the bar chart to complete the information. Each pupil in Class 3T chose only one favourite toy.

Favourite toys of pupils in Class 3T

(a) Complete the table.

Type	doll	puppy	toy car	dinosaur	toy aeroplane
Number of pupils					

(b) The toy favoured by the greatest number of pupils

is _____, and that by the fewest pupils

is _____.

The difference is ☐ pupils.

(c) Based on the data above, there are ☐ pupils in total in Class 3T.

What learning will pupils have achieved at the conclusion of Question 2?

● The composition of a bar chart will have been developed further and consolidated

● Pupils will have reinforced their ability to read bar charts with a scale interval of 2.

● Pupils will have practised completing bar charts with a scale interval of 2.

● The transfer of data from a bar chart to a frequency table and vice versa will have been practised.

● Pupils will have interrogated data in a bar chart to answer one-stage and two-stage questions.

Activities for whole-class instruction

● Give pupil pairs **Resource 3.4.2** Colour charts and look at the first bar chart together.

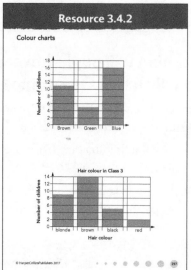

● Ask pupils to suggest what the data might represent. Their ideas may include statistics on recycling bins, paint colours, eye colours and so on. Let them enjoy the confusion that can arise when there is no title, so that they begin to appreciate how important it is to include one.

● Tell them that the correct title that has been omitted is 'Eye colour in Class 3'.

● Ask pairs of pupils to make up questions about the data and choose individual pupils to put their questions to the rest of the class. Listen to the answers, which should be given in whole sentences.

● If the question is not posed, ask them how many pupils are in Class 3 and how they work it out. Pupils answers should include: 'I found the total of the three colours, 11 + 3 + 16 = 30'.

All say... *To find the total number in a survey, add up the numbers in all bars.*

- Ask pupils to construct a frequency table from the bar chart on their mini whiteboards. Check that the top row has the title, 'Eye colour' and column headings 'brown', 'green' and 'blue' with the correct numbers in the row beneath.

- Pupils complete Question 2 in the Practice Book.

Same-day intervention

- Ask pupils to use the data from the bar chart on eye colour to draw a frequency table.

- Look at the second bar chart on **Resource 3.4.2** Colour charts showing hair colour in Class 3.

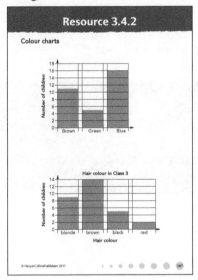

- In pairs, pupils make up five questions on the data. From the data draw a frequency table.

Same-day enrichment

- Survey the eye colour of pupils in the class and ask them to construct a frequency table and bar chart, using **Resource 3.4.1c** Blank bar chart. You can introduce a fourth category, 'Other', for those pupils whose eyes are not brown, green or blue but amber, hazel, grey, violet, black and so on.

Question 3

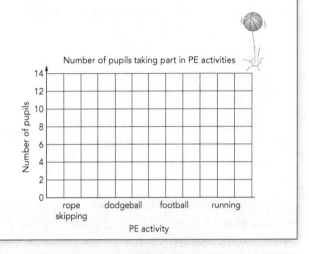

3 Jenna did a class survey on the number of pupils taking part in PE activities and presented the data in the table below.
Complete a bar chart based on the data.

Type	rope skipping	dodgeball	football	running
Number of pupils	6	11	8	13

What learning will pupils have achieved at the conclusion of Question 3?

- Pupils will have practised completing bar charts with a scale interval of 2.

- Pupils will have interrogated data in a bar chart to answer two stage questions.

- The transfer of data from a bar chart to a frequency table and vice versa will have been further consolidated.

Activities for whole-class instruction

- To introduce the subject, ask pupils to count how many letters they have in their first name. Invite them to put up their hand if they have four or fewer letters, five letters, six letters, seven or more letters.

- Explain that Year 3 in a school collected this data and made the following frequency table.

Letters in first name	4 or fewer	5	6	7 or greater
Number of pupils	11	14	19	16

- Challenge pupils to tell you how many pupils are in this Year 3 and explain how they carried out the calculation.

- Give them a copy of **Resource 3.4.1c** Blank bar chart, and work together to construct the bar chart from the frequency table.

- Ask pupils, in pairs, to make up questions about the data and choose individuals to put their questions to the rest of the class. Listen to the answers that should be given in whole sentences.
- Pupils complete Question 3 in the Practice Book.

Same-day intervention

- The frequency table shows the types of animals in a petting area at a children's farm.

Type of animal	Guinea pigs	Lambs	Piglets	Rabbits
Number of animals	8	3	5	6

- Using the blank bar chart, **Resource 3.4.1c**, invite pupils to draw a bar chart with intervals of 1.

Same-day enrichment

- At another children's farm, there are 25 animals in the petting area. Ten of the animals are rabbits. There are half as many lambs as rabbits. There is one more lamb than there are guinea pigs. The remaining animals are goats. Challenge pupils to determine the numbers of each animal and construct a frequency table. Using the blank bar chart, **Resource 3.4.1c**, invite them to draw a bar chart with intervals of 1.

Challenge and extension question

Question 4

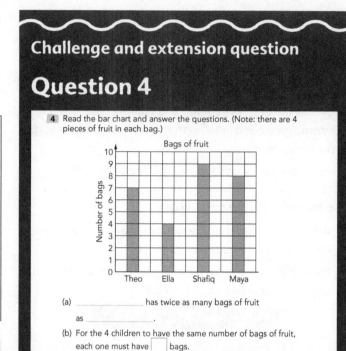

4 Read the bar chart and answer the questions. (Note: there are 4 pieces of fruit in each bag.)

(a) _____ has twice as many bags of fruit

as _____ .

(b) For the 4 children to have the same number of bags of fruit, each one must have ☐ bags.

(c) Theo has ☐ fewer pieces of fruit than Shafiq.

The bar chart in the extension question shows numbers of bags of sweets. Each bag contains four sweets. The challenge is to recognise when the question is asking about bags of sweets and when it is asking the actual numbers of sweets.

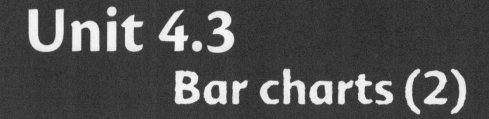

Unit 4.3
Bar charts (2)

Conceptual context

This is the final unit of the statistics chapter. Statistics, an important aspect of mathematics, involves gathering and recording information and then presenting it in a clear and relevant way.

Bar charts are a basic statistical tool. They present data in a way that enables at-a-glance comparison of a number of datasets simultaneously. Using information presented on axes and data labels it is also possible to infer and quantify a great deal of information from a single chart.

Pupils need to develop the skills both to construct charts correctly and to effectively interpret the information contained within them, looking for explanations.

The steps required to construct a bar chart are as follows:

- Determine what has to be written on each axis.
- Look at the numbers and identify the largest count or measure.
- Work out the best scale for the axis that will show how many or how much (usually the vertical axis) so that the largest count or measure fits on the bar chart.
- Work out the width and spacing of the bars, checking that the bars are of the same width and evenly spaced.
- Complete the labels on the axes and construct the bars.
- Give the bar chart an appropriate explanatory title.

At this stage in their learning pupils are not expected to be able to complete the entire process and are provided with a grid to populate.

Learning pupils will have achieved at the end of the unit

- Pupils will have reinforced their ability to interrogate data in pictograms to answer one-step and two-step questions (Q1)
- Pupils will have explored transferring data from a pictogram into a bar chart (Q1)
- Pupils will have continued to practise constructing bar charts with different scale intervals (Q1, Q2)
- Pupils will have practised choosing suitable scale intervals (Q2)
- Pupils will have reinforced their ability to interrogate data in bar charts to answer two-step questions (Q2)

Resources

mini whiteboards; **Resource 3.4.1c** Blank bar chart; **Resource 3.4.3** Spring flowerbed

Vocabulary

pictogram, bar, bar chart, table, cell, vertical axis, horizontal axis, interval, scale

Question 1

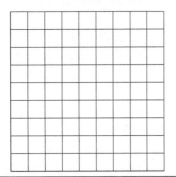

1 The pictogram shows information about the ages of the pupils who took part in a school activity.

Age	Number of pupils

Each 🧍 stands for 2 pupils.

(a) (i)   How many pupils are aged 9? ☐

   (ii) How many are younger than 9? ☐

   (iii) How many are older than 9? ☐

(b) How many pupils in total took part in the school activity? ☐

(c) Complete a bar chart based on the information in the pictogram on page 99.

What learning will pupils have achieved at the conclusion of Question 1?

- Pupils will have reinforced their ability to interrogate data in pictograms to answer one-step and two-step questions.
- Pupils will have reviewed pictograms in which a picture symbol represents two items/people and half the symbol represents one item/person.
- Pupils will have explored transferring data from a pictogram into a bar chart.
- Pupils will have practised completing bar charts with a scale interval of 1.

Activities for whole-class instruction

- Give pupils **Resource 3.4.3** Spring flowerbed and work together to complete it, judging how much support is appropriate to provide.

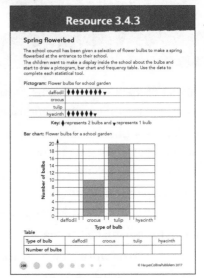

- Pupils complete Question 1, answering questions on a pictogram and using the same data to complete a bar chart.

Same-day intervention

- Give pupils **Resource 3.4.1c** Blank bar chart and this table showing the goals scored by four of the teams.

Football teams	United	City	Rovers	Athletic
Number of goals	19	9	17	10

- Label the bar chart and complete the bars, using the data.
- Which team scored the most goals?
- Which team scored the fewest goals?
- What is the difference in goals between the most and fewest?

Same-day enrichment

- Complete the Same-day intervention task.
- In the next two games the following number of goals are scored:
    - United – 1; City – 4; Rovers – 2; Athletic – 3
- Update the table and the bar chart.
- Which two teams have now scored the same number of goals?
- Show pupils this chart, which lists the number of games won, drawn and lost for each team.

Teams	Won	Drawn	Lost
United	10	1	4
City	4	3	8
Rovers	13	0	2
Athletic	5	2	8

Teams score 3 points for a win, 2 points for a draw and 1 point for a loss.

- Ask pupils to calculate the number of points each team has and, using a new blank bar chart, draw a bar chart to show the results. Ask pupils to compare this chart with the one showing the goals scored. Are the teams in the same order?

Question 2

2 During PE, Asif, Tom, Joe, Sarah and Lila had a competition to see who could bounce a ball the most times in one minute. Their results are recorded in the table below.

Results of ball bouncing

Name	Asif	Tom	Joe	Sarah	Lila
Number of times	40	25	50	45	35

(a) Look at the table and put the children's results in order. Write the names, starting with the child with the highest score.

_____

(b) Show the results in a bar chart.

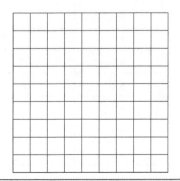

What learning will pupils have achieved at the conclusion of Question 2?

- Pupils will have been introduced to choosing suitable scale intervals.
- Pupils will have practised constructing bar charts with a scale interval of 5.
- Pupils will have reinforced their ability to interrogate data in bar charts to answer one-step and two-step questions.

Activities for whole-class instruction

- Ask pupils to look at Question 2 and find the greatest number of bounces. Agree that it is Joe's score of 50. Ask them whether there are enough squares on the page to use a scale of 1 cell for each bounce. Agree that there are not and that they will need to choose a different scale.
- Give pupils **Resource 3.4.1c** Blank bar chart and explain that they are going to learn how to choose the best scale for a bar chart.
- The choice of scale depends on the highest number.
    - If the highest number in the table is 10, count the cells and agree that a scale with intervals of one will fit on the bar chart.
    - Ask: *If the highest number in the frequency table is 20, what would the best scale interval be?* A scale with intervals of two will fill the bar chart. Look at the bar chart and count in twos together.
    - Move on to thinking about the best scale interval for a highest number of between 40 and 50. Establish that now intervals of five will be best. Look at the bar chart and count in fives together.
    - Finally, consider the best interval for a highest number of between 80 and 100. Numbers of this magnitude would require intervals of 10.
- Discuss what happens if you choose the 'wrong' scale. Establish that if the highest number is large, choosing a small interval means the columns will be very high – and probably would not fit on the page. If the highest number is small, choosing a large interval means the columns will be very short and there would be very little difference in the height of the columns. Show some examples.
- Return to Question 2 and ask pupils which they think is the best scale and why. Ask them to explain in a whole sentence.

Intervals of five are best because 50 will fill the bar chart.

- Pupils complete Question 2 in the Practice Book.

Same-day intervention

- The table shows the different soups eaten during one week in a school. Use the data and **Resource 3.4.1c** Blank bar chart, to complete a bar chart.

Soup	Chicken	Mushroom	Tomato	Vegetable
Number	35	15	45	30

- While pupils are carrying out the exercise, ask them to explain their choice of scale.

Same-day enrichment

- The table shows the different soups eaten during one week in a school, broken down day by day. The data and **Resource 3.4.1c** Blank bar chart, can be used to complete a number of bar charts, looking at the popularity of the soups day-by-day or over the week.

	Monday	Tuesday	Wednesday	Thursday	Friday
Chicken	5	13	8	7	12
Mushroom	5	4	4	5	2
Tomato	8	11	10	9	12
Vegetable	7	4	8	10	6

Adapt the task depending on the learning needs of pupils. Pupils can also be asked to make up questions on the data.

Challenge and extension question

Question 3

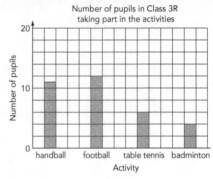

3 Use the bar chart to complete the sentences. Each pupil takes part in only one activity.

(a) In the graph, for the number of pupils, 1 unit represents ☐ pupil(s).

(b) The activity with the greatest number of participants is _____ and that with the fewest participants is _____ .

(c) There are ☐ pupils taking part in all the activities in Class 3R.

(d) The difference between the number of pupils taking part in handball and the number taking part in football is ☐ .

The Challenge and extension question requires pupils to answer questions on data from a bar chart showing sports activities. The scale is marked in intervals of 10, with each cell representing two children.

Chapter 4 test (Practice Book 3A, pages 103–106)

Test Question number	Relevant Unit	Relevant questions within unit
1	4.1	3
	4.2	1
	4.3	1, 2
2	4.1	3
	4.2	1, 2
	4.3	1, 2
3	4.1	2, 3
	4.2	1, 2
	4.3	1, 2
4	4.1	3
	4.2	1, 2, 3
	4.3	1, 2

Chapter 5
Introduction to time (III)

Area of mathematics	National Curriculum statutory requirements for Key Stage 2	Shanghai Maths Project reference
Measurement	Year 3 Programme of study: Pupils should be taught to:	
	■ tell and write the time from an analogue clock, including using Roman numerals from I to XII, and 12-hour and 24-hour clocks	Year 3, Unit 5.2
	■ estimate and read time with increasing accuracy to the nearest minute; record and compare time in terms of seconds, minutes and hours; use vocabulary such as o'clock, a.m./p.m., morning, afternoon, noon and midnight	Year 3, Units 5.1; 5.2
	■ know the number of seconds in a minute	Year 3, Unit 5.1
	■ know the number of days in each month, year and leap year	Year 3, Units 5.3; 5.4
	■ compare durations of events (for example to calculate the time taken by particular events of tasks)	Year 3, Units 5.2; 5.4

Unit 5.1
Second and minute

Conceptual context

This chapter looks at equivalent units of time, extending this concept from pupils' learning in KS1 where they mastered the language of years, months, days, hours and minutes and how they relate to each other. In Book 3 the focus is on seconds and minutes and equivalence between these. Pupils will use their experience of time duration to reason about the units needed to carry out various activities. They also extend their understanding of how to tell the time from to the nearest five minutes learned in Book 2 to the nearest minute in Book 3.

Learning pupils will have achieved at the end of the unit

- Pupils will have extended their knowledge of units of time to include seconds and consolidated their knowledge about relationships between units of time (seconds, minutes and hours) (Q1, Q5)
- Pupils will have extended their understanding of reading the time to the nearest minute (Q2)
- Pupils will have developed a sense of how long it takes to do something (Q3, Q4)

Resources

individual clocks; stopwatch per group; interlocking cubes; timers

Vocabulary

second, minute, hour, analogue time, digital time

Question 1

> **1** Complete the sentences.
> (a) There are ☐ seconds in a minute.
> (b) There are ☐ minutes in an hour.
> (c) There are ☐ seconds in half a minute.
> (d) There are ☐ seconds in 10 minutes.

What learning will pupils have achieved at the conclusion of Question 1?

- Pupils will have extended their knowledge of units of time to include seconds and consolidated their knowledge about relationships between units of time (seconds, minutes).
- Pupils will have used seconds, minutes and hours to find equivalent units of time.

Activities for whole-class instruction

- Ask pupils to tell you how many days there are in one week, then two weeks, four weeks and eight weeks. Ask: *How did you work these out? Did you use doubling?*

- Ask: *How many seconds are there in one minute?*

 All say... *There are 60 seconds in one minute.*

- Repeat the above activity so that pupils practise using known facts. For example, 120 seconds = two minutes, 240 seconds = four minutes, 480 seconds = eight minutes, 600 seconds = 10 minutes, 1 200 seconds = 20 minutes. Ask pupils to tell you how many minutes there are in half a minute, quarter of a minute and three-quarters of a minute.

 All say...
- *There are 30 seconds in half a minute*
- *There are 15 seconds in one quarter of a minute*
- *There are 45 seconds in three-quarters of a minute*

Ask, for example *How many seconds in, for example:*
- $1\frac{1}{2}$ minutes
- $3\frac{1}{2}$ minutes

- Pupils complete Question 1 in the Practice Book.

Same-day intervention

- Work with pupils who have difficulty using known facts to make new ones. Focus on finding the number of seconds in different minutes. If, for example, they know that there are 60 seconds in one minute, they can work out how many in two minutes by doubling 60 using their knowledge that double six is 12.

Same-day enrichment

- Pupils combine the different numbers of equivalent seconds and minutes that they found during the main activities to find, for example, how many seconds are equivalent to three $\frac{1}{4}$ minutes (one minute + two minutes + $\frac{1}{4}$ of a minute), seven $\frac{3}{4}$ minutes (four minutes + two minutes + one minute + $\frac{3}{4}$ of a minute). Repeat for other combinations.

Question 2

> **2** Draw lines to match the times.
> 03:05 08:26 04:59 11:11

What learning will pupils have achieved at the conclusion of Question 2?

- Pupils will have extended their understanding of reading the time to the nearest minute.

Activities for whole-class instruction

- Give each pupil a clock. Ask them to tell you all they can about their clock, for example, around the outside of the clock, between one whole hour and the next there are five marks. As the minute hand moves past each of these marks it shows that one minute has passed. At the same time, the hour hand moves a very, very small part of a turn that we can hardly see. If possible, with a geared clock, move through a minute, and another, and another, and another and so on. Pupils should do the same. Ask: *Can you see how slowly the hour hand moves?*

- Ask pupils to revise different o'clock, half past, quarter past and quarter to times and then find times that are multiples of five minutes (also revision). Now move on to times to the nearest minute. Ask pupils to show you 13 minutes past 10. Ask: *How do you know you have found 13 minutes past?* (Find 10 minutes past and then count on another three minutes.) Ask them to find 54 minutes past eight and to tell you how many minutes until the time is nine o'clock. Repeat with different times.

- Ask pupils to show you 45 minutes past seven. Ask: *How many minutes to 8 o'clock?* Ask them to write down what this time would be as a 12-hour digital time (7:45). Ask about the similarities and differences between the representations of these times. Establish the time is the same, but the representation is different. The first is an analogue time and the second is digital. When talking in analogue times we say the minutes to or past first and then the hour. When expressing time in digital terms we say the hour first and then the minutes past. Repeat for times to the nearest five minutes.

- Ask pupils to show you 16 minutes past nine. How do they think this would look as a digital time? The hour is 9 and the minutes past is 16, the digital time is therefore 9:16. Ask: *Is this a morning or evening time?* Agree that it could be either. Ask: *Do you know the symbols to show morning and afternoon/evening times?* Agree a.m. and p.m.

- Ask pupils to show six minutes past five. Write the digital time like this on the board: 5:6. Ask: *Is this correct or not?* Write 5:16 beside 5:6 and ask pupils to tell you the value of the minutes in each, for example 16 minutes past is made from one ten and six ones. Six minutes past is six ones. Ensure they see that in your representation you have positioned six in the tens position. Establish that there are no tens and so 5:6 should be written with a place holder in the tens position: 5:06. Repeat for other digital times to the nearest minute. Tell pupils whether these are morning/afternoon/evening times and ask them what symbols should be written to show this.

- Ask pupils to draw their own clock faces. They write 12, 6, 3 and 9 first and then position the other numbers between these. They draw a time, then pass their clock to a partner who writes the equivalent digital time. They decide what time of day the time is and write a.m. or p.m. accordingly. Invite pupils to share some of their digital times for the class to read as an analogue time with morning/afternoon/evening.

Look out for … pupils who write digital times with single minutes in the tens position. Show examples

of how single minutes are written: 01, 02, 03 and so on to 10 minutes, explaining the place value of each digit.

- Ask pupils to complete Question 2 in the Practice Book.

Same-day intervention

- Work with pupils who have difficulty making the link between analogue and digital time. Give each pupil a clock. Ask them to point to the hour numbers and count around the clock in intervals of five minutes beginning at 12 (zero minutes). Ask pupils to show 20 minutes past 10. Ask pupils to tell you how many minutes past 10 their clocks show. They write the digital time, the hour number first followed by the minutes number. Repeat for other times focusing on minutes past, for example 40 minutes past seven.

Same-day enrichment

- Give pupils four digits, for example 2, 4, 3 and 7. They make as many digital times as they can using the digits, for example 2:34, 2:43, 2:37, 2:47, 3:24, 3:27. They draw these times on analogue clocks and write the digital times.

Questions 3 and 4

3 Write a suitable unit of time in each space.

(a) It took Tom 48 _____ to swim 50 metres.

(b) Anya practises the piano 1 _____ every day.

(c) Ahmed's father works 8 _____ a day.

(d) The pupils have a 50 _____ lunch break.

4 Try the following activities and record the results.

(a) I can read ☐ English words in 1 minute.

(b) I can skip with a rope ☐ times in 1 minute.

(c) It takes me ☐ seconds to walk 50 metres.

(d) My heart beats ☐ times in 10 seconds.

What learning will pupils have achieved at the conclusion of Questions 3 and 4?

- Pupils will have developed an understanding of how to select suitable units of time to measure events.

- Pupils will have developed an appreciation of the length of time it takes to carry out tasks.

Activities for whole-class instruction

- Show pupils a stopwatch and ask what it is used for. Agree that it can be used to measure short time intervals. Ask pupils what unit a short time interval could be measured in. Agree seconds and minutes.

- Ask pupils to close their eyes. Tell them that you are going to put the stopwatch on for 30 seconds. They put their hands up when they think that amount of time has elapsed. Repeat this for 15 and then 45 seconds. Demonstrate how to use the stopwatch and then invite pupils to time the class for different numbers of seconds.

- Ask pupils what they think they could do in a minute. Try some of their ideas out. For example, can they stand on one leg for a minute? Can they throw and catch a ball without dropping it for one minute?

- Ask pupils what they think would take an hour, for example a maths lesson; walking to town and back. Ask: *How many hours are you in school every week?* Tell pupils that they spend around six hours in school every day – that's 30 hours every week.

- Ask pupils to work in groups. Give each group a stopwatch and ask them to write down a list of five quick activities, for example building a tower of ten interlocking cubes; writing their full name. They use the stopwatch to time each other as they carry out each task. They record the number of seconds each pupil took in a table. For example:

Name	Build a tower of cubes	Write full name	10 hops	Draw a stickman	Read three sentences
Joe					
Sophia					
Yukesh					
Gemma					

- Ask pupils to compare results, who was the fastest/slowest, what was the difference between pairs of results, what was the total time spent on one activity? What is that number of seconds converted to minutes and seconds?

- Pupils complete Questions 3 and 4 in the Practice Book.

Same-day intervention

- Work with pupils with a poor sense of time, particularly seconds and minutes. Give each pupil one-minute timers. Ask them to see how many numbers they can write in one minute. They turn their timers over and start writing. After a minute ask them to count the numbers that they wrote.

Same-day enrichment

- Pupils work with a partner to design two or three tasks that they think will take between one and two minutes to carry out. They carry out their chosen tasks, timing themselves with a stopwatch.

Question 5

5 Convert the times into different units.

(a) $\frac{1}{2}$ hour = ☐ minutes

(b) 5 minutes = ☐ seconds

(c) $\frac{3}{4}$ hour = ☐ minutes

(d) 120 seconds = ☐ minutes

(e) 1 hour 40 minutes = ☐ minutes

(f) 2 minutes 30 seconds = ☐ seconds

(g) 90 seconds = ☐ minute ☐ seconds

(h) 100 minutes = ☐ hour ☐ minutes

What learning will pupils have achieved at the conclusion of Question 5?

- Pupils will have consolidated their understanding of converting between units of time.

Activities for whole-class instruction

- Ask: *How many minutes there are in one hour?*

All say... *There are 60 minutes in one hour.*

- Write 1 hour = 60 minutes on the board. Ask pupils to double to find the number of minutes in two hours. They double again for four hours and then again for eight. As in Question 1, they use this information to find other equivalences by adding, subtracting and multiplying by 10.

- Ask pupils how many minutes are equivalent to half an hour, then one quarter. Ask them to explain how they found 30 minutes and 15 minutes. Can they find any other fractions of an hour, for example one third, one tenth and one sixth? Once they have unit fraction equivalents by dividing 60 by the denominator of the fraction, ask them to find non-unit fractions, for example $\frac{1}{6}$ = 10 minutes, $\frac{2}{5}$ = 20 minutes, $\frac{3}{6}$ = 30 minutes and $\frac{4}{6}$ = 40 minutes.

- Give pupils a mixture of hours and minutes to change to minutes, for example 1 hour 35 minutes, 2 hours 54 minutes, 3 hours 36 minutes and 4 hours 12 minutes. Observe how they change the hours. For example, do any pupils double one hour to find the minutes in two hours, double this for four hours, add one and two for three hours? Encourage them to use what they already know to find equivalences.

- Ask: *How many seconds there are in one minute and 30 seconds?* Elicit whether pupils recognise that they should use the same strategies with a mixture of minutes and seconds as they did with hours and minutes. Ask another similar question, for example: *How many seconds are there in three minutes and 20 seconds?*

- Pupils complete Question 5 in the Practice Book.

Same-day intervention

- Give each pupil a clock. Ask them how many minutes will pass between one o'clock time and the next o'clock time. Skip count in fives from the 12. Ask pupils to point to the seven and to tell you how many minutes past the hour it is (35 minutes past). Repeat for other hour numbers.

Same-day enrichment

- Give pupils a list of numbers of minutes above 60 and below 120, for example 64 minutes, 75 minutes. They convert these to one hour and however many minutes are left, 64 minutes is one hour and four minutes, 75 minutes is one hour and 15 minutes.

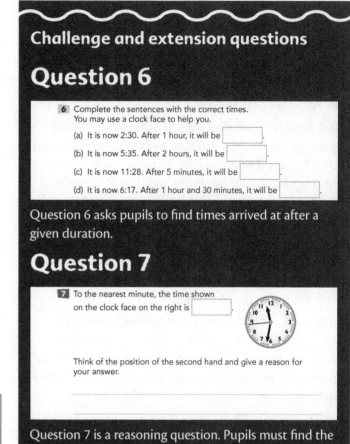

Challenge and extension questions

Question 6

> 6 Complete the sentences with the correct times. You may use a clock face to help you.
>
> (a) It is now 2:30. After 1 hour, it will be ▢.
>
> (b) It is now 5:35. After 2 hours, it will be ▢.
>
> (c) It is now 11:28. After 5 minutes, it will be ▢.
>
> (d) It is now 6:17. After 1 hour and 30 minutes, it will be ▢.

Question 6 asks pupils to find times arrived at after a given duration.

Question 7

> 7 To the nearest minute, the time shown on the clock face on the right is ▢.
>
> Think of the position of the second hand and give a reason for your answer.

Question 7 is a reasoning question. Pupils must find the time to the nearest minute using the second hand as their clue. You may need to spend some time looking at a clock with a second hand so that they can see its movement past a minute and to the next one.

Unit 5.2
Times on 12-hour and 24-hour clocks and in Roman numerals

Conceptual context

This chapter looks at making links between 12-hour and 24-hour times so that pupils will be able to convert from one to the other. For pupils this is the first introduction to the idea of 24-hour time. However, they are likely to be familiar with these times from their experiences outside school. They will also be introduced to Roman numerals for the first time. Roman numerals are still used today, for example book pages, dates, clocks.

Learning pupils will have achieved at the end of the unit

- Pupils will have consolidated their understanding of hours and days as units of time and equivalences between them (Q1)
- Pupils will have consolidated the use of a.m. and p.m. as notation for morning and afternoon/evening (Q1)
- Pupils will have developed their understanding of how to write Roman numerals to 12 and how to read an analogue clock with Roman numerals (Q2, Q3)
- Pupils will have begun to develop their understanding of how to convert from a 12-hour clock time to a 24-hour clock time (Q4, Q5)
- Pupils will have developed their understanding of how to find time durations (Q6)

Resources

individual clocks; pictures of real-life examples of Roman numerals (e.g. the clock tower of Big Ben, pages from the beginning of a book, names of kings and queens of the UK); flight timetables

Vocabulary

second, minute, hour, analogue time, digital time, a.m., p.m., 12-hour time, 24-hour time, Roman numerals

Question 1

> **1** Fill in the boxes.
>
> (a) There are ☐ hours in a day.
>
> (b) There are ☐ hours in half a day.
>
> (c) The time at noon is ☐ o'clock.
>
> (d) In a day, the time from midnight to noon is
> called _____ or a.m.
>
> (e) The time from noon to midnight is called _____ or p.m.

What learning will pupils have achieved at the conclusion of Question 1?

- Pupils will have used hours and days to find equivalent units of time.
- Pupils will have understood the use of a.m. and p.m. to indicate whether a time is in the morning or afternoon/evening.

Activities for whole-class instruction

- Ask pupils to tell you how many hours there are in one day, then two days, four days and eight days. Encourage them to use doubling to do this. Do they partition, double each part and recombine? Find out if any other strategies were used. Ask them to use this information to work out how many hours there are in six, 12, 14, 20, 40 and 80 days. Ask: *Did you add and multiply the appropriate days to find the new ones?*

 24 hours are equivalent to one day.

- Recap when morning times occur and then afternoon and evening times. Midnight to noon is the 'morning'. The 'afternoon' is after noon, as the word suggests, and goes on until it starts to get dark. When it gets dark we use the term 'evening'. Ask pupils to describe some of the things that they do in the morning, afternoon and then evening. Discuss the fact that any time after midnight and before noon is known as the morning. The morning lasts for the first 12 hours and the afternoon/evening is the second 12 hours.

- Remind pupils that 'noon' is another way of saying midday. This time is in the middle of the day, as the word midday suggests. Noon happens at 12 o'clock. It is the time that separates the morning and afternoon. On the board write: afternoon, morning, evening and noon.

- Ask pupils what abbreviations are used to describe morning and afternoon/evening times. Agree that the 12-hour clock is when the 24 hours of the day are divided

into two periods: a.m. and p.m. These abbreviations come from an old language called Latin. The first 12-hour period is called a.m., which means ante meridiem, before midday. The second 12-hour period is called p.m., which means post meridiem, after midday.

All say ... *a.m. means ante meridiem, before midday; p.m. means post meridiem, after midday.*

- Pupils complete Question 1 in the Practice Book.

Same-day intervention

- Work with pupils who have difficulty identifying a.m. and p.m. Give them two pieces of card. They write a.m. on one and p.m. on the other. Ask them to find different times on a clock, focussing on minutes past times, for example 20 minutes past nine, 50 minutes past three. For each time, tell them whether it is morning, afternoon or evening. They hold up the appropriate card.

Same-day enrichment

- Ask pupils to make an information sheet to show the meaning of seconds, minutes, hours, days, noon, midday, midnight, a.m. and p.m.

Questions 2 and 3

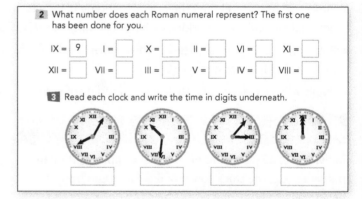

> **2** What number does each Roman numeral represent? The first one has been done for you.
>
> IX = 9   I = ☐   X = ☐   II = ☐   VI = ☐   XI = ☐
>
> XII = ☐   VII = ☐   III = ☐   V = ☐   IV = ☐   VIII = ☐
>
> **3** Read each clock and write the time in digits underneath.

What learning will pupils have achieved at the conclusion of Questions 2 and 3?

- Pupils will have developed their understanding of how to write Roman numerals to 12 and how to read an analogue clock with Roman numerals.

Activities for whole-class instruction

- Show the pictures that you have collected of Roman numerals in real life. Ask pupils what they think these are. Give some background information of the history of Roman numerals – see Information Point.

  (i) Roman numerals appear to have started out as notches on tally sticks about 3000 years ago in Rome. They continued to be used by some Italian shepherds into the 19th century. Ask pupils: *What years did the 19th century include?* (1800 to 1899). *How long does a century last? Do you know our current century?* I is a notch on the stick to represent one of something, II is 2, III is 3 and early IIII was 4. Λ or V represented a hand, and represented 5 because of the fingers; × represents two hands, one inverted and, therefore, 10 for ten fingers. So, the first Roman numerals would have looked something like this: IIIIΛIIIIXIIIIΛIIIIXII and so on, a bit like European tally marks today. Numbers got very large to write so abbreviations were used, for example IV is one before 5, so 4, VI is one after 5, so 6, IX is one before 10, so 9. Symbols were also introduced for specific larger numbers (L = 50, C = 100, D = 500 and M = 1000).

- Write the first Roman numerals on the board: IIIIΛIIIIXIIIIΛIIIIXII. Ask: *What are these similar to?* (They are similar to a tally chart.) *Can you remember what a tally looks like from your work in Book 2?* Ask pupils to draw a tally for the numbers to 20. Discuss the similarities and differences between the tally and early Roman numerals.

  卌 卌 卌 卌

- Ask pupils to write what they think would be the numbers 1 to 12 in Roman numerals, using the information given about 4 being one before 5, so IV and 6 being one after 5, so VI. Ask pairs of pupils to share their thinking. Ask: *Do you agree?* Write the numbers 1 to 12 on the board, explaining each one as you do (I, II, III, IV, V, VI, VII, VIII, IX, X, XI, XII). Pupils check their numerals with yours.

- Call out different numbers to 12, pupils write these in Roman numerals. Ask them to write down five Roman numerals and give them to a partner who writes the equivalent number using our number system. Repeat this so that each pupil has had the opportunity to convert numbers to Roman numerals and vice versa.

- Give pairs 24 small pieces of paper. On 12, they write our numbers from 1 to 12. On the other 12 they write the equivalent Roman numerals. They lay their numbers face down on the table and take it in turns to pick two

at a time. They turn them over and if they are a pair, one Arabic and one Roman of the same number, they keep them. If not, they put them back where they took them from. The winner of the game is the player with the most pairs.

- Give pupils plain paper and ask them to draw clock faces with Roman numerals instead of the usual Arabic numerals. They write XII, VI, III and IX first and then position the other numbers between these. They draw a time, then pass their clock to a partner who writes the analogue time.

- Pupils complete Questions 2 and 3 in the Practice Book.

Same-day intervention

- Work with pupils who have difficulty converting from Arabic to Roman numerals. Explain and demonstrate that one line is 1. Ask pupils to copy you. Continue with two lines for 2 and three lines for 3. Explain that the Romans used to draw four lines for 4, but then abbreviated it to one line in front of the symbol for 5, to show one less than 5 which is 4. Continue in this way to 12. Draw the numerals on different pieces of paper. Hold up one at a time, in order first and then randomly. Ask pupils to write down our equivalent number.

Same-day enrichment

- Ask pupils to make a conversion table with two columns, our numbers in one column and the equivalent Roman numerals in the other. Ask: *Can you use what you know about numbers to 12 to reason about how to write the other numbers to 20?* They add these to their conversion table. For example:

Arabic	Roman
1	I
2	II
3	III
4	IV

Questions 4 and 5

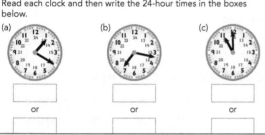

4 Complete the table for converting 12-hour time to 24-hour time. The first one has been done for you.

12-hour time	24-hour time
3:03 a.m.	03:03
8:00 a.m.	
	13:36
	23:58
12:00 midnight	

5 Read each clock and then write the 24-hour times in the boxes below.

(a)　　　　(b)　　　　(c)

or　　　　or　　　　or

What learning will pupils have achieved at the conclusion of Questions 4 and 5?

- Pupils will have developed an understanding of how to convert between 12-hour and 24-hour times.

Activities for whole-class instruction

- Write these two times on the board: 8:16 a.m. and 08:16. Ask pupils to show you this time on a clock. Ask: *What is the same/different about them?* Elicit that they both show 16 minutes past eight and they are both morning times. They are different representations of the same time. One is a 12-hour time and the other is 24-hour time. Ask: *Where have you seen times shown using a 24-hour representation?* Pupils are likely to mention mobile phones, computers and computer games. They may also mention bus and train timetables. Show examples as appropriate.

- Draw a time number line on the board similar to the one below:

24-hour time

0 1 2 3 4 5 6 7 8 9 10 11 12 13 14 15 16 17 18 19 20 21 22 23

a.m.　　　　　　　　　p.m.

12 1 2 3 4 5 6 7 8 9 10 11 12 1 2 3 4 5 6 7 8 9 10 11

12-hour time

- Ask: *What do you notice?* Elicit that the 12-hour time shows two sets of numbers from 12 to 11. The first set begins at midnight and ends at 11:59 a.m. The second set begins at noon and ends at 11:59 p.m. The 24-hour time begins at midnight and goes through to 23:59 at the end of the day.

- Call out some 12-hour o'clock times and ask pupils to write down the 24-hour equivalent, for example seven o'clock in the evening, six o'clock in the evening.

- Give each pupil a clock. Ask them to make a time to the nearest five minutes and to show it to a partner. They say the time as an analogue time, for example 10 minutes to five in the morning. Their partner writes the time as a 12-hour time with either a.m. or p.m. and then the 24-hour equivalent. They take it in turns to do this several times. Invite pupils to share some of their times. Elicit the fact that the minutes are the same on both representations.

- Ask pupils to look at the clocks in Question 5 of the Practice Book. Discuss the similarities/differences between these clocks and the number line you drew earlier. Point out that the hour numbers from 12 to 11 could indicate 12-hour and 24-hour times between midnight and noon. The numbers inside the main hour numbers show the equivalent 24-hour times for the hours after noon. Ask pupils what needs adding to the a.m. hour numbers. Agree that a zero as a place holder for the tens numbers needs adding to the hours from one to nine.

- Pupils complete Questions 4 and 5 in the Practice Book.

Same-day intervention

- Work with pupils who have difficulty converting from 12-hour to 24-hour times. Ask them to draw their own number line on a strip of paper beginning and ending at midnight. They write on the 12-hour numbers from 1 to 11, then noon at the 12 o'clock midday position, then the hour numbers 1 to 11 again. They write a.m. under the first set of numbers and p.m. under the second. Above these 12-hour numbers they write the 24-hour numbers from 01 to 23. Ask them to put their finger on different 12-hour positions and to tell you the 24-hour equivalent.

Same-day enrichment

- Pupils show clock times to the nearest minute to a partner. Their partner writes down the three possible times: analogue, 12-hour and 24-hour digital.

Question 6

> | 6 | [road sign: 07:30 – 10:30 / 16:30 – 19:30] | The traffic road sign shows that the road is closed to all traffic from ☐ to ☐ in the morning and from ☐ to ☐ in the afternoon. |
>
> The road is closed to traffic for ☐ hours each day.

What learning will pupils have achieved at the conclusion of Question 6?

- Pupils will have begun to develop their understanding of how to find time durations.

Activities for whole-class instruction

- Give each pupil a clock and ask them to find different times, for example nine minutes past eight, 18 minutes to four. Say whether these times are morning, afternoon or evening. They find them and then write the equivalent 12-hour and 24-hour digital times.

- Ask pupils to find 12:50 on their clocks. They tell you what this is as an analogue time (10 minutes to one). Ask pupils what the time would be one and a half hours later than 12:50. Discuss how they could do this. Encourage them to move the hour hand on one hour or, if using geared clocks, the minute hand round for one hour. They then move the minute hand another 30 minutes to give the time 2:20. Repeat this for other starting times, keeping the time one and a half hours later. Repeat for other lengths, for example one hour 15 minutes, two hours 25 minutes.

- Ask this problem: *The café opened at 07:45 and closed at 10:00. For how long was it open?* Pupils find 07:45 on their clocks. They then move the minute hand round twice for two hours. If you are not using geared clocks, ensure pupils move the hour hand round so that the time shows 09:45. Ask: *How many minutes from this time to 10:00?* Agree 15 minutes. The café is open for two and a half hours. Tell them that it opens again at 16:00 and closes at 19:30. Pupils work out how long It is open for this time in the same way. Repeat these questions but varying the opening times.

- (i) For problems with time, such as the ones for Question 5, use clocks, geared if possible. Time number lines are not a good method to use for the problems in Question 5 because pupils might get confused and work with tens and hundreds. Time is base 60.

- Pupils complete Question 6 in the Practice Book.

Same-day intervention

- Work with pupils who have difficulty finding time durations. Begin by asking problems where the start time is an o'clock time and the finish time a half past time, for example: *Billy went to play football with his friends. He started playing at 16:00 and finished at 18:30. For how long was he playing football?*

Same-day enrichment

- Pupils make up their own problems that involve finding time durations to give to a partner to solve. Their starting or finishing time should be a time to the nearest minute, for example 10:37.

Challenge and extension question

Question 7

> 7 Draw a line to match each pair of times that would look the same on the clock face. One has been done for you.
>
> 22:30   half past ten   9:25
>
> 16:25   19:15
>
> 21:25   4:25
>
> a quarter past seven

Question 7 links the three ways of showing the time that pupils have been thinking about in this chapter. They need to find the equivalent 12-hour and 24-hour times and match to how it would appear on an analogue clock.

Unit 5.3
Leap years and common years

Conceptual context

This chapter extends pupils' understanding of units for measuring time to months and years. This is the first time that pupils have been introduced to the idea of a leap year. Some may be familiar with these if they have friends or family members that were born in a leap year. A leap year contains one additional day. This is added to synchronise the calendar with the actual time taken for the earth to orbit the sun, which is $365\frac{1}{4}$ days. All other years are referred to as common years. Pupils will be exploring calendars so ensure that you have plenty of calendar pages available for this year.

Learning pupils will have achieved at the end of the unit

- Pupils will have consolidated their understanding of hours, days, months and years as units of time and the equivalences between them (Q1)
- Pupils will have practised interrogating data (in this case the tabular calendar form) (Q2, Q3)
- Pupils will have developed their understanding of calendars and the number of days in each month (Q2, Q3)
- Pupils will have begun to develop their understanding of leap years and common years (Q4, Q5)

Resources

calendar pages for this year; A4 sheets of squared paper

Vocabulary

day, month, year, leap year, common year, calendar

Question 1

> **1** True or false? (Put a ✓ for true and a ✗ for false in each box.)
>   (a) One day always has 24 hours. . . . . ☐
>   (b) One week always has 7 days. . . . . ☐
>   (c) One month always has 30 days. . . . . ☐
>   (d) One year always has 12 months. . . . . ☐
>   (e) One year always has 365 days. . . . . ☐

What learning will pupils have achieved at the conclusion of Question 1?

- Pupils will have consolidated their understanding of equivalent units of time (hours, days, weeks, months and years).

Activities for whole-class instruction

- Give pairs a calendar page that shows the 12 months of this year. Ask them to talk to their partner about what they can see. Discuss with the whole class. Elicit that pupils understand that there are 12 months and each month has either 30, 31 or 28/29 days. Ask questions such as: *On which day is 4th August? … 21st January? How many weekends are there in May? … February? On which day is the 1st of (each month)? Which day is the last of (each month)?*

- Write 'A month has 31 days' on the board. Ask: *Is this always, sometimes or never true?* Pupils should give examples as proof. Let them use the calendar page to help. Agree that the statement is sometimes true. Ask: *How many days are in the other months? What do you know about the number of days in February?* Elicit that every four years there is an extra day. Establish that February has 28 days except during a leap year when it has 29. Ask: *Do you know what a leap year is?* Explain that it is a year with an extra day. We have a leap year to keep the calendar year in time with the solar year (the time it takes for the earth to go around the sun).

- Write 'One year always has 366 days' on the board. Ask: *Is this always, sometimes or never true?* Pupils should give examples as proof. Again, let them use the calendar page to help. You could give pupils the calendar page for a leap year to compare. Together as a class, you should add the number of days in each month together using a calculator, to find the total.

- Tell pupils that the year 2000 was a leap year. Using their knowledge that these occur every four years, ask them to work out some of the leap years before 2000. Ask: *When will the next leap year be? What effect does that have on the number of days in February?* Check that pupils know that there is one extra day in February during a leap year,

making 29 instead of the usual 28. Ask: *Do any of you know of anyone who has a birthday on February 29th?*

 (All say …) *A leap year occurs every four years.*

- Pupils complete Question 1 in the Practice Book.

Same-day intervention

- Work with pupils who have difficulty understanding that although the number of hours in a day, days in a week and months in the year are always the same, the number of days in each month varies from one month to the next. (And the number of days in February also changes if it is a leap year.) Ask them to use their calendar page and write down the 12 months of the year. They then look to see how many days each has and write the number beside the month.

Same-day enrichment

- Ask pupils to use a calendar page showing the whole year and find all the months that have five Fridays. They then find the months with five Mondays. Ask: *Are they the same months?* Ask *Why is this not possible?* (For a month to have two Fridays and two Mondays, it would need 32 days.)

Questions 2 and 3

> **2** Use a calendar for this year to complete the table.
>
> | Month | Jan | Feb | Mar | Apr | May | Jun |
> |---|---|---|---|---|---|---|
> | Number of days | | | | | | |
> | Month | Jul | Aug | Sept | Oct | Nov | Dec |
> | Number of days | | | | | | |
>
> **3** Use the information in the table in Question 2 to complete the facts.
>
>   (a) There are ☐ days in April. The months that have the same number of days as April are _____
>   _____ .
>
>   (b) Christmas Day is on the ☐ of _____ .
>   The months that have the same number of days as that month are _____
>   _____ .
>
>   (c) The month that has the fewest days is _____ .
>   There are ☐ days in that month.
>
>   (d) There are ☐ days in this year.

What learning will pupils have achieved at the conclusion of Questions 2 and 3?

- Pupils will have practised interrogating data (in this case the tabular calendar form).
- Pupils will have consolidated their understanding of the number of days in each month.

Activities for whole-class instruction

- Give pupils a calendar page for all of the months in this year. Ask them to circle the last day of each month.

 30 days have September, April, June and November. All the rest have 31 except February, which has 28 or in a leap year 29.

- Pupils work out how many weeks and days each month has, for example April has 30 days, which is equivalent to four weeks and two days.
- Recap what a leap year is and how that affects the number of days in February.
- Ask pupils to follow your instructions as they move around the calendar page. For example, put your finger on the 2nd February, move forward three days, move forward 14 days, move back 24 hours, move back one day. Find out where pupils are. Can they recognise that 14 days is equivalent to two weeks and, therefore, jump forward two weeks instead of counting 14 days? Ask: *Do you know that 24 hours is the same as one day?* Repeat this with other instructions that involve converting hours to days and days to weeks.
- Ask pupils to think of special occasions and celebrations, for example their birthday, Christmas. Ask them to circle their date of birth. They then pass their calendar page to a partner who identifies when their birthday is and on what day of the month it is. Their partner then circles their birthday and works out how many weeks and days are between the two birthdays.
- Ask them if their birthday will always be on the same day. Agree that it won't, because the months of the year are not all exactly four weeks in length. Show a calendar from last year or next year so that pupils can see the same date does not fall on the same day in consecutive years.
- Pupils complete Questions 2 and 3 in the Practice Book.

Same-day intervention

- Work with pupils who have difficulty identifying the number of days in a week and use this to move around a calendar. Ask them to tell you how many days there are in one week, then double to make two weeks and double to make 4. They use this information to find the days in 3 weeks (add one and two weeks), 5 weeks (add one and 4 weeks) and other combinations. They write this information on paper. Give them a date and they work out the date after 1, 2, 3 and 4 weeks.

Same-day enrichment

- Ask pupils to make a set of instructions to move around the calendar to give to a partner to follow. Encourage them to include numbers of days that can be changed into weeks and hours that can be changed into days.

Questions 4 and 5

4 Using a print or online calendar, find the number of days in each year and the number of days in February for any 10 consecutive years.

Year	Number of days in the year	Number of days in February

5 Use the information in the table in Question 4 to fill in the answers.

(a) There are [ ] days in most of the months of February. These years are called common years.

(b) There are [ ] days in some of the months of February. These years are called leap years.

(c) A common year has [ ] days. A leap year has [ ] days.

(d) 2016 was a leap year. The next two leap years will be [ ] and [ ].

What learning will pupils have achieved at the conclusion of Questions 4 and 5?

- Pupils will have consolidated their understanding of leap years and how these affect the number of days in February.

Activities for whole-class instruction

- Remind pupils that leap years occur every four years and that during a leap year February has 29 days. Write 2000, 2004, 2008, 2012 and 2016 on the board. Tell pupils that these were all leap years. Do they notice that these years end with multiples of 4 and, because 20 is a multiple of 4, 2000 must be, so these years are multiples of 4? Say: *Write down when the next five leap years will be by counting on in fours.* (For example, 2020, 2024, 2028.)

- Give each pupil a sheet of squared paper and a calendar page for this year. Ask them to copy the dates and days for February. Tell them that the first of February in 2016 was a Monday. They make a calendar page for February 2016. Check that they remember to add the extra day. Ask them to make another for 2017 when the first was a Wednesday. Repeat for other years ensuring that you tell pupils the day that the 1st falls on.

- Point out that the word common means ordinary or something that happens often. Ask: *What do you think that a common year is?* Agree that all the years that are not leap years are common years. They happen most often – leap years are quite **UN**common.

- Pupils complete Questions 4 and 5 in the Practice Book.

Same-day intervention

- Work with pupils who having difficulty knowing that leap years are those with 29 days in February. Give them a selection of calendar pages for February in several leap and common years. Ask them to tell you similarities and differences between them. Ask them to sort them into two piles, one in which February has 29 days and the other in which February has 28 days. Ask them to explain what is special about the pile where February has 29 days. Establish that the extra day is there because it is a leap year.

Same-day enrichment

- Ask pupils to work out when all the leap years in this century after 2020 will occur. They then work out how many leap years there will be in total this century.

Challenge and extension question

Question 6

> 6 Alvin is at primary school. He said: "I have only had three birthdays, including the day I was born."
> On what day of the year do you think he was born?

Question 6 asks pupils to use and apply their knowledge of leap years to solve a problem that involves finding the birth day of a child in primary school, who was born in a leap year.

Unit 5.4
Calculating the duration of time

Conceptual context

This chapter recaps and extends equivalent units of time. The main focus is calculating time durations. Time is base 60, which means times cannot be added or subtracted using the usual column written methods because our number system is base 10. If pupils, for example, used a column written method to add 2 hours and 25 minutes to 7:45, they are likely to reach an answer of 9:70 if they choose to use a written method, because units of time are not decimal, (base 10) they are base 60.

Pupils will learn to use timelines to calculate time durations. For example, to find the duration from 12:30 to 15:45, pupils might count on 30 minutes to 13:00 then two hours to get to 15:00 and then another 45 minutes. The parts would be added together to give a duration of three hours 15 minutes. In the first example above, two hours would be added to 7:45 to give 9:45. Fifteen minutes would be added to give 10:00 and then the remaining 10 minutes, giving a final time of 10:10.

Learning pupils will have achieved at the end of the unit

- Pupils will have extended their understanding of hours, days, months and years as units of time and the equivalences between them, while practising use of inequality symbols (Q1)
- Pupils will have consolidated their understanding of calculating time durations (Q2, Q3, Q4, Q5)

Resources

calendar pages for this year; clocks

Vocabulary

seconds, minutes, hours, days, weeks, years, greater than, fewer than, timeline, duration

Question 1

> **1** Write >, < or = in each ( ).
>
> (a) 1 hour ( ) 60 seconds
> (b) 2 minutes ( ) 100 seconds
> (c) 2 days ( ) 20 hours
> (d) 3 hours ( ) 200 minutes
> (e) 1 month ( ) 27 days
> (f) 52 weeks ( ) 1 year

What learning will pupils have achieved at the conclusion of Question 1?

- Pupils will have extended their understanding of equivalent units of time (hours, days, weeks, months and years).

Activities for whole-class instruction

- Recap the number of seconds in a minute, minutes in an hour, hours in a day, days in a week, month and year. Ask: *Can you tell me the number of days in each month, including February in a leap year and a common year?*

- Ask questions that involve finding the seconds in different numbers of minutes. For example, ask: *How many seconds in two/three/four minutes?* Expect pupils to explain their strategies for finding the totals. Repeat for minutes in different numbers of hours and hours in days.

- Ask pupils to explain the difference between a common year and a leap year, including the numbers of days in each.

- Write >, < and = on the board. Ask pupils to explain what each symbol means. Practise comparing periods of times – years, months, weeks, days, hours, minutes and seconds.

- Ask: *How many minutes and seconds are there in 75 seconds?* Encourage pupils to partition the number of seconds into 60 for a minute and 15 for the number of seconds left. Repeat for other numbers of seconds, for example 90 (60 and 30 seconds), 130 (120 and 10 seconds), 200 (180 and 20 seconds). Ask: How many hours and minutes are there in 140 hours? (120 for two hours and 20 minutes). Repeat for other numbers of minutes, for example 260 (four hours and 20 minutes), 305 (five hours and five minutes), 320 (five hours and 20 minutes).

- Pupils complete Question 1 in the Practice Book.

Same-day intervention

- Work with pupils who have difficulty comparing units of time. Ask pupils to write down the number of seconds in one, then two, three, four and five minutes. As a group, create a table to display in class. Pupils use this information to write some comparison statements, for example: 2 minutes > 60 seconds, 180 seconds = 3 minutes, 180 seconds < 4 minutes. At another time, ask pupils to use the information to make up comparison statements for minutes and hours.

Same-day enrichment

- Ask pupils to make up their own comparison statements for seconds, minutes, hours, days and weeks. They make up five for >, < and =. Can they compare seconds and hours, hours and weeks?

Questions 2 and 3

> **2** Find the answers.
>
> (a) May left home at 9:00 a.m. and came back at 1:00 p.m.
>
> She was away for [ ] hours.
>
> (b) Lottie began her homework at 6:00 p.m. and finished at 6:45 p.m.
>
> She spent [ ] minutes doing her homework.
>
> (c) A party started at 19:30 and ended at 21:40.
>
> It lasted [ ] hours and [ ] minutes.
>
> (d) A volleyball match started at 19:30 and lasted 155 minutes.
>
> It ended at [ ].
>
> **3** ABC Superstore's opening hours on weekdays are shown below.
>
> ABC Superstore
> Opening hours
> 8:30 to 21:00
>
> (a) The superstore opens at [ ] in the morning and closes at [ ] in the evening.
>
> (b) The superstore is open for [ ] hours and [ ] minutes on weekdays.
>
> (c) It takes Mum 20 minutes to go from work to the superstore. If she leaves work for the superstore at 8 p.m., does she still have time to do shopping? _____.
>
> If so, for how long? _____.

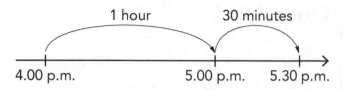

1 hour   30 minutes

4.00 p.m.   5.00 p.m.   5.30 p.m.

What learning will pupils have achieved at the conclusion of Questions 2 and 3?

- Pupils will have developed their understanding of finding time durations using clocks.

Activities for whole-class instruction

- Ask pupils what they think a duration of time is. Agree that it is the amount of time that passes between one thing happening and another. Give pupils mini whiteboards and small clocks. Ask: *How much time passes:*

  - *between six o'clock and seven o'clock?*

  - *between half past eight and ten o'clock?*

  - *from 10 past three to 20 to four?*

- For each example, give pupils plenty of time to discuss ideas with a partner and to write or draw anything that helps them to think about their answer.

- Ask the following questions:

  - *What do you think about to help you work it out?*

  - *What did you draw?*

  - *What did you imagine in your head?*

- Discuss how these time durations can be worked out. Agree that clocks could be used.

- Show pupils how to draw a timeline:

6.00 p.m.   7.00 p.m.   8.00 p.m.

- Tell pupils that this is a simple timeline showing two hours between 2 o'clock in the afternoon and 4 o'clock in the afternoon. Ask: *Where on this line is half past three?* Point to quarter to four and ask: *What time does this represent?*

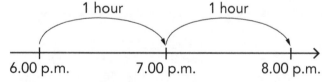

1 hour   1 hour

6.00 p.m.   7.00 p.m.   8.00 p.m.

- Set pupils some problems in context that involve finding the difference between two times, for example ask: *Sam began his swimming lesson at four o'clock, he finished at half past five. For how long did his swimming lesson last?*

- Build up the timeline together on the board, discussing what the start and finish times are, what the interim time markers should be and what the 'jumps' should be labelled. Ask: *How long did Sam's lesson last?*

- Repeat by asking the same problem but vary the finishing time of the lesson, for example: *It began at four o'clock and finished at $\frac{1}{4}$ to five, $\frac{1}{4}$ past five, $\frac{1}{4}$ to six.* Also vary the problem by giving digital finishing times, for example: 4:50, 5:10, 5:25. Pupils use clocks and timelines to find the time differences. They record their answers, discuss their preferred method and reasons for their choices.

- Make up other problems that have different digital starting times.

- Set pupils some problems that involve finding a finishing time, for example: *Jade went to play at her friend's house. She arrived at her friend's at two o'clock and stayed for two hours. When did she leave her friend's house?* Again, vary the problem by changing the length of time Jade stayed. Include whole hour times, hours and minutes and minutes only, for example: *Jade stayed three hours, one hour 35 minutes, 75 minutes.* Pupils use their clocks to find the finishing times and then record them by drawing their own clock and labelling it with the 12-hour and 24-hour digital times.

- Pupils complete Questions 2 and 3 in the Practice Book.

Same-day intervention

- Work with pupils who have difficulty finding time durations. Give each pupil a clock. You will need one to model with. Ask pupils to find seven o'clock. Tell them that this is the start time of a party and that the party finishes at nine o'clock. Ask: *Can you explain how you could find the length of the party?* Agree that they could use their fingers and count on from seven to nine. They could also move the minute hand around for two hours, ensuring that they move the hour hand appropriately. Repeat this for other o'clock times. Repeat for other times where the duration is a whole number of hours, for example half past six to half past nine.

Same-day enrichment

- Ask pupils to make up their own problems that involve finding finishing times in hours and minutes. They give these to a partner to solve.

Questions 4 and 5

> **4** The 2012 Summer Olympics was held in London from 27 July 2012 to 12 August 2012.
>
> What was the duration of the event? _____
>
> **5** A local computer store had three promotional sales in the first half of 2015. The dates are given below.
>
> | First promotion | 23/02/2015 until 02/03/2015 |
> |---|---|
> | Second promotion | 23/03/2015 until 01/04/2015 |
> | Third promotion | 25/04/2015 until 03/05/2015 |
>
> (a) Which promotion lasted for the longest period of time?
>
> _____
>
> How long was it? _____
>
> (b) Which promotion lasted for the shortest period of time?
>
> _____
>
> How long was it? _____
>
> (c) For how many days did the store have a promotion in the first half of 2015? _____

What learning will pupils have achieved at the conclusion of Questions 4 and 5?

- Pupils will have developed their understanding of finding time durations using calendars.

Activities for whole-class instruction

- Give pupils a year's calendar page showing dates for the current year. Ask pupils to point to different dates, for example: September 17th, January 30th, March 3rd. For each date, pupils tell you on which days of the week they are. Have pupils call out some dates for their peers to find and identify the day.

- Ask pupils to point to 13th February. Ask: *What date is eight days later? How did you find this? Did you count on one week and one day?* Repeat for other starting dates, encouraging them to find the date eight days later in the most efficient way. Repeat for adding nine, 12 and 14 days.

- Ask pupils to put a cross on 12th April and then May 4th. They work out the length of time from one date to the other, including both dates, in days. (23) They then work out how long this is in weeks and days. (Three weeks and two days.) Repeat for other dates.

- Set some problems that involve finding time durations in months. As they answer these, pupils need to work out if they count the first date or if they count on from it. For example: *Today is the 15th June, Sadie's birthday is on 1st July. In how many days is her birthday?* (Pupils count on from 15th June.) *Ben arrived in Spain on 29th July and left on 15th August. How long was he in Spain?* (Pupils count the first date.) Pupils use their calendar page to work out the solutions.

- Pupils complete Questions 4 and 5 in the Practice Book.

Same-day intervention

- Work with pupils who having difficulty finding time durations in days. Give them a calendar page and ask them to find different dates. They put their finger on each date and then identify the date one day later. Repeat this gradually increasing the number of days to count on to two, three, four, five, six and seven. When they count on seven days, ensure they understand this is equivalent to a week.

Same-day enrichment

- Ask pupils to make up their own problems that involve finding durations of time in weeks. They give these to a partner, who solves them using a calendar.

Challenge and extension question

Question 6

> **6** A popular TV series was broadcast from Thursday 7 March 2013 to 4 April 2013. Two episodes of the series were broadcast every day from Monday to Thursday and one episode was broadcast on Saturdays and Sundays. It was not shown on Fridays.
>
> The series had ⬚ episodes and was broadcast for ⬚ days.

Question 6 asks pupils to use and apply their understanding of finding time durations in days to solve a complex problem involving episodes of a drama series. Ensure that they are given the calendar months of March and April to help them out should they need it.

Chapter 5 test (Practice Book 3A, pages 119–121)

Test Question number	Relevant Unit	Relevant questions within unit
1	5.1	2
	5.2	3
2	5.2	5
3	5.1	1, 4
	5.2	1
	5.3	1
4	5.4	1
5	5.3	1, 2, 3, 4, 5
6	5.4	2, 3, 4
7	5.4	4
8	5.4	3, 5

Chapter 6
Consolidation and enhancement

Chapter overview

Area of mathematics	National Curriculum statutory requirements for Key Stage 2	Shanghai Maths Project reference
Number – number and place value	Year 3 Programme of study: Pupils should be taught to:	
	■ write and calculate mathematical statements for multiplication and division using the multiplication tables that they know, including for two-digit numbers times one-digit numbers, using mental and progressing to formal written methods	Year 3, Units 6.1, 6.2, 6.3
	■ solve problems, including missing number problems, involving multiplication and division, including positive integer scaling problems and correspondence problems in which n objects are connected to m objects.	Year 3, Units 6.1, 6.2, 6.3
	■ recognise the place value of each digit in a three-digit number (hundreds, tens, ones)	Year 3, Units 6.6
	■ compare and order numbers up to 1000	Year 3, Unit 6.7
	■ identify, represent and estimate numbers using different representations	Year 3, Unit 6.6
	■ read and write numbers up to 1000 in numerals and in words	Year 3, Units 6.6, 6.7
	■ solve number problems and practical problems involving these ideas.	Year 3, Units 6.6, 6.7
	Year 4 Programme of study: Pupils should be taught to:	
	■ find 1000 more or less than a given number	Year 4, Unit 6.6
	■ recognise the place value of each digit in a four-digit number (thousands, hundreds, tens, and ones)	Year 4, Units 6.6, 6.7
	■ order and compare numbers beyond 1000	Year 4, Unit 6.7

Number – addition and subtraction (extract only)	Year 3 Programme of study: Pupils should be taught to: ■ solve problems, including missing number problems, using number facts, place value, and more complex addition and subtraction.	Year 3, Units 6.4, 6.5
Number – multiplication and division	Year 3 Programme of study: Pupils should be taught to: ■ recall and use multiplication and division facts for the 3, 4 and 8 multiplication tables	Year 3, Units 6.6, 6.7
	■ solve problems, including missing number problems, involving multiplication and division, including positive integer scaling problems and correspondence problems in which n objects are connected to m objects.	Year 3, Unit 6.4, 6.6, 6.7

Unit 6.1
5 threes plus 3 threes equals 8 threes

Conceptual context

Pupils will now have a good knowledge and understanding of multiplication facts and will have a rich experience of different representations of arrays that they can visualise or draw. This deep conceptual understanding will be important in this unit as pupils will begin to explore the distributive property of multiplication.

(i) The distributive property means that a multiplication fact can be partitioned into the sum of two other multiplication facts. So far, pupils have learned that a product has two factors; now they will learn that one of those factors can be partitioned into two parts. For example:

This 6 × 5 array has been partitioned so that it represents 4 × 5 + 2 × 5 = 6 × 5 = 30.

Learning pupils will have achieved at the end of the unit

- Using concrete equipment and pictorial representations pupils will have added related multiplication facts together (Q1)
- Pupils will have explored relationships and multiplicative structures between the 2, 4 and 6 times table (Q2)
- Partitioning one axis in an array to create two part products will have been practised and pupils will have recognised that, together, the part products are equivalent to the product of the whole array (Q2)
- Pupils will have applied their understanding of the distributive property of multiplication to solve multiplication and addition problems (Q3, Q4, Q5)
- Pupils will have applied their knowledge of the distributive property of multiplication to solve problems (Q6)

Resources

mini whiteboards and pens; interlocking cubes; red and blue counters; paper plates; 0–9 dice (two per group); red and blue toy cars; **Resource 3.6.1a** Out of the frying pan; **Resource 3.6.1b** Multiplication tables; **Resource 3.6.1c** What's the same? What's different?

Vocabulary

multiplication, multiply, times, factor, product, part product, add, addition, plus

Question 1

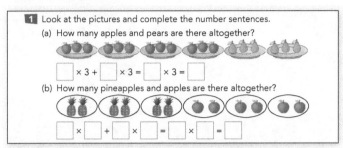

1 Look at the pictures and complete the number sentences.

(a) How many apples and pears are there altogether?

☐ × 3 + ☐ × 3 = ☐ × 3 = ☐

(b) How many pineapples and apples are there altogether?

☐ × ☐ + ☐ × ☐ = ☐ × ☐ = ☐

What learning will pupils have achieved at the conclusion of Question 1?

- Using concrete equipment and pictorial representations pupils will have added related multiplication facts together.

Activities for whole-class instruction

- Invite two pupils to the front of the class. Give Pupil A three sticks of four interlocking cubes. Give Pupil B five sticks of four interlocking cubes. Ask pupils to use a multiplication sentence to describe how many cubes Pupil A has. (3 × 4) Repeat for Pupil B. (5 × 4) Ask: *How many cubes do Pupil A and Pupil B have altogether? Can you write this as a number sentence?* (3 × 4 + 5 × 4 = 8 × 4 = 32)

- Using toy cars, show pupils three blue cars. Ask: *How many wheels are there on each car? Can you write a multiplication sentence to represent the number of wheels altogether?* (3 × 4 = 12)

Now show pupils two red cars. Ask: *How many wheels are there on each car? Can you write a multiplication sentence to represent the number of wheels altogether?* (2 × 4 = 8)

- Combine the blue and red cars. Say: *I know that 3 × 4 + 2 × 4 is the same as 5 × 4.* Record this as a number

sentence on the board. Ask: *True or false? How do you know?* Ask: *How many wheels are there altogether? Add this to your number sentence.* (3 × 4 + 2 × 4 = 5 × 4 = 20)

- Show pupils the image above of daffodils and tulips. Ask: *What do you notice?* Ask pupils to speak to their partner.

ⓘ The term used to describe the whole group (daffodils and tulips) is extremely important. Pupils must understand that the whole group Is partitioned into two sub-groups (daffodils becoming one group and the tulips becoming another). This will help pupils to find the difference between sub-groups in later units.

- Ask: *How many daffodils are there? Can you write this as a multiplication sentence?* (4 × 2 = 8). Ask: *How many tulips are there? Can you write this as multiplication sentence?* (2 × 2 = 4). Ask: *How many flowers are there altogether?* (12). *How do you know?*

- On the board write:

  4 × 2 + 2 × 2 = ☐ × 2

- Pupils to talk to their partner about the number sentence and then fill in the missing value (6). Ask pupils to explain how they came to that answer.

- Show pupils the images and multiplication sentences below to represent each part of the multiplication sentence.

4 × 2

2 × 2

6 × 2

- Pupils should complete Question 1 in the Practice Book.

Same-day intervention

- Put five paper plates on a table. Put two interlocking cubes on three of the plates and two counters on each of the remaining plates.

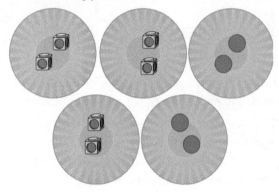

- Ask pupils to talk to their partner about what's the same and what's different.
- When describing the plates, it is important that pupils use precise language about the names of the groups and sub-groups they are referring to. Ask pupils to write a multiplication sentence to represent the number of interlocking cubes (3 × 2) and the number of counters (2 × 2). Ask: *How many objects are there altogether in the whole group? Can you write a multiplication sentence to represent the number of groups of two objects altogether?* (5 × 2). Ask: *Is adding 3 × 2 and 2 × 2 is the same as 5 × 2? How do you know?*
- Using cubes and counters, repeat with groups of 4.

Same-day enrichment

- Give pupils **Resource 3.6.1a** Out of the frying pan. Ask them to write their own number sentence to represent the image.

- Give pupils two 0–9 dice. Ask pupils to roll them to get two factors. How many different ways can they partition the multiplication? Pupils should record this as a number sentence.

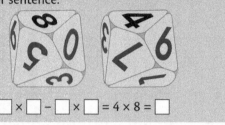

□ × □ − □ × □ = 4 × 8 = □

Question 2

2 Complete the table and then fill in the answers below.

	1	2	3	4	5	6	7	8	9	10
2 times										
4 times										
6 times										

2 times a number plus 4 times the same number equals □ times this number.

2 × □ + 4 × □ = □ × □

What learning will pupils have achieved at the conclusion of Question 2?

- Pupils will have explored relationships and multiplicative structures between the 2, 4 and 6 times table.
- Partitioning one axis in an array to create two part products will have been practised and pupils will have recognised that, together, the part products are equivalent to the product of the whole array.

Activities for whole-class instruction

- Using red counters, ask pupils to represent the multiplication fact 2 × 5 as an array.
- Using blue counters, ask pupils to add another group of five (this might be an addition row or an additional column depending on how pupils have represented the array).
- Repeat this instruction one group at a time until pupils have added four groups of 5.

- Ask: *Can you use your array to give me any more multiplication facts?* Pupils should see that 6×5 is now represented.

- Say: *I think that 2×5 plus 4×5 equals 6×5. True or false? Convince me!*

- Pupils should complete Question 2 in the Practice Book.

Same-day intervention

- Give pupils **Resource 3.6.1b** Multiplication tables.

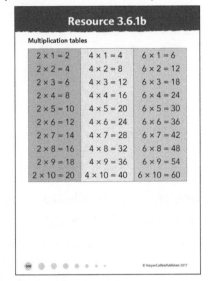

- Ask pupils what they recognise about the relationship between the multiplication tables.

- Give pupils the following number sentences. Ask them to represent the sentences using interlocking cubes.

  $2 \times 3 + 4 \times 3 = 6 \times 3 = 18$

  $2 \times 8 + 4 \times 8 = 6 \times 8 = 48$

Same-day enrichment

- Ask: *If you know that 2 times a number plus 4 times the same number equals 6 times this number, what other statements can you write about other numbers?*

  - $\boxed{}$ times a number plus $\boxed{}$ times the same number equals $\boxed{}$ times this number.

  - Prove it!

Questions 3, 4 and 5

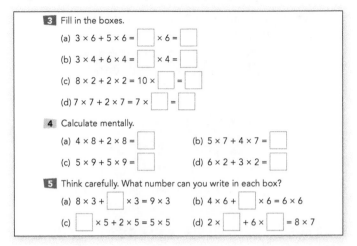

3 Fill in the boxes.

(a) $3 \times 6 + 5 \times 6 = \boxed{} \times 6 = \boxed{}$

(b) $3 \times 4 + 6 \times 4 = \boxed{} \times 4 = \boxed{}$

(c) $8 \times 2 + 2 \times 2 = 10 \times \boxed{} = \boxed{}$

(d) $7 \times 7 + 2 \times 7 = 7 \times \boxed{} = \boxed{}$

4 Calculate mentally.

(a) $4 \times 8 + 2 \times 8 = \boxed{}$ (b) $5 \times 7 + 4 \times 7 = \boxed{}$

(c) $5 \times 9 + 5 \times 9 = \boxed{}$ (d) $6 \times 2 + 3 \times 2 = \boxed{}$

5 Think carefully. What number can you write in each box?

(a) $8 \times 3 + \boxed{} \times 3 = 9 \times 3$ (b) $4 \times 6 + \boxed{} \times 6 = 6 \times 6$

(c) $\boxed{} \times 5 + 2 \times 5 = 5 \times 5$ (d) $2 \times \boxed{} + 6 \times \boxed{} = 8 \times 7$

What learning will pupils have achieved at the conclusion of Questions 3, 4 and 5?

- Pupils will have applied their understanding of the distributive property of multiplication to solve multiplication and addition problems.

Activities for whole-class instruction

- Show pupils the bar model. Ask: *What do you notice?*

6	6	6	6	6	6	6	6
6	6	6	6	6	6	6	6
48							

- Say: *The top bar represents $3 \times 6 + 5 \times 6$. True or false?*

- Ask: *What does the second bar represent?*

- Ask pupils to use the bars to complete the number sentence:

  $\boxed{} \times 6 + \boxed{} \times 6 = \boxed{} \times 6 = \boxed{}$

- On the board, write the list of calculations below. Ask: *What's the same and what's different?*

  $1 \times 8 + 7 \times 8 = 8 \times 8 = 64$

  $2 \times 8 + 6 \times 8 = 8 \times 8 = 64$

  $3 \times 8 + 5 \times 8 = 8 \times 8 = 64$

- Ask pupils to work in pairs to find the remaining number multiplication sentences where 8×8 had been partitioned into part products. Ask: *How do you know that you have them all?*

- Pupils should complete Questions 3, 4 and 5 In the Practice Book.

Same-day intervention

- Give pupils **Resource 3.2.1c** What's the same? What's different?

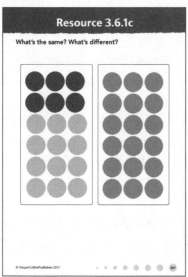

- Ask: *What's the same? What's different?* Pupils should be given time to discuss what the arrays represent.

 All say … *2 times 3 plus 4 times 3 equals 6 times 3.*

 Look out for … pupils who may not have a secure understanding of order of operations and attempt to calculate $(2 \times 3 + 4) \times 3$.

- Ask pupils to talk to a partner; can they complete the number sentence?

$2 \times \boxed{} + 4 \times \boxed{} = \boxed{} \times \boxed{}$

- Draw two arrays of six columns of five rows. Partition one array so that four columns are one colour and two columns are a different colour.

- Ask: *What's the same? What's different?* Pupils should be given time to discuss what the arrays represent.

All say … *2 times 5 plus 4 times 5 equals 6 times 5.*

- Ask pupils to talk to a partner; can they complete the number sentence?

$2 \times \boxed{} + 4 \times \boxed{} = \boxed{} \times \boxed{}$

- Ask: *What did you notice about the relationship between 2x, 4x and 6x? How did this help you to complete the number sentence? Can you think of a different way of completing the number sentence?*

Same-day enrichment

- Ask pupils to complete the following number sentences. Ask: *How many ways of doing each are there?*

$\boxed{} \times 4 + \boxed{} \times 4 = \boxed{} \times 4 = 48$

$\boxed{} \times 5 + \boxed{} \times 5 = \boxed{} \times 5 = 66$

Question 6

> **6** Each pen costs £9. Tom bought 2 pens and Joana bought 3 pens. How much did they pay in total?

What learning will pupils have achieved at the conclusion of Question 6?

- Pupils will have applied their knowledge of the distributive property of multiplication to solve problems.

Activities for whole-class instruction

- Write the following problem on the board:

  In a toy shop, toy cars cost £3. Louis bought 4 cars and Jo bought 5 cars. How much did they pay in total?

- Ask pupils to draw a picture to represent the problem. Ask: *Using the picture, can you now write a number sentence?* ($3 \times £3 + 5 \times £3 = 8 \times £3 = £24$)

Same-day intervention

- On the board, write the following problem. Ask pupils to use concrete equipment to represent the problem and then solve it.

  Tennis balls cost £2 each. Kelly bought 7 tennis balls and Sean bought 3 tennis balls. How much did they pay in total?

Same-day enrichment

● Ask pupils to solve the following problems.

- A bag of sweets cost £2. Katie bought 3 bags and Jake bought 7 bags. How much did they pay altogether?

- Each book costs £7. Pauline bought 3 books and Michael bought 5. How much did they pay in total?

Challenge and extension question

Question 7

7 Think carefully and then fill in the boxes.

(a) $4 \times 8 + 8 = \boxed{} \times 8$ (b) $6 \times 5 + \boxed{} = 7 \times 5$

Pupils will need to understand that $1 \times 8 = 8$. There is no need to write this fact as a multiplication as there is only one group.

Question 8

8 Show different ways to express 9×5.

(a) $9 \times 5 = \boxed{} \times \boxed{} + \boxed{} \times \boxed{}$

(b) $9 \times 5 = \boxed{} \times \boxed{} + \boxed{} \times \boxed{}$

(c) $9 \times 5 = \boxed{} \times \boxed{} + \boxed{} \times \boxed{}$

(d) $9 \times 5 = \boxed{} \times \boxed{} + \boxed{} \times \boxed{}$

(e) $9 \times 5 = \boxed{} \times \boxed{} + \boxed{} \times \boxed{} + \boxed{} \times \boxed{}$

Finding 'different ways' to express a fact provides further challenge because there are more possibilities. Pupils might choose to work systematically to answer this question.

Unit 6.2
5 threes minus 3 threes equals 2 threes

Conceptual context

From the previous unit, pupils will be able to apply their understanding of the distributive property of multiplication to subtract one part product from another.

Learning pupils will have achieved at the end of the unit

- Using concrete equipment and pictorial representations pupils will have built strong mental images of relationships within products (Q1, Q2)
- Pupils will have explored relationships and identify patterns between the 9, 5 and 4 times table (Q2)
- Pupils will have used their understanding of the distributive property of multiplication to solve multiplication and subtraction problems (Q3, Q4, Q5)
- Pupils will have solved problems in the context of money (Q6)

Resources

mini whiteboards and pens; interlocking cubes; counters; **Resource 3.6.2a** More multiplication tables; **Resource 3.6.2b** Multiplication and subtraction puzzles

Vocabulary

multiplication, multiply, times, factor, product, how many more to make…?, how many more… subtract, subtraction, take (away), minus, how many are left/left over?

Question 1

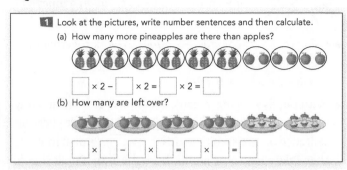

1 Look at the pictures, write number sentences and then calculate.
(a) How many more pineapples are there than apples?

☐ × 2 – ☐ × 2 = ☐ × 2 = ☐

(b) How many are left over?

☐ × ☐ – ☐ × ☐ = ☐ × ☐ = ☐

What learning will pupils have achieved at the conclusion of Question 1?

- Using concrete equipment and pictorial representations pupils have built strong mental images of relationships within products.

Activities for whole-class instruction

- Show pupils that you have eight sticks of four interlocking cubes. Ask: *How many groups of 4 do I have altogether?* Record the multiplication sentence on the board. (8 × 4)

- Invite a pupil to the front of the class. Give them three sticks of interlocking cubes. Ask: *How many groups of 4 does Pupil A have?* Record the multiplication sentence on the board. (3 × 4)

- Ask: *How many groups of 4 do I have left?* (5 × 4) *How many cubes do I have left?* (20)

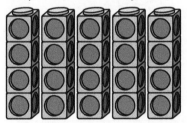

- Ask pupils to write a number sentence to represent what has happened. (8 × 4 – 3 × 4 = 5 × 4 = 20)

- Show pupils the image below. Ask: *What's the same and what's different?*

- Ask pupils to work out how many chocolate chip cookies there are (4 × 3) and how many jammy biscuits there are

(2 × 3). Pupils should share how they worked it out. Ask: *What did you do?* Record the multiplication sentences on the board.

- Ask: *What can you say about the number of cookies, compared to the number of jammy biscuits? How many more groups of cookies are there?* (2 × 3)

- Model writing this as a number sentence on the board:

  4 × 3 - 2 × 3 = 2 × 3 = 6

- Ask pupils to talk to a partner about the image below. Using mini whiteboards and pens, have pupils write a number sentence to represent how many more wheels there are on the black bikes than the number of wheels there are on the white bikes. (6 × 2 - 4 × 2 = 2 × 2 = 4)

- Pupils should complete Question 1 in the Practice Book.

Same-day intervention

- Give pupils 20 interlocking cubes. Ask: *Can you put your cubes into groups of 4? Can you write a multiplication sentence to represent this?* (5 × 4 = 20).

- Take two groups of 4 away from each pupil. Ask: *How many groups of 4 did I take? How many groups of 4 do you have left? Can you write this as a number sentence?* (5 × 4 – 2 × 4 = 3 × 4 = 12)

- Repeat for:

  6 × 5 – 3 × 5 = 3 × 5 = 15

  4 × 7 – 2 × 7 = 2 × 7 = 14

Same-day enrichment

- Write the following numbers and number sentence template on the board. Ask pupils to complete the number sentence using the given numbers. They can use a number more than once.

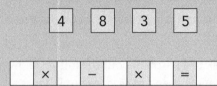

Question 2

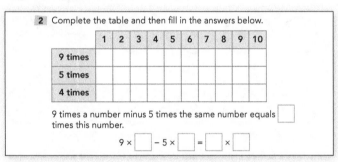

2 Complete the table and then fill in the answers below.

	1	2	3	4	5	6	7	8	9	10
9 times										
5 times										
4 times										

9 times a number minus 5 times the same number equals ☐ times this number.

$9 \times \boxed{} - 5 \times \boxed{} = \boxed{} \times \boxed{}$

What learning will pupils have achieved at the conclusion of Question 2?

- Using concrete equipment and pictorial representations pupils have built strong mental images of relationships within products.
- Pupils will have explored relationships and identify patterns between the 9, 5 and 4 times table.

Activities for whole-class instruction

- Put pupils into pairs. Ask Pupil A to represent the multiplication $9 \times 6 = 54$ using counters. Ask pupil B to take away 5 groups of 6 counters. Ask: *How many groups of 6 are left?*

All say ... *9 times 6, minus 5 times 6, equals 4 times 6.*

- Using mini whiteboards, ask pupils to work together in their pairs to write this as a number sentence. $(9 \times 6 - 5 \times 6 = 4 \times 6 = 24)$.

- With pupils working in pairs, ask Pupil A to build an array with a product of 48 or less and record the multiplication sentence on their whiteboards ($\boxed{} \times \boxed{}$). Pupil B must then take away some groups from the array. Pupils should continue their number sentence ($\boxed{} \times \boxed{} - \boxed{} \times \boxed{}$). Ask: *Using the array that is left, can you finish the number sentence?* ($\boxed{} \times \boxed{} - \boxed{} \times \boxed{} = \boxed{} \times \boxed{} = \boxed{}$)

- Repeat with further arrays in which one of the factors is 9 and this becomes partitioned.

- Pupils should complete Question 2 in the Practice Book.

Same-day intervention

- Give pupils **Resource 3.6.2a** More multiplication tables.

Resource 3.6.2a

More multiplication tables

$9 \times 1 = 9$	$5 \times 1 = 5$	$4 \times 1 = 4$
$9 \times 2 = 18$	$5 \times 2 = 10$	$4 \times 2 = 8$
$9 \times 3 = 27$	$5 \times 3 = 15$	$4 \times 3 = 12$
$9 \times 4 = 36$	$5 \times 4 = 20$	$4 \times 4 = 16$
$9 \times 5 = 45$	$5 \times 5 = 25$	$4 \times 5 = 20$
$9 \times 6 = 54$	$5 \times 6 = 30$	$4 \times 6 = 24$
$9 \times 7 = 63$	$5 \times 7 = 35$	$4 \times 7 = 28$
$9 \times 8 = 72$	$5 \times 8 = 40$	$4 \times 8 = 32$
$9 \times 9 = 81$	$5 \times 9 = 45$	$4 \times 9 = 36$
$9 \times 10 = 90$	$5 \times 10 = 50$	$4 \times 10 = 40$

102   © HarperCollinsPublishers 2017

- Ask them to use squared paper to make the arrays. Discuss what they recognise about the relationship between the multiplication tables.

- On the board, write the following number sentences. Ask pupils to work with a partner and represent the sentences, using interlocking cubes.

$9 \times 3 - 5 \times 3 = 4 \times 3 = 12$

$9 \times 8 - 5 \times 8 = 4 \times 8 = 32$

Same-day enrichment

- Ask: *If you know that 9 times a number minus 5 times the same number equals 4 times this number, what other statements can you write about other numbers?*
  - ☐ *times a number minus* ☐ *times the same number equals* ☐ *times this number.*
  - *Prove it!*

Questions 3, 4 and 5

3 Fill in the boxes.

(a) $8 \times 6 - 5 \times 6 = \boxed{} \times 6 = \boxed{}$

(b) $7 \times 4 - 6 \times 4 = 4 \times \boxed{} = \boxed{}$

(c) $5 \times 9 - 5 \times 6 = 5 \times \boxed{} = \boxed{}$

(d) $5 \times 7 - 2 \times 7 = \boxed{} \times \boxed{} = \boxed{}$

(e) $9 \times 6 - 4 \times \boxed{} = \boxed{} \times 6 = \boxed{}$

4 Calculate mentally.

(a) $12 \times 8 - 10 \times 8 = \boxed{}$

(b) $15 \times 7 - 9 \times 7 = \boxed{}$

(c) $18 \times 9 - 9 \times 9 = \boxed{}$

(d) $16 \times 2 - 7 \times 2 = \boxed{}$

5 Think carefully and fill in the boxes.

(a) $8 \times 3 - \boxed{} \times 3 = 6 \times 3$

(b) $12 \times 6 - \boxed{} \times 6 = 9 \times 6$

(c) $\boxed{} \times 5 - 6 \times 5 = 4 \times 5$

(d) $10 \times \boxed{} - 6 \times \boxed{} = \boxed{} \times 7$

What learning will pupils have achieved at the conclusion of Questions 3, 4 and 5?

- Pupils will have used their understanding of the distributive property of multiplication to solve multiplication and subtraction problems.

Activities for whole-class instruction

- On the board, write:

Lindy says:

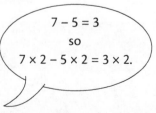

$7 - 5 = 3$
so
$7 \times 2 - 5 \times 2 = 3 \times 2.$

- Ask: *Is Lindy correct? Can you explain what she means?*

- Show pupil the model below. Ask: *What does the bar represent?*

6	6	6	6	6	6	6	6

6	6	6	6	6	6	~~6~~	~~6~~

- Ask pupils to write a number sentence that is represented by the bar. ($8 \times 6 - 6 \times 6 = 2 \times 6 = 12$)

- On the board, copy the bar below. Ask pupils to copy it, then ask: *What does it represent?* ($4 \times 4 = 16$)

4	4	4	4

- Ask pupils to draw another bar underneath that represents the number sentence $4 \times 4 - 2 \times 4 = 2 \times 4 = 8$.

- Pupils should complete Questions 3, 4 and 5 in the Practice Book.

Same-day intervention

- Pupils should complete **Resource 3.6.2b** Multiplication and subtraction puzzles.

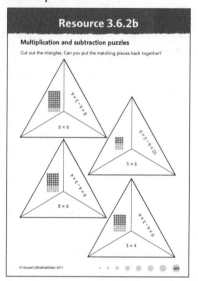

Resource 3.6.2b

Multiplication and subtraction puzzles

Cut out the triangles. Can you put the matching pieces back together?

© HarperCollinsPublishers 2017

Same-day enrichment

- Ask: *How many ways can you complete the number sentences? Have you found all possibilities?*

$\boxed{} \times \boxed{} - \boxed{} \times \boxed{} = 5 \times 6 = 30$

$\boxed{} \times \boxed{} - \boxed{} \times \boxed{} = 3 \times 9 = 27$

$\boxed{} \times \boxed{} - \boxed{} \times \boxed{} = 4 \times 3 = 12$

Question 6

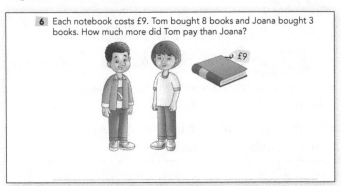

> 6 Each notebook costs £9. Tom bought 8 books and Joana bought 3 books. How much more did Tom pay than Joana?

What learning will pupils have achieved at the conclusion of Question 6?

- Pupils will have solved problems in the context of money.

Activities for whole-class instruction

- Each tin of chocolate costs £2. Callum bought 5 tins of chocolate. Kerry bought 3 tins of chocolate. How much more will Callum pay than Kerry?

- Write the problem above on the board and ask pupils to consider it. Now copy the model below. Ask pupils: *Will the model help you to solve the problem?*

£2	£2	£2	£2	£2
£2	£2	£2		

- Ask pupils to talk to a partner and discuss the questions: *What do we know? What calculation(s) do we need to do?*

- Tell pupils each teddy costs £5. In pairs, ask them if they can write their own problem using the image below.

- Pupils should share their problem with another pair and ask them to write a number sentence to solve the problem.

- Share some of the problems with the class.

- Pupils should complete Question 6 in the Practice Book.

Same-day intervention

- On the board, write the following problem. Ask: *Can you draw a picture to help you solve it?*

  Each box of building bricks costs £6. Felix bought 7 boxes of building bricks and Annie bought 2 boxes. How much more did Felix spend than Annie?

Same-day enrichment

- Write the following problem on the board and ask pupils to solve it.

  Each picture costs £7. Pete bought 3 pictures and Sandra bought 5. How much more did Sandra pay then Pete?

- Ask pupils to write their own problem for this number sentence.

  8 × £4 – 3 × £4 = 5 × £4

Challenge and extension question

Question 7

> 7 Think carefully and then fill in the boxes.
>
> (a) 8 × 8 – 8 = ☐ × ☐
>
> (b) 8 × 5 = 12 × 5 – ☐ × ☐
>
> (c) 8 × 7 – 2 × 7 – 4 × 7 = 7 × ☐
>
> (d) 2 × 2 = 9 × 2 – 4 × ☐ – ☐ × ☐

Pupils are required to apply a greater depth of knowledge to complete the questions. The variation in the question requires pupils to think deeply about the mathematics and, therefore, they cannot rely on following a procedure alone.

Question 8

> 8 Use different ways to express 3 × 6.
>
> (a) 3 × 6 = ☐ × ☐ – ☐ × ☐
>
> (b) 3 × 6 = ☐ × ☐ – ☐ × ☐
>
> (c) 3 × 6 = ☐ × ☐ – ☐ × ☐
>
> (d) 3 × 6 = ☐ × ☐ – ☐ × ☐ – ☐ × ☐

Finding 'different ways' to express a fact provides further challenge because there are more possibilities. Pupils might choose to work systematically to answer this question.

Unit 6.3
Multiplication and division

Conceptual context

From previous units, pupils will now have a secure understanding of multiplication and division. Using a range of models and images, they will understand the relationship between multiplication and division and known that they are inverse operations. In this unit, pupils will further develop their knowledge of conditions in which multiplication and division situations arise. Therefore, this unit will provide further practice with real-life problems involving multiplication and division so that pupils deepen their conceptual knowledge.

Learning pupils will have achieved at the end of the unit

- Pupils will have practised writing related multiplication and division facts (Q1)
- Pupils will have practised rapid recall of multiplication facts and used reasoning to complete number sentences (Q2)
- Pupils will have developed their understanding of the conditions in which multiplication and division situations arise (Q3)
- Pupils will have applied their knowledge of multiplication and division facts to solve a range of problems (Q4)
- Pupils will have interpreted mathematical language associated with multiplication and division to solve problems (Q4)

Resources

mini whiteboards and pens; green and yellow interlocking cubes; counters; **Resource 3.6.3a** Classroom; **Resource 3.6.3b** Books and shelves; **Resource 3.6.3c** Posing a question

Vocabulary

multiplication, multiply, times, times as many, factor, product, divide, division, divided by, inverse, dividend, divisor, quotient, linked fact

Question 1

<table>
<tr><td colspan="2">**1** Calculate mentally.</td></tr>
<tr><td>(a) 4 × 6 = ▢</td><td>(b) 7 × 8 = ▢</td></tr>
<tr><td>(c) 5 × 9 = ▢</td><td>(d) 10 × 10 = ▢</td></tr>
<tr><td>(e) 24 ÷ 4 = ▢</td><td>(f) 56 ÷ 7 = ▢</td></tr>
<tr><td>(g) 45 ÷ 9 = ▢</td><td>(h) 100 ÷ 10 = ▢</td></tr>
<tr><td>(i) 24 ÷ 6 = ▢</td><td>(j) 56 ÷ 8 = ▢</td></tr>
<tr><td>(k) 45 ÷ 5 = ▢</td><td>(l) 0 × 7 = ▢</td></tr>
</table>

What learning will pupils have achieved at the conclusion of Question 1?

- Pupils will have practised writing related multiplication and division facts.

Activities for whole-class instruction

- Tell pupils that Sammy says: *I know that 7 × 3 = 21 so I know that 21 ÷ 7 = 3 and 21 ÷ 3 = 7*

- Ask: *Is what Sammy says true or false? Convince me!* Give pupils counters, and ask whether they can use the counters to prove that this statement is correct. Representing these number sentences as an array is a good visual image for pupils to refer to.

- Show pupils the bar model below, by copying it on the board. Ask: *What facts can you write using the bar model?*

7	7	7	7	7
35				

  I know that 5 times 7 equals 35 so I also know that 35 divided by 7 is 5 and 35 divided by 5 is 7.

- Write the following facts on the board. Ask pupils to draw a bar model like the one above and write the two related division facts.

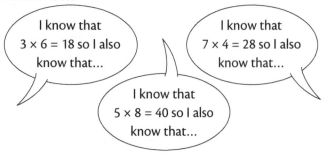

I know that 3 × 6 = 18 so I also know that...

I know that 7 × 4 = 28 so I also know that...

I know that 5 × 8 = 40 so I also know that...

- Pupils should complete Question 1 in the Practice Book.

Same-day intervention

- Give pupils some counters. Ask them to first use the multiplication sentence to create an array. Using their array, they should then write two division facts. Ask: *What do you notice about the relationship between the multiplication and division sentences?*

a) 2 × 7 = ▢

b) 5 × 8 = ▢

c) 3 × 9 = ▢

d) 3 × 10 = ▢

Same-day enrichment

- Write the following statement on the board: *I know that 3 × 6 = 18 so I also know that 180 ÷ 30 = 60 and 180 ÷ 60 = 30.*

- Ask: *Is this statement true or false? How do you know? Draw something that proves it. Can you use write any other facts using the multiplication fact 3 × 6 = 18?*

Question 2

<table>
<tr><td colspan="2">**2** What is the greatest number you can write in each box?</td></tr>
<tr><td>(a) ▢ × 5 < 42</td><td>(b) ▢ × 6 < 37</td></tr>
<tr><td>(c) ▢ × 8 < 80</td><td>(d) 7 × ▢ < 56</td></tr>
<tr><td>(e) 48 > 9 × ▢</td><td>(f) 58 > 6 × ▢</td></tr>
</table>

What learning will pupils have achieved at the conclusion of Question 2?

- Pupils will have practised rapid recall of multiplication facts and use their reasoning skills to complete number sentences.

Activities for whole-class instruction

- Write ▢ × 4 = 24 on the board. Give pupils time to discuss what the missing number is and what they did to work it out.

- Write ▢ × 7 < 30 on the board. Give pairs of pupils 30 counters. Ask them to create an array with groups of 7.

- Using their array, they should complete the number sentence finding the greatest number.

- Ask: *What if the number sentence was* ☐ × 7 < 36?

- Repeat for ☐ × 5 < 27.

- Pupils should be encouraged to draw upon known multiplications facts and visualise an array if necessary to help them answer these questions. Give pupils time to reflect on what they know about multiples of five. Ask: *Would this help you find the greatest factor without having to make the array?*

- Write 10 × 6 < 60 on the board. Ask: *Is this number sentence correct? Why not? Can you change the 10 to make it correct?*

- Pupils should complete Question 2 in the Practice Book.

Same-day intervention

- Continue to use arrays to help pupils identify the greatest number to complete the number sentence.

Same-day enrichment

- Give pupils the following problems:

  - A basket will hold 8 muffins. If there are 7 full baskets and 1 part-filled basket, how many muffins might there be?

  - Each car holds 5 people. There are 7 cars – 6 are full and 1 has some people in it. How many people might there be altogether?

  - Children in a class were put into groups of 6. There were 6 full groups and 1 part group. How many children might be in the whole class?

- Pupils should list all possible answers for each problem and be able to explain how they know they are right. Ask them to represent what they know about each problem using a picture or diagram.

Question 3

3 Draw lines to match the conditions to the questions. Then write the number sentences and answer the questions.

Each set consists of one desk and one chair.

| There are 6 rows of desks and chairs in the classroom. There are 7 sets in each row. | | How many rows are there? |

| There are 42 sets of desks and chairs. There are 7 sets in each row. | | How many sets are there in each row? |

| There are 42 sets of desks and chairs. They are put into 6 rows equally. | | How many sets of desks and chairs are there? |

(a) Number sentence: _____

There are ☐ rows.

(b) Number sentence: _____

There are ☐ sets in each row.

(c) Number sentence: _____

There are ☐ sets of desks and chairs.

What learning will pupils have achieved at the conclusion of Question 3?

- Pupils will have developed their understanding of the conditions in which multiplication and division situations arise.

Activities for whole-class instruction

- Give pupils **Resource 3.6.3a** Classroom.

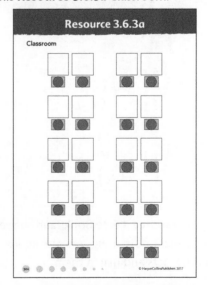

- Tell pupils that the image represents a classroom layout.

- Ask: *How many rows of desks and chairs are there?*

- Ask: *How many sets of rows and chairs are there each row?*

- Write the number sentence $5 \times 4 = 20$. Give pupils time to discuss what the number sentence represents. (There are five rows of four tables, so there are 20 tables altogether.)
- Give pupil pairs some interlocking cubes. Explain that each cube represents a set of one desk and one chair. Ask pupils the following questions:
- *If there are six desk and chair sets in each row, how many sets would there be altogether?* ($5 \times 6 = 30$).
- *There is an equal number of desks and chairs in each row. If there are 40 desk and chair sets arranged into five rows, how many sets are there in each row?*
- Give pupils the linked facts below. Ask them to work in pairs to think of a question.
  - In a classroom, children are standing in rows of 8. There are 4 rows.
  - There are 32 children in a classroom. They are standing in 8 equal rows.
  - There are 32 children in a classroom. They are standing in 4 equal rows.
- Ask pupils to continue to work together to solve the question. They must write a number sentence first.
- Pupils should complete Question 3 in the Practice Book.

Same-day intervention

Resource 3.6.3b

Books and shelves

- Give pupils **Resource 3.6.3b** Books and shelves. Pupils will benefit from using concrete equipment to represent the facts and match them.

Same-day enrichment

Resource 3.6.3c

Posing a question

- Give pupils **Resource 3.6.3c** Posing a question and ask them to work through the activity, creating a multiplication and a division question having thought of some linked facts about the picture.

Question 4

4 Application problems.
  (a) There are 6 sheep on the hillside.
       There are 6 times as many deer as sheep.
       How many deer are there? ☐
  (b) 10 boys made some paper models. Each of them made 4 paper models. A group of girls made 32 paper models in total. Who made more, the boys or the girls?
  (c) A taxi can seat 4 passengers.
       There are 27 passengers.
       What is the smallest number of taxis needed to carry all the passengers? ☐
  (d) The month of July has ☐ weeks and ☐ days.
       The division sentence that shows this is:

What learning will pupils have achieved at the conclusion of Question 4?

- Pupils will have applied their knowledge of multiplication and division facts to solve a range of problems.
- Pupils will have interpreted mathematical language associated with multiplication and division to solve problems.

Activities for whole-class instruction

- Write the following problem on the board:

  There are 7 boys in a class. There are 3 times as many girls as boys. How many girls are there?

- Talk about the language with pupils. Ask: *What does '3 times as many' mean?*

- Give pupils some interlocking cubes. Tell them that the green cubes represent the number of boys and the yellow cubes represent the number of girls. Ask: *Can you use the cubes to help you solve the problem?*

- Ask a pupil to come to the front of the class to explain what they did and why.

- Write the following problem on the board:

  There are 36 cookies. Each tray holds 8 cookies. How many full trays are there?

- Ask pupils to work with a partner. Ask: *Can you solve the problem by drawing a picture or diagram to represent the problem?*

- Ask: *How many cookies are left over?* Ask pupils: *If there were four trays only, how many cookies would need to be on each tray to make sure that there were none left over? Can you write a division sentence to represent this?*

- Pupils should complete Question 4 in the Practice Book.

Same-day intervention

- It is important that pupils understand the vocabulary in a problem to be able to interpret and solve it. Spend time discussing what the language in the problems below means and then ask pupils to represent the problem using counters and/or by drawing an image. Ask them to write a number sentence to find the answer.

  - There are 3 bees on a tree. There are 6 times as many butterflies as bees. How many butterflies are there?

  - 5 boys ate 3 chocolates each. A group of girls ate 18 chocolates in total. Who ate the most chocolates, the boys or the girls?

  - A kennel holds 4 dogs. There are 26 dogs. At least how many kennels are needed to hold all of the dogs?

Same-day enrichment

- Ask pupils to write their own multiplication and division problems for a partner to solve.

Challenge and extension question

Question 5

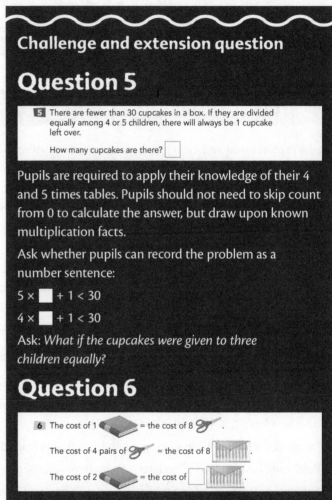

5   There are fewer than 30 cupcakes in a box. If they are divided equally among 4 or 5 children, there will always be 1 cupcake left over.

How many cupcakes are there? ☐

Pupils are required to apply their knowledge of their 4 and 5 times tables. Pupils should not need to skip count from 0 to calculate the answer, but draw upon known multiplication facts.

Ask whether pupils can record the problem as a number sentence:

$5 \times \blacksquare + 1 < 30$

$4 \times \blacksquare + 1 < 30$

Ask: *What if the cupcakes were given to three children equally?*

Question 6

6   The cost of 1 📕 = the cost of 8 ✂.

The cost of 4 pairs of ✂ = the cost of 8 🎹.

The cost of 2 📕 = the cost of ☐ 🎹.

Encourage pupils to record what they know in a table. Ask: *Can you use this information to find other facts?* This will help pupils to find the cost of the items.

Unit 6.4
Mathematics plaza — dots and patterns

Conceptual context

Unit 4 explores odd and even numbers and patterns using them. Pupils should be given opportunities to investigate whether a number is odd or even, using a variety of objects.

One way to do this is by pairing objects. If they can be paired with none left unpaired, then the number is even. If an object is left without a 'partner', the number is odd.

Another way is to use two sorting rings and share the objects one by one into alternate rings. If the groups have an equal number of objects, the number of objects is even. In this way, pupils can see that an even number can be divided into two equal groups and an odd number cannot. Even numbers have 0, 2, 4, 6 or 8 in the ones place while odd numbers have 1, 3, 5, 7 or 9.

Learning pupils will have achieved at the end of the unit

- Pupils will have reinforced conceptual understanding of odd and even numbers (Q1, Q2, Q3)
- Pupils will have recalled that even numbers have 0, 2, 4, 6 or 8 in the ones place while odd numbers have 1, 3, 5, 7 or 9 (Q1, Q3, Q4, Q5)
- Pupils will have practised writing number sentences from dot diagrams (Q2)
- Pupils will have revised using letters to represent numbers (Q2)
- Number patterns involving odd and even numbers will have been explored (Q3, Q4, Q5, Q6)
- Pupils will have revisited division as grouping (Q4)
- Pupils will have considered square numbers and their properties (Q6)

Resources

interlocking cubes; 0–20 number line; 0–100 number cards; mini whiteboards; number plates; double-sided counters; 1–6 dice; multiplication square; five blank cards; **Resource 3.6.4a** Investigate 12; **Resource 3.6.4b** Odd and even number patterns; **Resource 3.6.4c** True, false or sometimes true

Vocabulary

even number, odd number, number pattern, difference, remainder, square number

Question 1

> **1** Look at the dot diagrams and write 'even' or 'odd' in each box.
> (Remember: even numbers end in 0, 2, 4, 6 and 8, and odd numbers end in 1, 3, 5, 7 and 9.)
>
> (a) ☐ (b) ☐
>
> (c) ☐ (d) ☐

What learning will pupils have achieved at the conclusion of Question 1?

- Pupils will have reinforced conceptual understanding of odd and even numbers.
- Pupils will have recalled that even numbers have 0, 2, 4, 6 or 8 in the ones place while odd numbers have 1, 3, 5, 7 or 9.

Activities for whole-class instruction

- Show pupils ten interlocking cubes randomly arranged and invite a pupil to demonstrate and explain whether there is an even number or odd number of cubes. Ask all pupils: *How could you use these cubes to show that 10 is an even number?*
- Pupils may suggest pairing the cubes to make five pairs and explain that because there are none left over, the number is even.

- Pupils may also arrange the cubes to show two sets of five cubes, illustrating that the number can be divided exactly by 2 with no remainder. They can explain that because there are two equal groups, the number is even.

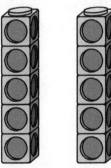

- Repeat with 13 cubes and confirm that 13 is an odd number because after pairing, one cube is left over. If the cubes are shared into two groups, they will not be even. One group will have six cubes and the other, seven.

- Give pairs of pupils a handful of interlocking cubes (between four and 12) and ask them to arrange the cubes to show whether their number is odd or even and to explain how they know. For example: *Our number is 7. 7 is an odd number because after the cubes are paired, there is 1 left over.*
- Using a 0–20 number line as support, establish that the pattern of numbers is even, odd, even, odd and so on.
- Look at the values of the ones digit for even numbers and establish that even numbers always end 0, 2, 4, 6 or 8.
- Look at the values of the ones digit for odd numbers and establish that odd numbers always end 1, 3, 5, 7 or 9.

 Even numbers can be divided exactly by 2. Odd numbers cannot be divided exactly by 2, they have a remainder of 1.

(i) Zero is an even number. It fits the definition of an even number: an integer that can be divided into two equal groups. There is no remainder when it is divided by 2.

- Pupils complete Question 1 in the Practice Book.

Same-day intervention

- Ask pupils to say whether there is an even or odd number of pupils in the intervention group. Tell them to get into pairs to check their answer.
- Using a pack of 0–50 number cards, select and show cards at random and ask pupils to jump up if the number is odd and clap their hands if the number is even. Ask individual pupils to explain how they know. Use counters to settle any disputes.

Same-day enrichment

- Shuffle a pack of 0–100 number cards and deal out 10 cards to each pair of pupils. Ask them to sort the cards into odd and even numbers. Whether a number is odd or even is a property of every number. Ask pupils to think of another property for each of the numbers.

Question 2

> 2 Look at the dot diagrams and write the number sentences.
>
> (a) ▢ + ▢ = ▢
>
> (b) ▢ + ▢ = ▢
>
> (c) ▢ + ▢ = ▢

What learning will pupils have achieved at the conclusion of Question 2?

- Pupils will have reinforced conceptual understanding of odd and even numbers.
- Pupils will have practised writing number sentences from dot diagrams.
- Pupils will have revised using letters to represent numbers.

Activities for whole-class instruction

- Give pupils mini whiteboards and ask pupils to write and answer the following number sentences:

  $1 + 1 =$ $3 + 3 =$ $5 + 5 =$ $7 + 7 =$ $9 + 9 =$

- Ask: *Can you see a pattern of odd and even numbers?* Establish that for every number sentence the addends are all odd numbers and the sum is an even number. Draw a dot diagram to show this, for example $5 + 5$.

- Ask pupils to draw a dot diagram for another number sentence and draw attention to the fact that each odd number has a bit sticking out. Explain that when two odd numbers are added, we can put the two sticking out bits together, so the sum has no bits sticking out.

- Explain that if O stands for an odd number and E stands for an even number, this can be written as

  $O + O = E$

- Draw the following dot diagram and ask them to write the number sentence that it represents.

- Confirm that the number sentence is $6 + 8 = 14$.

- Ask: *Are the addends in this number sentence odd or even? What about the sum?*

- Establish that they are all even.

- Ask pupils to draw a dot diagram with any two small even numbers and see if the result is the same.

- Point out the visual image of the dots in generic plastic or card number plates of even numbers. These even numbers have straight vertical edges. When these two straight edges are added, a new straight edge is formed.

  Hence the sum of any two even numbers is always even. Ask: *How can this be expressed?* ($E + E = E$)

- Tell pupils there is a combination left to try – an odd number and an even number ($O + E$). Ask them to discuss with a partner what they think will happen when one odd and one even number are added together. The even number has a straight edge, while the odd number has a bit sticking out that will still be sticking out when the number plates are put together. Thus, the sum is always odd. The unpaired odd number remains unpaired. Try a few examples and share results. This can be expressed as $O + E = O$.

- Pupils complete Question 2 in the Practice Book.

Same-day intervention

- Show pupils the following number statements, where O represents an odd number and E represents an even number.

  $E + E = E$ $O + O = E$ $O + E = O$

- Encourage pupils to run their finger around the perimeter of number plates to actually feel the 'one' that 'sticks out' when the number is odd and compare this with the outline of number plates that represent even numbers.

- Let pupils see and feel why adding two even numbers will result in an even sum ($E + E = E$).

- Let them see and feel why two odd numbers 'interlock', creating an outline with no protrusions. ($O + O = E$).

- When an odd and an even number are added, use number plates to show that $O + E = O$.

- Ask pupils to write three examples to illustrate each number statement.

 All say... $E + E = E$ $O + O = E$ $O + E = O$

Same-day enrichment

- Give pupil pairs a copy of **Resource 3.6.4a** Investigate 12 and 12 double-sided counters.

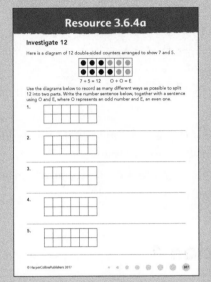

Answers:

1 + 11 = 12   O + O = E

2 + 10 = 12   E + E = E

3 + 9 = 12   O + O = E

4 + 8 = 12   E + E = E

5 + 7 = 12   O + O = E

6 + 6 = 12   E + E = E

Question 3

> **3** Find the pattern and then fill in the boxes.
> (a) 1, 3, 5, 7, 9, ☐ , ☐ , ☐
> (b) 2, 4, 6, 8, 10, ☐ , ☐ , ☐
> (c) 30, 28, 26, 24, 22, ☐ , ☐ , ☐
> (d) 2, 5, 4, 7, 6, 9, 8, 11, ☐ , ☐ , ☐ , ☐

What learning will pupils have achieved at the conclusion of Question 3?

- Pupils will have reinforced conceptual understanding of odd and even numbers.
- Pupils will have recalled that even numbers have 0, 2, 4, 6 or 8 in the ones place while odd numbers have 1, 3, 5, 7 or 9.
- Number patterns involving odd and even numbers will have been explored.

Activities for whole-class instruction

- Using a class pack of 0–50 number cards, select one, for example 37, and ask whether the number is odd or even. Ask pupils to count from 37 back to 1.

  Repeat with other starting numbers, counting on and counting back.

- On the board, write the following number pattern and ask pupils to discuss with a partner what the next numbers are.

  1, 24, 3, 22, 5, 20, 7, 18, 9, ☐ , ☐ , ☐ , ☐

- Agree that the next numbers are 11, 16, 13, 14. The pattern is composed of two patterns with alternate numbers from each sequence, 1, 3, 5, 7, 9, *11*, *13* and 24, 22, 20, 18, *16*, *14*.

- Ask pairs of pupils to write their own two-pattern sequences of odd and even numbers on mini whiteboards and challenge another pair to write the next four numbers.

- Pupils should now complete Question 3 in the Practice Book.

Same-day intervention

- Give pupils **Resource 3.6.4b** Odd and even number patterns.

Resource 3.6.4b

Odd and even number patterns

- The following odd and even number patterns have an error in them.
- In some there is an incorrect number and in others a number is missing.
- Cross out the incorrect number and add an arrow to show where a missing number is required.
- Write the correct number pattern on the line below and add the next three numbers. The first one has been done for you.

1. 1, 3, 5, 7, 8, 11, 13, 15, 19, ____, ____, ____
   1, 3, 5, 7, 9, 11, 13, 15, 19, 21, 23, 25

2. 50, 48, 46, 42, 40, 38, 36, ____, ____, ____

3. 31, 29, 26, 25, 23, 21, 19, ____, ____, ____

4. 102, 104, 106, 108, 110, 121, 114, 116, ____, ____, ____

5. 467, 465, 463, 461, 459, 457, 455, 451, ____, ____, ____

6. 220, 222, 224, 226, 228, 232, 234, 236, ____, ____, ____

© HarperCollinsPublishers 2017

Answers:

1. 8 incorrect/9; 21, 23, 25
2. 44 missing; 34, 32, 30
3. 26 incorrect/27; 17, 15, 13
4. 121 incorrect/112; 118, 120, 122
5. 453 missing; 449, 447, 445
6. 230 missing; 238, 240, 242

Same-day enrichment

- Give pupils **Resource 3.6.4c** True, false or sometimes true.

Resource 3.6.4c

True, false or sometimes true

Decide whether the statements are true, false or sometimes true and underline your answer.
Give three examples as evidence.

1. When you double a number, the answer is always even.
   True False Sometimes true

2. When you halve an even number, the answer is always even.
   True False Sometimes true

3. If you count on 2 from an odd number, the answer is even.
   True False Sometimes true

4. The number that comes after an even number is always odd.
   True False Sometimes true

5. An odd number added to odd number gives an odd number.
   True False Sometimes true

© HarperCollinsPublishers 2017

- Question 2 is sometimes true. When even numbers that are multiples of four are halved, the two halves are even numbers.
- Answers: Pupils' answers will be individual.
  1. True, for example: 3 + 3 = 6, 4 + 4 = 8
  2. Sometimes, for example: Half of 8 is 4, which is even; half of 6 is 3, which is odd.
  3. False, for example: 5, 6, 7 – 7 is odd; 9, 10, 11 – 11 is odd.
  4. True, for example: 7 comes after 6 and is odd; 3 comes after 2, which is odd.
  5. False, for example: 7 + 7 = 14, which is an even number.

Question 4

4 Look at the diagram and then fill in the boxes.

○○○ ○○○ ○○○ ○○○ ○○○ ○

(a) 16 ÷ 3 = ☐ (groups) with a remainder of ☐ (circle)

(b) 16 ÷ 5 = ☐ (circles) with a remainder of ☐ (circle)

What learning will pupils have achieved at the conclusion of Question 4?

- Pupils will have recalled that even numbers have 0, 2, 4, 6 or 8 in the ones place while odd numbers have 1, 3, 5, 7 or 9.
- Number patterns involving odd and even numbers will have been explored.
- Pupils will have revisited division as grouping.

Activities for whole-class instruction

- Give pupils 13 counters and ask them to put them in groups of 2. Remind pupils that when we put a set of objects into groups, there may be some left over. This is called the *remainder*.
- Check that they have six groups of 2 and 1 left over. (13 is an odd number).
- Ask: *How can we write this as a number sentence?*
  6 × 2 + 1 = 13.
- Now ask pupils to put the cubes into groups of 3 and record the results in the same way (4 × 3 + 1 = 13).

- Ask pupils to record what happens if the counters are put into groups of 4, 5 or 6. Check that pupils are able to record the correct number sentences ($3 \times 4 + 1 = 13$; $2 \times 5 + 3 = 13$; $2 \times 6 + 1 = 13$).
- Pupils should now complete Question 4 in the Practice Book.

Same-day intervention

- Write the general statement number sentence $\Box \times \Box + \Box =$ (Number of cubes)
- Give pupil pairs a number to explore, for example 11, 17 or 19. Give them counters and ask them to try making groups of 2, 3, 4 and 5 and to record the number sentences.

Same-day enrichment

- Complete each calculation with a different remainder.

  $\Box \div 4 = 6 \text{ r} \Box$ $\Box \div 5 = 3 \text{ r} \Box$ $\Box \div 6 = 2 \text{ r} \Box$

  Find three solutions   Find four solutions   Find five solutions
- Ask pupils to explain why there are three possible answers when the divisor is 4; four possible answers when the divisor is 5 and five possible answers when the divisor is 6.

Question 5

> 5 Write the numbers.
> (a) 5 odd numbers: ☐☐☐☐☐
> (b) 5 even numbers: ☐☐☐☐☐
> (c) All 2-digit odd numbers with a 3 in the tens place:
> _____
> (d) All 2-digit even numbers with a 6 in the tens place:
> _____
> (e) Four 2-digit even numbers after 19: ☐☐☐☐

What learning will pupils have achieved at the conclusion of Question 5?

- Pupils will have recalled that even numbers have 0, 2, 4, 6 or 8 in the ones place while odd numbers have 1, 3, 5, 7 or 9.
- Number patterns involving odd and even numbers will have been explored.

Activities for whole-class instruction

- Shuffle a pack of 0–100 number cards and turn them over one by one. Ask pupils to say whether the card is an odd or even number.
- Give pupils mini whiteboards and ask them to write as many two-digit odd numbers as they can with a 7 in the tens place. Share their results and establish that there are five possible numbers, because odd numbers can only have 1, 3, 5, 7 or 9 in the ones place.

  71   73   75   77   79
- Ask: *How many even numbers with a 7 in the tens place do you think there are? What are they?*
- Confirm that again there are five possible numbers, 70, 72, 74, 76 and 78, because even numbers have 0, 2, 4, 6 or 8 in the ones place.
- Pupils are ready to complete Question 5 in the Practice Book.

Same-day intervention

- Ask pupils to investigate what happens when you add:
  - 2 consecutive numbers (for example 2 + 3, 5 + 6)
  - 3 consecutive numbers (for example 2 + 3 + 4, 5 + 6 + 7)
- Ask: *Are the answers odd or even numbers?*

Same-day enrichment

- Carry out the Same-day intervention task and explain **why** the answers are odd or even numbers.
- Ask pupils to find all three-digit odd numbers with 6 in the tens place and an odd hundreds digit.

  161 163 165 167 169, 361 363 365 367 369, 561 563 565 567 569, 761 763 765 767 769, 961 963 965 967 969
- Challenge pupils in pairs to write and answer a similar problem of their own.

Question 6

> 6 Think carefully and fill in the boxes.
> (a) $1 + 3 = 2 \times 2 = \Box$
> (b) $1 + 3 + 5 = 3 \times 3 = \Box$
> (c) $1 + 3 + 5 + 7 = 4 \times \Box = \Box$
> (d) $1 + 3 + 5 + 7 + 9 = \Box \times \Box = \Box$
> (e) $1 + 3 + 5 + 7 + \Box + \Box = \Box \times \Box = \Box$

What learning will pupils have achieved at the conclusion of Question 6?

- Number patterns involving odd and even numbers will have been explored.
- Pupils will have considered square numbers and their properties.

Activities for whole-class instruction

- Ask pupils to write an addition sentence and a multiplication sentence for 18:

  10 + 8 = 18

  3 × 6 = 18.

- Remind pupils that, since both equal 18, these expressions can be written as an equivalent calculation:

  10 + 8 = 3 × 6.

- Ask pupils to write addition and multiplication sentences for other two-digit even numbers.

- Display a multiplication square.

x	1	2	3	4	5	6	7	8	9	10
1	1	2	3	4	5	6	7	8	9	10
2	2	4	6	8	10	12	14	16	18	20
3	3	6	9	12	15	17	21	24	27	30
4	4	8	12	16	20	24	28	32	36	40
5	5	10	15	20	25	30	35	40	45	50
6	6	12	18	24	30	36	42	48	54	60
7	7	14	21	27	35	42	49	56	63	70
8	8	16	24	32	40	48	56	64	81	80
9	9	18	27	36	45	54	63	72	81	90
10	10	20	30	40	50	60	70	80	90	100

- Look at the numbers along the diagonal, 1, 4, 9, 16 and so on. Tell or remind pupils that these numbers are known as square numbers. Explain that these are the numbers produced when a number is multiplied by itself.

  (i) A square number is the number obtained when a number is multiplied by itself.

  1 × 1 = 1; 2 × 2 = 4; 3 × 3 = 9 and so on. Therefore, 1, 4, 9 … are square numbers.

4 is 2 × 2

9 is 3 × 3

- Pupils should complete Question 6 after reading it carefully.

Same-day intervention

- Shuffle a pack of 0–50 number cards. Ask pupils to sort the cards into odd and even numbers and then put them into the pattern of odd and even numbers: 1, 3, 5 … and 0, 2, 4, 6 … and so on.

Same-day enrichment

- Show pupils the following diagram and challenge them to discuss with a partner and explain how this diagram relates to Question 6.

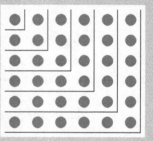

Challenge and extension questions

Questions 7 and 8

7 Complete the calculations.

(a) 8 + 1 = ☐   (b) 8 + 2 = ☐   (c) 9 + 1 = ☐

(d) 8 + 3 = ☐   (e) 8 + 4 = ☐   (f) 9 + 3 = ☐

(g) 8 + 5 = ☐   (h) 8 + 6 = ☐   (i) 9 + 5 = ☐

(j) 8 + 7 = ☐   (k) 8 + 8 = ☐   (l) 9 + 7 = ☐

8 Think carefully. Then write 'odd number' or 'even number' in each answer space.

(a) even number + odd number = _____

(b) even number + even number = _____

(c) odd number + odd number = _____

The challenge and extension questions examine the effect of subtracting odd and even numbers.

In Question 8 they examined what happens when they are added. Encourage pupils to draw diagrams or use generic plastic or card number plates to support their reasoning.

Unit 6.5
Mathematics Plaza — magic square

Conceptual context

In a magic square every row, column and diagonal adds up to the same number. The simplest magic square is a 3 × 3 grid and contains the numbers 1–9, giving totals of 15.

Magic squares are a fun way to practise problem solving and addition.

 A 3 × 3 magic square remains magic:

- if each number is multiplied by the same number
- if the same number is added to each number
- if the outside two rows or columns are interchanged

Using these transformations, many different 3 × 3 magic squares can be constructed. Not all totals are possible, only multiples of 3 from 15. The central number is always the fifth number when the numbers are ordered and the magic number for any square is 3 times the central number.

The order of a magic square is determined by the number of rows or columns, so a 3 × 3 square is Order 3. Order 4 magic squares use a 4 × 4 grid and so on.

Magic squares were known by Chinese mathematicians more than 2000 years ago. An ancient legend describes a huge flood and a turtle emerging from the water with the pattern of a magic square on its shell.

This is known as the Lo Shu square and since that time the pattern has been considered to have magical properties.

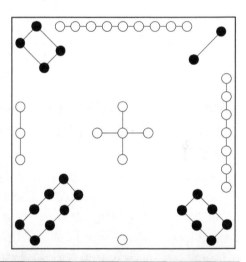

Learning pupils will have achieved at the end of the unit

- Pupils will have been introduced to 3 × 3 magic squares, in which each row, column and diagonal adds up to the same number (Q1, Q2, Q3)
- Pupils will have developed skills to fill in partially-completed magic squares in a logical way (Q1, Q2, Q3, Q4)
- The use of letters to represent numbers will have been revised (Q4)
- Pupils will have explored solving addition and subtraction problems presented in different ways (Q1, Q2, Q4, Q5)

Resources

squared paper; mini whiteboards; five small blank cards;
Resource 3.6.5 Magic square

Vocabulary

magic square, total, row, column, diagonal

Questions 1, 2 and 3

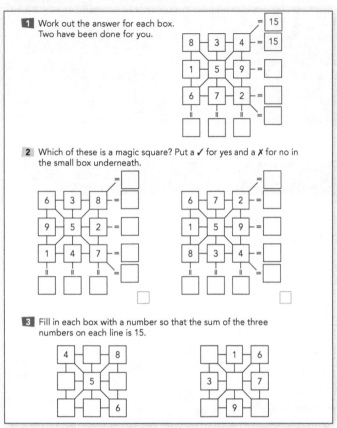

1 Work out the answer for each box.
Two have been done for you.

2 Which of these is a magic square? Put a ✓ for yes and a ✗ for no in the small box underneath.

3 Fill in each box with a number so that the sum of the three numbers on each line is 15.

What learning will pupils have achieved at the conclusion of Questions 1, 2 and 3?

- Pupils will have been introduced to 3 × 3 magic squares, in which each row, column and diagonal adds up to the same number.
- Pupils will have developed skills to fill in partially-completed magic squares in a logical way.
- Pupils will have explored solving addition and subtraction problems presented in different ways.

Activities for whole-class instruction

Look at Question 1 in the Practice Book. Discuss the layout of the 3 × 3 grid and establish that the numbers are 1–9. Revise the terms *row*, *column* and *diagonal*, then work together to add each row, column and diagonal, noting that each one adds up to 15.

- Ask pupils to discuss with a partner whether they think that if the numbers were placed in any position, they would still add up to 15. Listen to some answers. Pupils may recognise that swapping numbers will destroy the 'magic'.

- Pupils complete Question 2 in the Practice Book to establish that the numbers cannot be placed randomly.
- Show pupils the following partially-completed magic square, which has a magic number of 24.

		5
6	8	

- Ask: *Which empty square can be completed?* The square to the right of the 8 can be completed, because two of the numbers are already known.

  6 + 8 + ? = 24

- The missing number is 10. Once the 10 is placed, the right-hand row can be completed with 9 to make 24. The next step is to complete the diagonal 9 + 8 + ? = 24. Continue to fill in the remaining squares.
- Pupils complete Question 3 in the Practice Book.

Same-day intervention

- Copy the magic square below on the board. Ask pupils to add 4 to each number and check whether the square is still magic.

4	9	2
3	5	7
8	1	6

- Establish that the square is still magic, with a magic number of 27.

Same-day enrichment

- Ask pupils to look at the layout of the numbers in the magic squares they have created in Question 3 and the magic squares in Questions 1 and 2 to see if the arrangements of the numbers are related.

Question 4

4 Fill in the cells with different numbers so that all three numbers in each row, each column and each diagonal add up to the number in the circle above.

What learning will pupils have achieved at the conclusion of Question 4?

- Pupils will have developed skills to fill in partially-completed magic squares in a logical way.
- The use of letters to represent numbers will have been revised.
- Pupils will have explored solving addition and subtraction problems presented in different ways.

Activities for whole-class instruction

- On the board, copy the following partially-completed magic square, which has a magic number of 30.

8	A	4
B	10	C
D	E	F

- Ask pupils which letters they can solve immediately. Confirm that they can find A in the top row and D and F from the diagonals. Once these have been calculated the other letters can be solved to give the left-hand magic square below.

8	18	4
6	10	14
16	2	12

- Write the second magic square beside this one and ask pupils to discuss with a partner what they notice about the two squares. They may notice that the left-hand square is composed of the first nine even numbers. They may also spot that the numbers in the left-hand square are double the numbers in the right-hand one.

- Give pupils a blank 3 × 3 square and ask them to try adding the same small number (less than 10) to every number in the right-hand magic square and then to see whether the square is still magic. Confirm that it is.

- Pupils complete Question 4 in the Practice Book.

Same-day intervention

- On the board, copy the 1-9 magic square below. Ask pupils to create a new magic square by multiplying each number by 3.

4	9	2
3	5	7
8	1	6

- Ask pupils what the magic number for this magic square is.

Same-day enrichment

- Give pupils **Resource 3.6.5** Magic square and ask pupils to complete the 4 × 4 magic square.

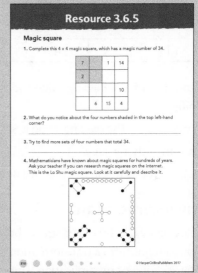

Answers:

7	12	1	14
2	13	8	11
16	3	10	5
9	6	15	4

The four numbers in each corner total 34.

The central four numbers also total 34.

Question 5

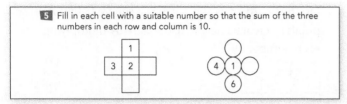

5 Fill in each cell with a suitable number so that the sum of the three numbers in each row and column is 10.

What learning will pupils have achieved at the conclusion of Question 5?

- Pupils will have explored solving addition and subtraction problems presented in different ways.

Activities for whole-class instruction

- On the board, write the following bar model, in which the whole has been divided into three parts.

20		
11	4	5

- Ask pupils to calculate the missing value and explain their reasoning.

  20 – 11 – 4 = 5   The missing value is 5.

- Repeat with some other values.
- Pupils complete Question 5 in the Practice Book.

Same-day intervention

- On the board, copy the 1-9 magic square below. Ask each pupil to perform a transformation on the numbers by addition or multiplication. They should then rub out five of the numbers (or copy the remaining numbers onto a new blank square) and ask a friend to solve the magic square and determine the magic number.

2	9	4
7	5	3
6	1	8

Same-day enrichment

- Ask pupils to look at all the 3 × 3 magic squares they have studied. Ask: *Can you find a relationship between the number in the centre of the square and the magic number?* (The magic number is always 3x the central number.)
- Ask pupils to create their own magic square.

Challenge and extension question

Question 6

6 Put these numbers in the circles so that the sum of the three numbers on each arm is the same.

2   4   6   8   10   12   14

The challenge and extension question asks pupils to fill in the first 7 even numbers so that each line has the same sum. From the investigations on magic squares, they may deduce that when the numbers are arranged in order, the fourth number should be in the middle position.

Pupils could also try filling in the first 7 odd numbers into the same diagram.

Unit 6.6
Numbers to 1000 and beyond

Conceptual context

Pupils are familiar with place value to 1000. This unit develops their ability to read and write four-digit and five-digit numbers, extending their understanding of place value to include the 'thousands' and 'Ten Thousands' columns.

Ten Thousands	Thousands	Hundreds	Tens	Ones
	3			

The 3 in the thousands column is 3 × 1000, 3000.

Ten Thousands	Thousands	Hundreds	Tens	Ones
2				

The 2 in the Ten Thousands column is 2 × 10 000, 20 000.

Zeros are used as placeholders in positions that have no value.

(i) Conventionally, four-digit numbers are written with no spaces, for example 1000, 2643, while five-digit numbers have a half-spacing between the hundreds and thousands places, for example 10 000, 23 157. This spacing makes large numbers easier to read.

Learning pupils will have achieved at the end of the unit

- Pupils will have revised addition, subtraction, multiplication and division calculations using mental methods (Q1)
- Pupils will have recognised the place value of each digit in three-digit, four-digit and occasional five-digit numbers (Ten Thousands, thousands, hundreds, tens, and ones) (Q2, Q4, Q5, Q6)
- Pupils will have explored different ways to represent four-digit numbers (Q2, Q5)
- Use of zero as a placeholder will have been practised (Q2, Q3, Q4, Q5, Q6)
- Pupils will have explored counting forwards and backwards in multiples of 1000 (Q3)
- Pupils will have begun to communicate about place value in abstract terms with understanding (Q6)

Resources

0–100 number cards; place value chart; sticky notes; small 0–9 digit cards; blank counting stick; A5 blank cards; base 10 blocks; counters; place value arrow cards; **Resource 3.6.6a** Number pairs 1; **Resource 3.6.6b** Number pairs 2; **Resource 3.6.6c** Number questions; **Resource 3.6.6d** Number values

Vocabulary

ones, tens, hundreds, thousands, Ten Thousands, placeholder, base 10 apparatus (one, rod, block), place value, place value chart, place value arrow cards, consecutive

Question 1

> **1** Calculate mentally.
>
> (a) $57 - 33 + 45 =$ ☐ (b) $1 \times 6 \times 6 =$ ☐
>
> (c) $64 \div 8 - 0 =$ ☐ (d) $23 + 82 + 92 =$ ☐
>
> (e) $6 \times 6 \times 0 =$ ☐ (f) $50 \div 5 + 72 =$ ☐

What learning will pupils have achieved at the conclusion of Question 1?

- Pupils will have revised addition, subtraction, multiplication and division calculations using mental methods.

Activities for whole-class instruction

- Shuffle a class pack of 0–100 number cards. Turn over the top card and ask pupils what has to be added to make 100. Ask individual pupils to describe their method. Repeat with the next card.

- Ask: *What effect does adding zero to a number sentence have?*

  For example, $11 + 23 = ?$ and $11 + 23 + 0 = ?$

- Ask: *What effect does subtracting zero from a number sentence have?*

  For example, $15 + 28 = ?$ and $15 + 28 - 0 = ?$

- Ask: *What effect does multiplying by zero have on a number sentence?*

  For example, $5 \times 8 = ?$ and $5 \times 8 \times 0 = ?$

   If pupils ask about division by zero, explain that it does not make sense.

  Division means splitting into equal parts and it is impossible to split a number into zero equal parts.

  We say that dividing by zero is undefined.

- Pupils complete Question 1, working mentally to revise the four operations.

Same-day intervention

- Write the following two-digit numbers on the board:

  16   45   50   54   32   33

- Challenge pupils to put the numbers in pairs, so that each pair shares a mathematical fact. For example, 45 and 33 are a pair because they are both odd; 16 and 32 are a pair because they are both in the 4x table; 50 and 54 are a pair because they both have a 5 in the tens column. Many other correct pairings are possible.

Same-day enrichment

- Write the following two-digit numbers on the board: 15, 25, 36.

- Challenge pupils to give a reason why each of the numbers is the odd one out. For example:

  - *15 is the odd one out because it is the only number that is not a square number.*

  - *25 is the odd one out because it is the only one not divisible by 3.*

  - *36 is the odd one out because it is the only even number.*

  - Repeat with another three random numbers.

Question 2

> **2** Complete the place value chart.
>
> | | | | tens | ones |
> |---|---|---|---|---|
> | place | place | place | place | place |

What learning will pupils have achieved at the conclusion of Question 2?

- Pupils will have recognised the place value of each digit in three-digit, four-digit and occasional five-digit numbers (Ten Thousands, thousands, hundreds, tens, and ones).

- Pupils will have explored different ways to represent four-digit numbers.

- Use of zero as a placeholder will have been practised.

Activities for whole-class instruction

- Show pupils a base 10 1000 block. Ask: *How many hundreds are needed to make this block?* Demonstrate that ten 100 blocks can be stacked to make 1000.

   10 times 100 makes 1000

- Give pupils a blank 5 columns × 2 rows table on thin card, to fill a landscape A4 sheet.

- Explain that they are going to make their own place value chart. The top row will have the place value titles. Ask: *What is the title of the right-hand column?* Establish that this is 'Ones'. Continue naming the columns one place at a time. Most pupils will know that the fourth column from the right is 'Thousands'. Ask the name of the final column and enter the title 'Ten Thousands'. Discuss with pupils the fact that each column is 10x the value of the column to its right. So, 10 is 10 × 1; 100 is 10 × 10; 1000 is 10 × 100 and 10 000 is 10 × 1000, so the title of the column is 'Ten Thousands.'

- Check that pupils' place value charts now look like this:

Ten Thousands	Thousands	Hundreds	Tens	Ones

- Ask: *How many placeholder zeros are required to write 1000?* Three zeros are required to ensure the 1 is in the correct place. Repeat the question for 100, 10 000 and 10.

- Challenge pupils to discuss with a partner the largest number that could be recorded on their place value chart and agree that it is 99 999. Ask whether this is this the largest possible number and elicit that it is not, the next number would be a 6-digit number 100 000, which is one hundred thousand (not a million, as they may suggest).

  However, explain that this unit will only be looking at four-digit and occasional five-digit numbers.

  Keep the place value chart to use throughout this unit and the next one, but put them aside for now.

- Pupils are ready to complete Question 2 in the Practice Book.

Same-day intervention

- Write sufficient pairs of sticky notes to match the number of pupils in the group. For example, for a group of ten pupils, write: 200, two hundred, 2000, two thousand, 20 000, twenty thousand, 4000, four thousand, 40, forty. Explain that there are pairs of identical numbers, one in words and one in numerals.

  Tell pupils to hold out their hands, with their eyes closed, and randomly place a sticky note on their hand. Ask them to open their eyes and challenge them to find their partner. Give 5 points to the first pair to find each other, 3 points for the second pair and 1 point for the third. Repeat.

Same-day enrichment

- Give pupil pairs set of 1–9 small digit cards and a separate pile of four digit cards all with zero. Ask pupils to select the cards required to make two thousand and place them correctly on their place value chart.

- Support pupils to explain the structure of the number in a whole sentence, for example: *Two thousand has a 2 in the thousands column and three zeros in the hundreds, tens and ones columns. If the zeros weren't there, the number would look like a 2.* Repeat, using the same sentence structure with other numbers, for example:

  six hundred

  five thousand

  Ten Thousand

  eighty

  nine hundred

  seven thousand

  twelve thousand.

Question 3

> **3** Complete the number pattern.
>
> 1000, 2000, ☐, 4000, ☐, 6000, ☐, 8000, ☐, 10 000, ☐, 12 000

What learning will pupils have achieved at the conclusion of Question 3?

- Use of zero as a placeholder will have been practised.
- Pupils will have explored counting forwards and backwards in multiples of 1000.

Activities for whole-class instruction

- Use a blank counting stick and work together as a group to label the stick in thousands from 0–10 000.

- Count forwards and backwards, pointing at the numbers.

- Ask: *Counting in thousands, what numbers comes after 10 000?* Confirm that the numbers are 11 000, 12 000, 13 000 and so on.

- Label nine small A5 cards with 1000–9000. Shuffle them and invite two pupils to choose a card each. Ask them to tell you the number on the card, to say whose card is bigger/smaller, to calculate the total of/difference between the two cards. Repeat with new cards.

- Pupils are ready to complete Question 3 in the Practice Book.

Same-day intervention

- Shuffle the small A5 cards labelled in thousands from the whole-class activities and ask pupils to put them in order. Count forwards and backwards, pointing at the numbers.

Same-day enrichment

- Ask pupils to fill in the missing numbers and add the next two numbers in the number pattern.
  - 100, 1100, 2100, 3100, ☐, 5100, 6100, ☐, ☐
  - 9080, 8080, 7080, ☐, 5080, 4080, 3080, ☐, ☐
  - 1567, 2567, 3567, 4567, ☐, 6567, 7567, 8567, ☐, ☐
- Now ask pupils to write a thousands number pattern of their own.

Question 4

4 Write the digits of each number in the place value chart. The first one has been done for you.

(a) 4208

TTh	Th	H	T	O
0	4	2	0	8

(b) 9990

TTh	Th	H	T	O

(c) 10 008

TTh	Th	H	T	O

(d) 2006

TTh	Th	H	T	O

What learning will pupils have achieved at the conclusion of Question 4?

- Pupils will have recognised the place value of each digit in three-digit, four-digit and occasional five-digit numbers (Ten Thousands, thousands, hundreds, tens, and ones).
- Use of zero as a placeholder will have been practised.

Activities for whole-class instruction

- Write the following table on the board:

Ten Thousands	Thousands	Hundreds	Tens	Ones
	5	2	7	6

- Ask pupils what number is shown and confirm that it is five thousand two hundred and seventy-six. Add a zero to the Ten Thousands column and ask: *What number is now shown?*

Ten Thousands	Thousands	Hundreds	Tens	Ones
0	5	2	7	6

- Elicit from pupils that the number is still five thousand two hundred and seventy-six, because there are no digits to the left of this zero. Therefore, it is not acting as a placeholder and is not increasing the value of the number. Ask: *How is this number written?* Confirm that it is 5276. Explain that zeros to the left of a number are not included in a number because they do not change its value.

- Write these two tables on the board:

Ten Thousands	Thousands	Hundreds	Tens	Ones
0	1	4	6	9

Ten Thousands	Thousands	Hundreds	Tens	Ones
1	0	4	6	9

- Ask pupils to discuss the two numbers with a partner. Agree that they are 'one thousand four hundred and sixty-nine' and 'Ten Thousand four hundred and sixty-nine'.

- Ask: *In which of the numbers is the zero acting as a placeholder?* (It is a placeholder in 'Ten Thousand four hundred and sixty-nine' because there is a positive digit to the left of the placeholder zero.) Are pupils able to tell you that without the zero, the number would look like 1469 – the value would be different?

- Show the number five hundred on a place value chart.

Ten Thousands	Thousands	Hundreds	Tens	Ones
		5	0	0

- Ask: *Will the value of the number change if zeros are placed in the thousands and Ten Thousands columns?* Agree that the number remains five hundred and does not change.

- Show the following on a place value chart on the board.

Ten Thousands	Thousands	Hundreds	Tens	Ones
0	6	0	3	0

- Establish that the chart shows six thousand and thirty. Ask pupils to discuss with a partner which zeros are acting as placeholders in the number. Share their opinions and agree that the zeros in the hundreds and ones columns are the ones that are placeholders (while the zero in the Ten Thousands place is not a placeholder).

- Look at the first example in Question 4 in the Practice Book with pupils. Establish that in this question they are expected to complete each box in the place value chart even where it does not affect the value of the number. Pupils are now ready to complete Question 4.

> **Look out for** … pupils who find four-digit numbers challenging. Check that they are completely secure with three-digit numbers. Provide lots of practice using base 10 blocks, introducing the 1000 block.

Same-day intervention

- Use base 10 blocks to help build pupils' conceptual understanding of thousands numbers. Show them a four-digit number in blocks, for example 1452 and ask them to identify the value of each digit. There is 'one' thousand block, 'four' hundred blocks, 'five' tens rods and 'two' ones. Build up to reading the number as 'one thousand four hundred and fifty-two'.

- Repeat with more numbers, including numbers with placeholder zeros.

- Reverse the process, showing pupils a number, for example 2451, and ask them to make the number using base 10 blocks. Repeat with different numbers.

Same-day enrichment

- Write the following numbers on the board and ask pupils to name the column that contains a placeholder zero.

  309, 4095, 13 480, 2103, 10 479, 12 045, 3180, 11 308.

  (Answers: tens, hundreds, ones, thousands, hundreds, ones, tens)

Question 5

5 Write the numbers represented in each place value chart in words and numerals.

(a)

TTh	Th	H	T	O
	●●	●●		●●

In words: _____

In numerals: _____

(b)

TTh	Th	H	T	O
●				

In words: _____

In numerals: _____

(c)

TTh	Th	H	T	O
	●●●●●	●●	●●●	

In words: _____

In numerals: _____

(d)

TTh	Th	H	T	O
	●●●●		●●●●●	●●●

In words: _____

In numerals: _____

What learning will pupils have achieved at the conclusion of Question 5?

- Pupils will have recognised the place value of each digit in three-digit, four-digit and occasional five-digit number (Ten Thousands, thousands, hundreds, tens, and ones).

- Pupils will have explored different ways to represent four-digit numbers.

- Use of zero as a placeholder will have been practised.

Activities for whole-class instruction

- Give pupils the place value charts prepared for Question 2, plus some small counters.

- Tell them to place three counters in the thousands column and ask them what number this represents. Confirm that it is three thousand, written in numerals as 3000.

- Tell them that they can move one counter to another column to make a new number. Ask them to discuss with a partner the different numbers that can be made. Ask individual pupils to say the numbers that they have found (the possibilities are 2001, 2010, 2100, 12 000) and to write down the numbers in numerals.

- Give pupils five counters and ask them to place them in at least three different columns. Collect some of the numbers that they have made, for example 2120, 1301 and so on.

- Pupils are ready to complete Question 5 in the Practice Book.

Same-day intervention

- Give pupils **Resource 3.6.6a** Number pairs 1 and **Resource 3.6.6b** Number pairs 2 and ask pupils to play the game in pairs.

Resource 3.6.6a	Resource 3.6.6b

Number pairs 1

- Cut out and shuffle the cards.
- Place them face down. Take turns to turn over two cards. If they are a matching pair, keep them and have another turn. If they are not a pair, turn them back over, keeping them in the same position.
- Continue playing until all the cards have been collected. The winner has the greater number of cards.

3000	1426	7089	2954
5600	3098	5309	4444
three thousand	one thousand four hundred and twenty-six	seven thousand and eighty-nine	two thousand nine hundred and fifty-four
five thousand six hundred	three thousand and ninety-eight	five thousand three hundred and nine	four thousand four hundred and forty-four

© HarperCollinsPublishers 2017

Number pairs 2

© HarperCollinsPublishers 2017

Same-day enrichment

- Give pupils **Resource 3.6.6a** and **Resource 3.6.6b** and ask pupils, in pairs, to play the game and then, using the blank grid on page 2 of the resource sheet, pupils should design a new set of four-digit numbers in words and numerals.

- They should cut out cards and play their game then swap to play another pair's set.

Question 6

6 Complete each sentence.

(a) 10 ones make ☐, 10 tens make ☐, 10 hundreds make ☐ and 10 thousands make ☐.

(b) Counting from the right, the first digit of a number is its _____ place. The third digit is its _____ place. The fifth digit is its _____ place.

(c) A number consisting of 7 thousands, 5 hundreds, 2 tens and 3 ones is ☐.

(d) There are ☐ thousands or ☐ hundreds in 6000. There are ☐ tens in 170.

(e) Three consecutive numbers after 9998 are ☐, ☐, and ☐.

What learning will pupils have achieved at the conclusion of Question 6?

- Pupils will have recognised the place value of each digit in three-digit, four-digit and occasional five-digit numbers (Ten Thousands, thousands, hundreds, tens, and ones).

- Use of zero as a placeholder will have been practised.

- Pupils will have begun to communicate about place value in abstract terms with understanding.

Activities for whole-class instruction

- Write the number 6732 on the board.

- Give pupils mini whiteboards and ask them to answer the following questions.

- Choose individual pupils to explain the answer in a complete sentence.

  - *What is the value of the 3?* (The value of the 3 is 30 or 3 tens.)

  - *What is the value of the 6?* (The value of the 6 is 6000 or 6 thousands.)

  - *What is the value of the 7?* (The value of the 7 is 700 or 7 hundreds.)

  - *What is the value of the 2?* (The value of the 2 is 2 or 2 ones.)

- Write the number 15 493 on the board.

- Give pupils mini whiteboards. Ask the following questions:

  - *What digit is in the hundreds column?* (4)

  - *What digit is in the Ten Thousands column?* (1)

  - *What digit is in the ones column?* (3)

  - *What digit is in the thousands column?* (5)

  - *What digit is in the tens column?* (9)

- Write the number 2000 on the board. Ask the following questions:

  - *How many thousands in this number?* (2 thousands)

  - *How many hundreds make up this number?* (20 hundreds)

  - *How many tens make up this number?* (200 tens)

  - *How many ones make up this number?* (2000 ones)

- Pupils are ready to complete Question 6 in the Practice Book.

Same-day intervention

● Give pupils **Resource 3.6.6c** Number questions.

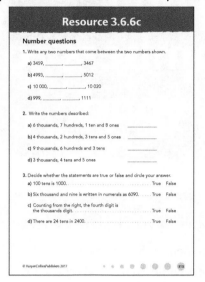

Answers:

1. Two numbers from 3460 – 4367; 4994 – 5011; 10 001 – 10 019; 1000 – 1110.

2. 6718; 4235; 9630; 3045

3. False; false; true; false

Challenge and extension question

Question 7

7 Complete the number patterns.

(a) 5078, 5079, ☐ , ☐ , 5082

(b) 2323, 3434, 4545, ☐ , ☐ , 7878

(c) 10 000, 9990, 9980, ☐ , ☐ , ☐

In the challenge and extension question pupils need to recognise different number patterns in four-digit numbers. Remind them to look at differences to see how the numbers are changing and whether they are increasing or decreasing. Ask pupils to describe the pattern in a whole sentence.

Same-day enrichment

● Give pupils **Resource 3.6.6d** Number values.

Answers: a) 20 tens b) 45 hundreds c) 56 tens
d) 710 tens e) 120 hundreds

Unit 6.7
Read, write and compare numbers to 1000 and beyond

Conceptual context

This unit, covering reading, writing and comparing numbers beyond 1000, deepens pupils' abstract understanding of the composition of larger numbers. The focus is on four-digit numbers, with occasional five-digit numbers. Each place value column is 10 times larger than its right-hand neighbour, so a thousand is 10× larger than a hundred and Ten Thousand is 10× larger than a thousand. Being able to recognise the magnitude of the values in each of the place value columns allows pupils to compare and order four-digit (and five-digit) numbers, using greater than and less than symbols appropriately.

Pupils need to continue to visualise pictorial representations, such as diagrams and number lines. Ensure that concrete and pictorial representations are always available and used frequently. Pupils vary a great deal in the time taken to be able answer questions simply through using abstract numbers.

(i) It is important for pupils' understanding that they practise reading numbers as words and do not fall into the habit of saying them as a string of digits; because pupils are most likely to comprehend the magnitude of a number through knowing 'the whole story' of the number. Strong number sense cannot develop unless pupils have good understanding about the numbers they are working with; referring to numbers by the 'word names' wherever possible, will enable pupils to develop strong number sense.

Learning pupils will have achieved at the end of the unit

- Pupils will have revised addition, subtraction, multiplication and division calculations using mental methods (Q1)
- Pupils' number world will have expanded to include Ten Thousands (Q2)
- Reading four-digit (and some five-digit numbers) will have been developed (Q2, Q3)
- Writing four-digit (and some five-digit numbers) in numerals and words will have been practised (Q2, Q3, Q4)
- Use of zero as a placeholder will have been reinforced (Q2, Q3, Q4, Q6)
- Pupils will have explored partitioning and combining four-digit numbers (Q3)
- Pupils will have practised writing four-digit numbers as w000 + x00 + y0 + z (expanded form) (Q3)
- Pupils will have compared pairs of three-digit and four-digit numbers using >, < and = symbols (Q5)
- Pupils will have developed their understanding that to compare and order four-digit numbers, they should begin by looking at the magnitude of the thousands digit, then the magnitude of the hundreds digit, then the magnitude of the tens digit and finally the magnitude of the ones digit (Q5, Q7)
- Overall fluency in number sense will have been developed, enabling pupils to read and write four-digit numbers with placeholders in numerals and words (Q6, Q7)
- Ordering of four-digit numbers will have been extended (Q7)

Resources

0–9 dice; mini whiteboards; place value arrow cards; place value stick abacus; base 10 apparatus; **Resource 3.6.7a** Blank tables square; **Resource 3.6.7b** Number grid; **Resource 3.6.7c** 4-digit numbers; **Resource 3.6.7d** 5-digit numbers; **Resource 3.6.7e** Greater than or less than; **Resource 3.6.7f** Place value; **Resource 3.6.7g** Making numbers

Vocabulary

ones, tens, hundreds, thousands, Ten Thousands, placeholder, base 10 blocks, place value, place value chart, place value arrow cards

Question 1

> **1** Calculate mentally.
>
> (a) 70 – 3 + 13 = ☐
> (b) 53 – 37 – 4 = ☐
> (c) 6 × 8 + 3 = ☐
> (d) 47 – 11 – 21 = ☐
> (e) 6 × 4 + 61 = ☐
> (f) 40 ÷ 4 – 5 = ☐
> (g) 90 ÷ 9 × 5 = ☐
> (h) 23 + 28 + 92 = ☐
> (i) 4 × 8 + 6 = ☐
> (j) 56 ÷ 8 × 6 = ☐

What learning will pupils have achieved at the conclusion of Question 1?

- Pupils will have revised addition, subtraction, multiplication and division calculations using mental methods.

Activities for whole-class instruction

- Choose a pupil to roll two 0–9 dice.
- Make both possible two-digit numbers and ask pupils to:
  - add the numbers
  - find the difference
  - multiply the two numbers
  - divide the two-digit numbers by each of the two dice roll numbers
- For example, here are the answers for dice rolls of 6 and 7.

  67 + 76 = 143   76 – 67 = 9   6 × 7 = 42

  67 ÷ 6 = 11 r 1   76 ÷ 6 = 12 r 4

  67 ÷ 7 = 9 r 4   76 ÷ 7 = 10 r 6

- Pupils are ready to complete Question 1 in the Practice Book.

Same-day intervention

- Give pupils **Resource 3.6.7a** Blank tables square to complete.

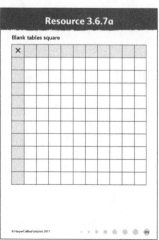

Activities for whole-class instruction

- Ask them to swap with a friend to check each other's answers and to make a note of any that they find tricky so that they can find strategies for better recall. For example, in the 7× table, if pupils know that 3 × 7 = 21, they can use this knowledge to find 6 × 7 by doubling, 6 × 7 = 42 and 9 × 7 = 63.

Same-day enrichment

- Give pupils a blank tables square with the tables filled in along the axes in a random order, for example:

x	5	8	3	12	7	1	2	10	4	11	6	9

- Numbers on the vertical axis can either be in the usual order or also mixed up. Ask pupils to complete the table and then swap with a friend to check each other's answers. Tell them to keep a note of any that are incorrect and discuss how to find the answer.

Question 2

> **2** Read and write the numbers in words and numerals to complete the table.
>
> | Words | Numerals |
> |---|---|
> | six thousand three hundred and forty-eight | |
> | five thousand and fifty | |
> | thirteen thousand and four | |
> | | 9008 |
> | | 4415 |
> | | 19006 |

What learning will pupils have achieved at the conclusion of Question 2?

- Pupils' number world will have expanded to include Ten Thousands.
- Reading four-digit (and some five-digit numbers) will have been developed.
- Writing four-digit (and some five-digit numbers) in numerals and words will have been practised.
- Use of zero as a placeholder will have been reinforced.

Activities for whole-class instruction

- Write this place value chart on the board.

Ten Thousands	Thousands	Hundreds	Tens	Ones
1	2	3	4	5

- Ask: *Which digit has the highest value?* Establish that it is the 1 in the Ten Thousands column.
- Ask pupils to read this number and agree that it is 12 345, twelve thousand three hundred and forty-five.
- Ask: *What will the number become if 1 is added to each of the five columns?* Here is the new place value chart:

Ten Thousands	Thousands	Hundreds	Tens	Ones
2	3	4	5	6

- The new number is 23 456, twenty-three thousand four hundred and fifty-six.
- Give pupils **Resource 3.6.7b** Number grid.

Resource 3.6.7b

Number grid

one thousand	seven hundred	and seventy	two
and thirty	two thousand	three	nine hundred
six hundred	six	three thousand	and twenty
eight	and fifty	four hundred	five thousand

318   © HarperCollinsPublishers 2017

- Read the numbers in the first column together.
- Ask: *Can you put the numbers in order to make a four-digit number?* Confirm that the correct order is one thousand six hundred and thirty-eight. Challenge them to write this number in numerals on mini whiteboards – 1638.
- Continue looking at each column in turn. Establish that they need to find the thousands number first, then the hundreds followed by the tens and finally the ones. (2756, 3473, 5922)
- Use the four numbers in each horizontal line to make four more four-digit numbers.
- Write each number in numerals (*1772, 2933, 3626, 5458*).
- invite pupils to look again at the numbers in the first column and look for the number in italics (thirty). Replace this digit with a zero and consider what the new four-digit number will be. Confirm that the new number is one thousand six hundred and eight, in numerals, 1608. The value in the tens column now is now zero.
- Look at the columns and rows again, making new numbers by replacing the number that is in italics with zero.
- Pupils are ready to complete Question 2 in the Practice Book.

Same-day intervention

Ask the following questions:
- *What is the value of the 4 in the following numbers?*
  1234   4321   1432   1342
- *What is the value of the 6 in the following numbers?*
  6592   5269   9652   2596
- Make place value arrow cards available for support.

Same-day enrichment

- Ask: *Without changing the order of the digits, how many different four-digit numbers can you make by adding a zero to the digits in the number 132?*
  (1032, 1302, 1320)
- Ask: *What numbers can you make when you add two zeros to the digits in the number 132?*
  (13 200, 13 020, 13 002, 10 320, 10 302, 10 032)

Question 3

3 Read and write the numbers in words and then fill in the boxes.

(a) 4632 _____

4632 = ☐ + ☐ + ☐ + ☐

(b) 2547 _____

2547 = ☐ + ☐ + ☐ + ☐

(c) 6003 _____

6003 = ☐ + ☐ + ☐ + ☐

(d) 2030 _____

2030 = ☐ + ☐ + ☐ + ☐

What learning will pupils have achieved at the conclusion of Question 3?

- Reading four-digit numbers will have been reinforced.
- Writing four-digit numbers in numerals and words will have been practised.
- Use of zero as a placeholder will have been reinforced.
- Pupils will have explored partitioning and combining four-digit numbers.
- Pupils will have practised writing four-digit numbers as w000 + x00 + y0 + z (expanded form).

Activities for whole-class instruction

- Using place value arrow cards, show pupils the following cards

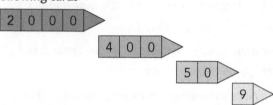

- Choose a pupil to overlap the cards and line up the arrows to form the four-digit number 2459.

- Read the number together, two thousand four hundred and fifty-nine. Expand the number again and write the number sentence 2459 = 2000 + 400 + 50 + 9.

- Ask: *What will the number be if I remove the 400 arrow card?* Confirm that the number becomes 2059 and the number sentence for this is 2059 = 2000 + 50 + 9.

- Replace the 400 and ask what the number will be if you remove the 50 arrow card. Confirm that the number becomes 2409 with a number sentence 2409 = 2000 + 400 + 9

- Invite individual pupils to select new arrow cards and repeat with the new numbers.

 When there is no value in a column, zero is required as a placeholder.

- When pupils demonstrate sufficient confidence, they should complete Question 3 in the Practice Book.

Same-day intervention

- Give pupils **Resource 3.6.7c** 4-digit numbers and ask them to complete the task.

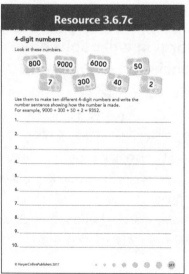

Same-day enrichment

- Give pupils **Resource 3.6.7d** 5-digit numbers and ask them to complete the task.

Question 4

4 Write the numbers in numerals.

(a) One thousand eight hundred and twelve: ☐

(b) Four thousand and fifty: ☐

(c) Six thousand five hundred: ☐

(d) Five thousand and six: ☐

What learning will pupils have achieved at the conclusion of Question 4?

- Writing four-digit numbers in numerals and words will have been practised.

- Use of zero as a placeholder will have been reinforced.

Activities for whole-class instruction

- Give pupils mini whiteboards.

- Tell them that you are thinking of a four-digit number. Write these clues on the board:

  – It is less than two thousand.

  – It has 3 in the hundreds place and the ones place

  – The value in the tens place is the sum of the values in the hundreds and ones places.

- Say that you are going to work together to find out the number from the clues.

- Write column headings on the board and short lines as shown. Ask pupils to do this on their whiteboards.

Th	H	T	O
—	—	—	—

- Look at the information given and add the numbers in the correct positions where you can. Deduce that the thousands column value must be 1 because the number is less than 2000 and it is a four-digit number.
- The hundreds digit and the ones digit are both 3.

Th	H	T	O
1	3	—	3

- The value of the tens digit is the sum of the hundreds and ones digits, 6 (3 + 3)

Th	H	T	O
1	3	6	3

- The number is 1363. Read this together as one thousand three hundred and sixty-three.
- Tell pupils that you are thinking of a four-digit number. Write the following clues on the board:
  - It has 7 in the thousands place
  - It has 7 in the ones place
  - The digits in the hundreds and tens places are the same and added together they total 4
- Challenge them to use the clues to work out the number on their whiteboards.
- Confirm that it is 7227, seven thousand two hundred and twenty-seven. Some pupils may use the column headings and lines as before, others may just use the column headings or just the lines, holding the other information in their heads.
- Tell pupils that you are thinking of a four-digit number and give them the following clues:
  - It is greater than 4000 but less than 5000
  - The digit in the ones column is one more than the digit in the thousands column.
  - The number contain two zeros.

    Challenge them to work out the number from the clues. Confirm that it is 4005, four thousand and five.
- Pupils are ready to complete Question 4 in the Practice Book.

Same-day intervention

- Write this speech bubble on the board and read it together.

> You write two thousand and seven like this – 20007

- Ask pupils to explain to a partner why this is wrong.
- Ask pupils to write the following numbers in numerals:
  - three thousand and six
  - four thousand and one
  - nine thousand and two
  - six thousand and five

Same-day enrichment

- Challenge pupils to write three clues for their own number riddle similar to the ones in the whole class activities to find a four-digit number.
- They should then try out their riddle on a friend.

Question 5

5 Write >, < or = in each ◯.

(a) 985 ◯ 895   (b) 1000 ◯ 999   (c) 7801 ◯ 7081

(d) 3877 ◯ 3787   (e) 5020 ◯ 2050   (f) 3456 ◯ 3546

(g) 5420 ◯ 5421   (h) 9887 ◯ 9987   (i) 4002 ◯ 4200

What learning will pupils have achieved at the conclusion of Question 5?

- Pupils will have compared pairs of three-digit and four-digit numbers using >, < and = symbols.
- Pupils will have developed their understanding that to compare and order four-digit numbers, they should begin by looking at the magnitude of the thousands digit, then the magnitude of the hundreds digit, then the magnitude of the tens digit and finally the magnitude of the ones digit.

Activities for whole-class instruction

- Write the following two numbers on the board.

  651   532

- Ask: *Which number is greater?* Agree 651 > 532 because 651 has 6 hundreds and 532 only 5 hundreds.
- Now look at these two numbers:

  516   532
- Ask: *Which number is greater?* Agree 532 > 516. Both numbers have 5 hundreds, so pupils need to look at the tens column – 3 tens is greater than 1 ten, so 532 > 516.
- Write the following two four-digit numbers on the board.

  3124   2319
- Ask: *How can these four-digit numbers be compared?* Agree 3124 > 2319 because a four-digit number with 3 thousands is greater than any number with only 2 thousands.
- Now consider these two four-digit numbers.

  5372   5749
- Ask: *How can these four-digit numbers be compared?* Agree that because they both have the same number of thousands, pupils need to compare the hundreds digits – 5372 < 5749 because 3 hundreds is less than 7 hundreds.
- Pupils are ready to complete Question 5 in the Practice Book.

Same-day intervention

- Give pupils **Resource 3.6.7e** Greater than or less than.

Resource 3.6.7e

Greater than or less than

Write these numbers in numerals and add > or < to compare the numbers.

1. four thousand six hundred and thirty-four ☐ three thousand nine hundred and twenty
2. six thousand and ninety ☐ six thousand four hundred and twenty-two
3. one thousand nine hundred and forty-eight ☐ one thousand seven hundred and forty-seven
4. three thousand and seventy ☐ four thousand
5. eight thousand eight hundred and eighty-eight ☐ eight thousand nine hundred

© HarperCollinsPublishers 2017

(Answers: 1. 4634 > 3920 2. 6090 < 6422
3. 1948 > 1747 4. 3070 < 4000 5. 8888 < 8900)

Same-day enrichment

- Shuffle a pack of 0–9 digit cards and lay out four cards, for example 3, 0, 6 and 8. Make two four-digit numbers
- from these numbers and use < to compare them, for example 3608 < 3806.
- Ask pupils to add two more numbers between the original two, for example 3608 < 3708 < 3800 < 3806.
- Repeat with new numbers.

Question 6

6 Multiple choice questions. (For each question, choose the correct answer and write the letter in the box.)

(a) Four thousand and five hundred is written as ☐ .
    A. 450   **B.** 4500   **C.** 4050

(b) Three thousand and seventeen is written as ☐ .
    A. 307   **B.** 3007   **C.** 3017

(c) Nine thousand and three is written as ☐ .
    A. 9030   **B.** 903   **C.** 9003

What learning will pupils have achieved at the conclusion of Questions 6?

- Use of zero as a placeholder will have been reinforced.
- Overall fluency in number sense will have been developed, enabling pupils to read and write four-digit numbers with placeholders in numerals and words.

Activities for whole-class instruction

- Give pupils mini whiteboards and write these numbers on the board.

  A. 2900   B. 290   C. 2090   D. 29 000
- Ask them to write down the letter of the largest number. Check that they choose D. Ask: *How do you know this is the largest number?* (It has the greatest number of digits. It has a value in the Ten Thousands column.)
- Ask them to write down the letter of the smallest number. Check that they choose B.
- Ask: *How do you know this is the smallest number?* (It is the only number with only three digits. It does not have any thousands.) Agree that the number is two hundred and ninety.
- Ask: *What is the same about the numbers A and C?* (They are both four-digit numbers. Both numbers contain the digits 2, 9 and two zeros.)

- Ask: *Which number is two thousand nine hundred?* Confirm that it is A, because it has 2 in the thousands column and 9 in the hundreds column.

- Ask: *What number is C?* Confirm that it is two thousand and ninety.

- Write these numbers on the board.

  A. 5060   B. 50 006   C. 5006   D. 506

- Ask: *Which number is five thousand and six?* Confirm that it is C.

- Ask pupils to read the other numbers with a partner and establish that they are:

  A. five thousand and sixty; B. fifty thousand and six; D. five hundred and six.

- If pupils cannot identify the correct number, provide additional support using base 10 blocks or place value stick abacus.

- Pupils are ready to complete Questions 6, 7 and 8 in the Practice Book.

Same-day intervention

- Give pupils **Resource 3.6.7f** Place value.

  Resource 3.6.7f

  Place value

  1. Make the number on the left with base 10 blocks or place value counters. How many tens or hundreds are needed to make each number? Circle the two correct answers.

     a) 4 hundreds are equal to   40 ones   400 tens   40 tens   400 ones

     b) 27 tens are equal to   270 tens   270 ones
     277 ones   2 hundreds and 7 tens

     c) 3 thousands are equal to   30 hundreds   30 tens
     3000 ones   30 thousands

  2. Draw the correct number of dots in each box to show the following numbers and write the number in numerals.

     a) three thousand two hundred and six

Thousands	Hundreds	Tens	Ones

     In numerals _____

     b) five thousand and twenty

Thousands	Hundreds	Tens	Ones

     In numerals _____

     c) six thousand and three

Thousands	Hundreds	Tens	Ones

     In numerals _____

     © HarperCollinsPublishers 2017

- Pupils should complete the tasks on the resource sheet, considering equivalences and representing numbers given in words as dot diagrams.

- Give pupils **Resource 3.6.7g** Making numbers.

(Answers: 1. 1007, 7001 2. 1070, 7010 3. 1007, 1070 4. 7010, 7100 5. 1007, 7001 6. Questions 1 & 5 have the same answer. The two numbers in 1 are the only possible odd numbers.)

Question 7

7 Use < to put the numbers in order, from the smallest to the greatest.

(a) 367 209 627 736

  ☐ < ☐ < ☐ < ☐

(b) 8070 8007 8700 7800

  ☐ < ☐ < ☐ < ☐

What learning will pupils have achieved at the conclusion of Question 7?

- Pupils will have developed their understanding that to order four-digit numbers, they should begin by looking at the magnitude of the thousands digit, then the magnitude of the hundreds digit, then the magnitude of the tens digit and finally the magnitude of the ones digit.

- Ordering of four-digit numbers will have been extended.

Activities for whole-class instruction

- Write the following three-digit numbers on the board.

  454   378   409   456   532

- Ask pairs of pupils them to put the numbers in order from the greatest to the least and to explain their reasoning.

- The order is 532 > 456 > 454 > 409 > 378.
- Pupils should explain that to find the largest number they need to look for the number with the largest hundreds digit, so 532 is the greatest number. There are three numbers with 4 in the hundreds place, so to order them they need to look at the digits in the tens column. Two numbers, 456 and 454, have the same tens digit so they need to look at the ones column, 456 > 454 and both these are greater than 409. The final number only has a 3 in the hundreds place so it is the smallest.
- Ask: *Having revised ordering three-digit numbers, how do you think you can order four-digit numbers?*
- Confirm that the process is the same, begin with the highest value digit - now thousands – and find the number with the largest number of thousands. If there is more than one number with the same number of thousands you will need to look at the digits in the hundreds column and so on.
- Write the following four-digit numbers on the board.

  2098   1356   2403   1540   1066

  Say: *Look at these four-digit numbers and put them in order from the greatest to the least.*

  There are two numbers with 2 in the thousands column: 2098 and 2403. 2403 > 2098 because 2403 has 4 in the hundreds column, while 2098 has 0 in the hundreds column. Continue ordering the remaining numbers in the same way to determine the final order: 2403 > 2098 > 1540 > 1356 > 1066.
- Try another set of five four-digit numbers, this time ordering them from the least to the greatest. Now, the process is reversed and pupils will need to look for the number with the smallest number of thousands and so on through the hundreds, tens and ones columns.
- Pupils are ready to complete Question 9 in the Practice Book.

Same-day intervention

- On the board, write these three-digit numbers: 253, 314, 272, 182 and ask pupils to make the numbers from base 10 blocks. (If necessary, remind them and demonstrate that 10 ones make a rod and 10 rods make a 100 block.)

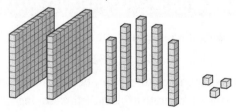

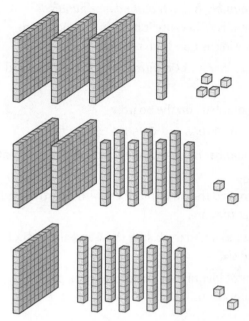

- Ask pupils to look at the numbers and order them from the greatest to least. Agree that the order is 314 > 272 > 253 > 182. To find the greatest number, explain that they need to look for the number with the most hundreds, so 314 is the greatest number.
- 272 and 253 have the same number of hundreds so they need to look at the tens column, 272 > 253. 182 is the smallest number because it has only 1 hundred.
- Try a new set of four three-digit numbers, using base 10 blocks for support as necessary.
- On the board, write these four-digit numbers: 1342, 2105, 1159, 2163 and ask pupils to make these numbers from base 10 blocks. (Remind them and demonstrate that ten 100 blocks make a block).

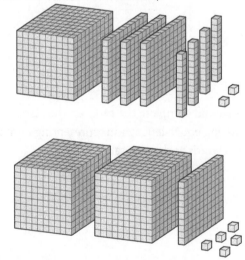

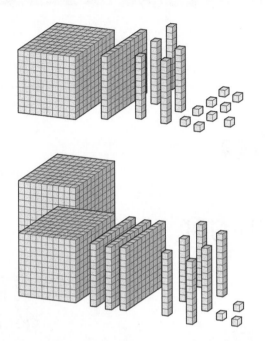

- Ask pupils to look at the numbers and order them from the least to the greatest. Agree that the order is 1159 < 1342 < 2105 < 2363. To find the smallest number, explain that they need to look for the number with the fewest thousands. There are two numbers with just 1 thousand, so they need to look at the value of the hundreds digit, 1159 < 1342. Repeat the process with the two numbers that have 2 in the thousands place.

- Try a new set of four four-digit numbers, using base 10 blocks for support as necessary.

Same-day enrichment

- Ask pupils to describe how they would order five-digit numbers. Listen to their explanations. The more digits a number has, the larger it is. These numbers have digits in the Ten Thousands column, so pupils must start by looking at the fifth digit, and then look at thousands, hundreds, tens and ones in turn.

- Order these numbers, from the least.

  20 324   16 782   17 980   21 354

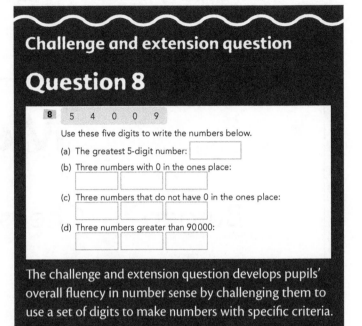

Challenge and extension question

Question 8

8   5   4   0   0   9

Use these five digits to write the numbers below.

(a) The greatest 5-digit number: ☐

(b) Three numbers with 0 in the ones place:
☐ ☐ ☐

(c) Three numbers that do not have 0 in the ones place:
☐ ☐ ☐

(d) Three numbers greater than 90 000:
☐ ☐ ☐

The challenge and extension question develops pupils' overall fluency in number sense by challenging them to use a set of digits to make numbers with specific criteria.

Chapter 6 test (Practice Book 3A, pages 144–149)

Test Question number	Relevant Unit	Relevant questions within unit
1	6.1	4
	6.2	4
	6.6	1
	6.7	1
2	6.3	2
3	6.3	1
4	6.1	3, 4, 5
	6.2	3, 4, 5
5	6.1	3, 4, 5
	6.2	3, 4, 5
6	6.5	1, 2, 3, 4
7	6.6	6
	6.7	2, 6, 7, C&E
8	6.3	4

Mia's Burger Bar

Menu

Burger

Cheeseburger

Fries

£3

£4

£1

Milkshake

Lemonade

£2

£1

Resource 3.2.1a

Related multiplication and division facts representations

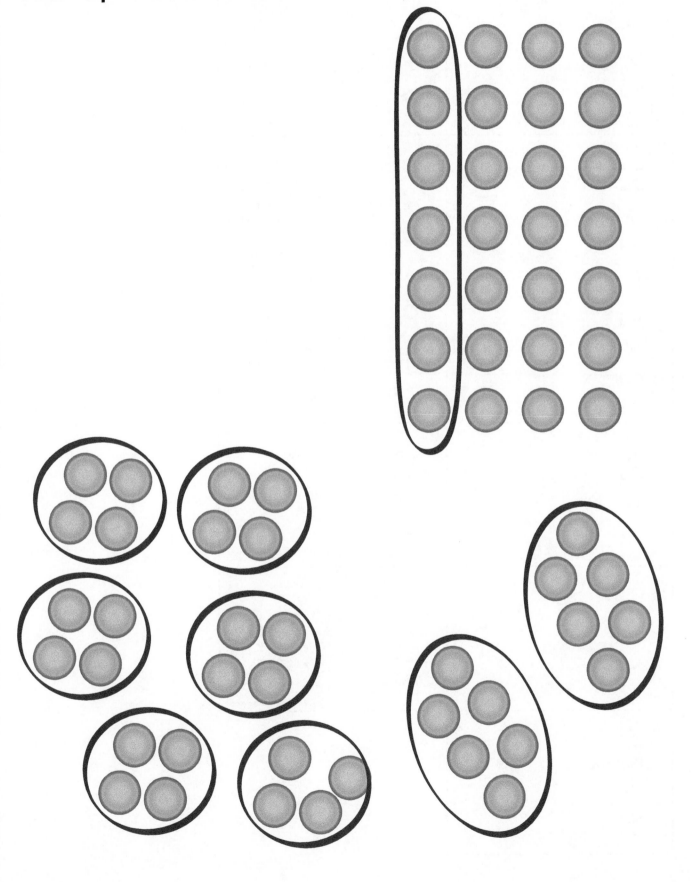

Domino cards

0 × 7	7	7 × 7	6
14 ÷ 7	21	7 × 8	9
7 × 3	7	63 ÷ 7	84
28 ÷ 4	2	11 × 7	0
5 × 7	49	7 × 12	35
42 ÷ 7	56		

Dividend, divisor, quotient

Dividend	Divisor	Quotient
6	3	
33	3	
15		5
	10	3
27	9	

Multiply and divide by 3

9	6	27
36	15	3

Resource 3.2.2c

Multiply and divide by 3 match up

What is the product of 3 groups of 4?	
How many groups of 3 are there in 18?	
What is 5 times 3?	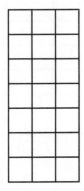
How many groups of 3 are there in 21?	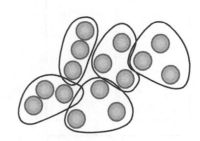

Making connections

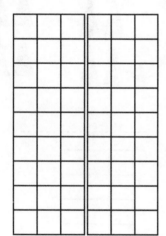

| 3 × 9 | 3 × 9 |

Multiplication squares

		42
		24
28	36	

		72
		18
24	54	

Problem matching and solving

There are 3 chocolate bars. Each bar has 9 pieces of chocolate. How many pieces are there altogether?

A bar of chocolate costs 12p. Jess buys 9 bars. How much does she pay?

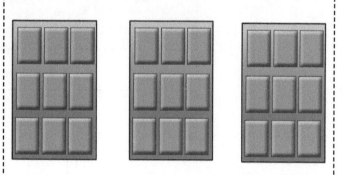

45 chocolates are shared equally between 5 people. How many chocolates will each person get?

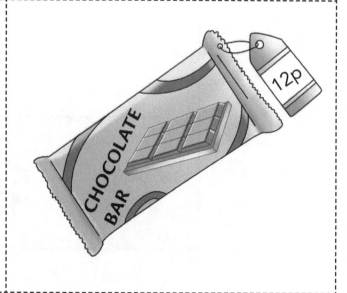

Scoring points

Kim threw red bean bags and Chris threw green bean bags.
Here is a diagram of where the bean bags landed.

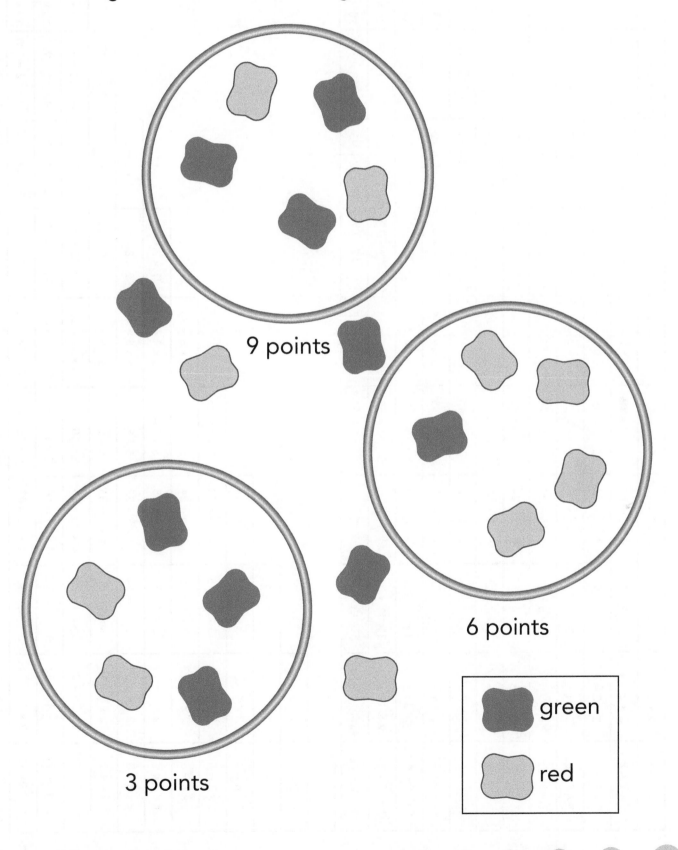

9 points

6 points

3 points

green	
red	

Multiplication grids

Grid 1

1×	2×	3×	4×	5×	6×	7×	8×	9×	10×
1 × 1 = 1	2 × 1 = 2	3 × 1 = 3	4 × 1 = 4	5 × 1 = 5	6 × 1 = 6	7 × 1 = 7	8 × 1 = 8	9 × 1 = 9	10 × 1 = 10
1 × 2 = 2	2 × 2 = 4	3 × 2 = 6	4 × 2 = 8	5 × 2 = 10	6 × 2 = 12	7 × 2 = 14	8 × 2 = 16	9 × 2 = 18	10 × 2 = 20
1 × 3 = 3	2 × 3 = 6	3 × 3 = 9	4 × 3 = 12	5 × 3 = 15	6 × 3 = 18	7 × 3 = 21	8 × 3 = 24	9 × 3 = 27	10 × 3 = 30
1 × 4 = 4	2 × 4 = 8	3 × 4 = 12	4 × 4 = 16	5 × 4 = 20	6 × 4 = 24	7 × 4 = 28	8 × 4 = 32	9 × 4 = 36	10 × 4 = 40
1 × 5 = 5	2 × 5 = 10	3 × 5 = 15	4 × 5 = 20	5 × 5 = 25	6 × 5 = 30	7 × 5 = 35	8 × 5 = 40	9 × 5 = 45	10 × 5 = 50
1 × 6 = 6	2 × 6 = 12	3 × 6 = 18	4 × 6 = 24	5 × 6 = 30	6 × 6 = 36	7 × 6 = 42	8 × 6 = 48	9 × 6 = 54	10 × 6 = 60
1 × 7 = 7	2 × 7 = 14	3 × 7 = 21	4 × 7 = 28	5 × 7 = 35	6 × 7 = 42	7 × 7 = 49	8 × 7 = 56	9 × 7 = 63	10 × 7 = 70
1 × 8 = 8	2 × 8 = 16	3 × 8 = 24	4 × 8 = 32	5 × 8 = 40	6 × 8 = 48	7 × 8 = 56	8 × 8 = 64	9 × 8 = 72	10 × 8 = 80
1 × 9 = 9	2 × 9 = 18	3 × 9 = 27	4 × 9 = 36	5 × 9 = 45	6 × 9 = 54	7 × 9 = 63	8 × 9 = 72	9 × 9 = 72	10 × 9 = 90
1 × 10 = 10	2 × 10 = 20	3 × 10 = 30	4 × 10 = 40	5 × 10 = 50	6 × 10 = 60	7 × 10 = 70	8 × 10 = 80	9 × 10 = 90	10 × 10 = 100
1 z 11 = 11	2 × 11 = 22	3 × 11 = 33	4 × 11 = 44	5 × 11 = 55	6 × 11 = 66	7 × 11 = 77	8 × 11 = 88	9 × 11 = 99	10 × 11 = 110
1 × 12 = 12	2 × 12 = 24	3 × 12 = 36	4 × 12 = 48	5 × 12 = 60	6 × 12 = 72	7 × 12 = 84	8 × 12 = 95	9 × 12 = 108	10 × 12 = 120

Grid 2

1×	2×	3×	4×	5×	6×	7×	8×	9×	10×
1 × 1 = 1									
1 × 2 = 2	2 × 2 = 4								
1 × 3 = 3	2 × 3 = 6	3 × 3 = 9							
1 × 4 = 4	2 × 4 = 8	3 × 4 = 12	4 × 4 = 16						
1 × 5 = 5	2 × 5 = 10	3 × 5 = 15	4 × 5 = 20	5 × 5 = 25					
1 × 6 = 6	2 × 6 = 12	3 × 6 = 18	4 × 6 = 24	5 × 6 = 30	6 × 6 = 36				
1 × 7 = 7	2 × 7 = 14	3 × 7 = 21	4 × 7 = 28	5 × 7 = 35	6 × 7 = 42	7 × 7 = 49			
1 × 8 = 8	2 × 8 = 16	3 × 8 = 24	4 × 8 = 32	5 × 8 = 40	6 × 8 = 48	7 × 8 = 56	8 × 8 = 64		
1 × 9 = 9	2 × 9 = 18	3 × 9 = 27	4 × 9 = 36	5 × 9 = 45	6 × 9 = 54	7 × 9 = 63	8 × 9 = 72	9 × 9 = 72	
1 × 10 = 10	2 × 10 = 20	3 × 10 = 30	4 × 10 = 40	5 × 10 = 50	6 × 10 = 60	7 × 10 = 70	8 × 10 = 80	9 × 10 = 90	10 × 10 = 100
1 z 11 = 11	2 × 11 = 22	3 × 11 = 33	4 × 11 = 44	5 × 11 = 55	6 × 11 = 66	7 × 11 = 77	8 × 11 = 88	9 × 11 = 99	10 × 11 = 110
1 × 12 = 12	2 × 12 = 24	3 × 12 = 36	4 × 12 = 48	5 × 12 = 60	6 × 12 = 72	7 × 12 = 84	8 × 12 = 95	9 × 12 = 108	10 × 12 = 120

Resource 3.2.6b

Growing multiplication

1	2	3	4	5	6	7	8	9	10
2	4	6	8	10	12	14	16	18	20
3	6	9	12	15	18	21	24	27	30
4	8	12	16	20	24	28	32	36	40
5	10	15	20	25	30	35	40	45	50
6	12	18	24	30	36	42	48	54	60
7	14	21	28	35	42	49	56	63	70
8	16	24	32	40	48	56	64	72	80
9	18	27	36	45	54	63	72	81	90
10	20	30	40	50	60	70	80	90	100

Resource 3.2.6c

Multiplication grid puzzle

×	3		5		7
	9			30	21
8		32			
	27		45		63
2		8		20	
		16			28
6		24		60	

Explain to a partner how you did it.

Can you create your own for a partner to solve?

Multiplication statements

	I use my 2 times table to help me with my fours.
	I use my 10 times table to help me with my fives.
	I use my 4 times table to help me with my eights.
	I use my 5 times table to help me with my sixes.
	I use my 10 times table to help me with my nines.

Representations

1

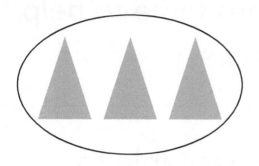

2

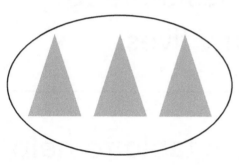

3

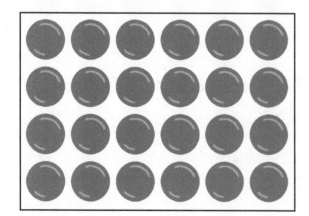

4

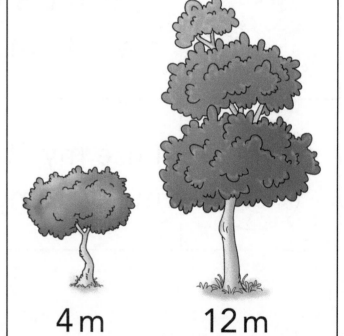

4 m   12 m

Resource 3.2.6f

Match up sentences

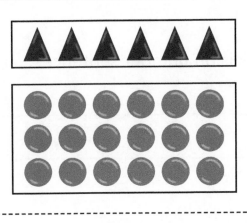

Dividing ☐ into ☐ equal groups.
Each group has ☐.

There are ☐ groups of ☐ in ☐.

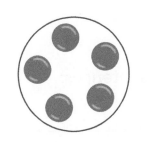 

☐ is ☐ times ☐.

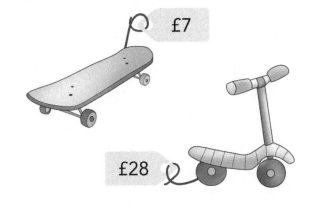

£7

£28

☐ times ☐ is ☐.

Missing numbers

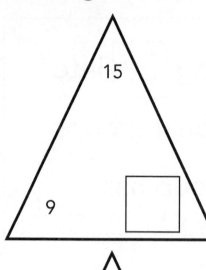

Multiplication and division number sentences:

_____ _____

_____ _____

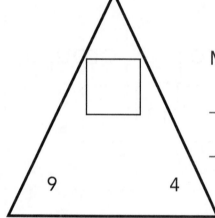

Multiplication and division number sentences:

_____ _____

_____ _____

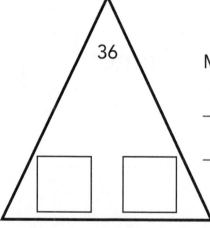

Multiplication and division number sentences:

_____ _____

_____ _____

Is there another way? And another?

Match up problems

There were 4 plates of cookies. Each plate had 10 cookies on it. How many cookies were there altogether?

$42 \div 6 = \boxed{}$

42 cookies were shared equally between 6 children. How many cookies did each child get?

$4 \times 10 = \boxed{}$

Muffins take 3 times as long to cook in the oven as cookies. Cookies take 12 minutes. How long do the muffins take?

$\boxed{} = 3 \times 12$

Resource 3.2.7b

What's the problem?

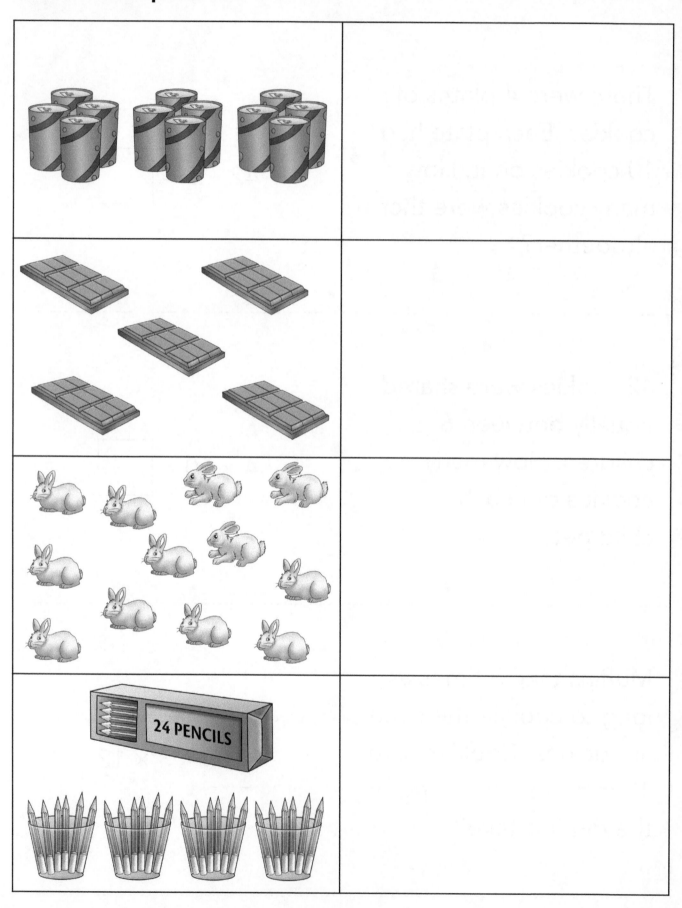

Resource 3.2.8a

Statement pairs

There are 5 cars.	There are 7 lamp-posts on each street.
There are 90 windows altogether.	Each house has 10 windows.
There are 4 streets.	There are 9 houses.
The cars each have 4 wheels.	There are 49 chairs.
There are 7 chairs around each table.	There are 3 pots of flowers outside each house.

Resource 3.2.8b

Linked facts

There is a lot of sports equipment on the school field.	There are 48 bibs of assorted colours.	
There are 10 bags of large balls, each bag contains 5 balls.	There are 56 small balls shared equally into 8 bags.	
There are 3 tubs of hockey sticks. Each tub contains 12 hockey sticks.	There are 6 empty bib bags.	

Resource 3.2.9

Multiplication and division

$13 = 2 \times \square + \square$

$13 = 3 \times \square + \square$

$13 = 4 \times \square + \square$

$13 = 5 \times \square + \square$

$13 = 6 \times \square + \square$

$13 = 7 \times \square + \square$

$13 = 8 \times \square + \square$

$13 = 9 \times \square + \square$

$13 = 10 \times \square + \square$

Cauliflowers

Coffee shop problems

A coffee shop stacks cups in towers of 3.

There are 28 cups altogether.

How many towers of 3 cups are there?

Another coffee shop stacks cups in towers of 4.

They also have 30 cups altogether.

How many cups are not in a tower of 4?

Cakes are arranged on plates of 5.

There are 27 cakes altogether.

How many plates of 5 cakes are there?

Cakes are arranged on plates of 5.

There are 23 cakes altogether.

How many cakes are not on a plate of 5?

Bite-size cookies are sold in bags of 6.

40 cookies are baked on a tray.

How many bags of 6 cookies can be made?

Bite-size cookies are sold in bags of 6.

40 cookies are baked on a tray.

How many cookies are not in a bag of 6?

How many more cookies are needed to make another bag of 6?

Coin arrays

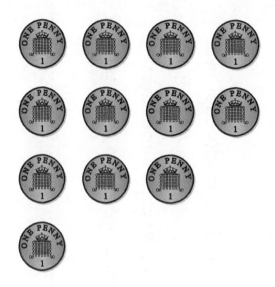

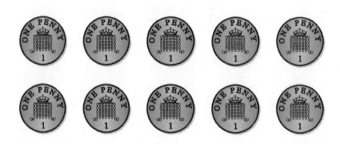

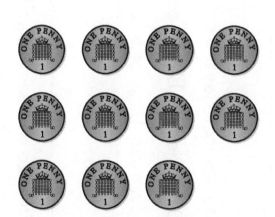

Arrays

Image 1

Image 2

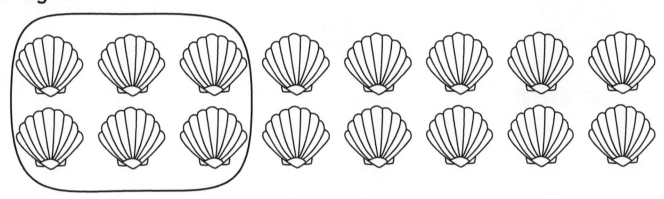

Image 3

Image 4

Division match up

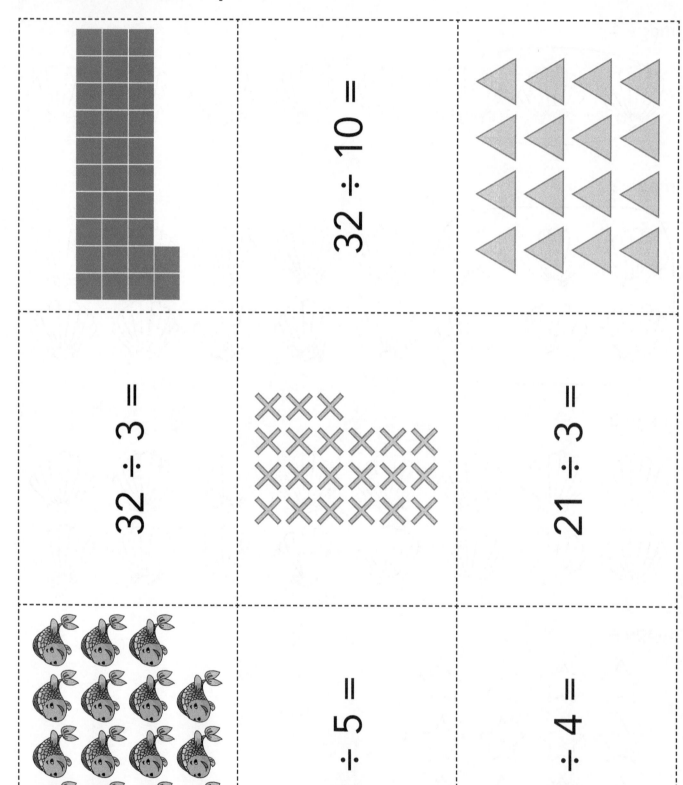

Plants and flower beds

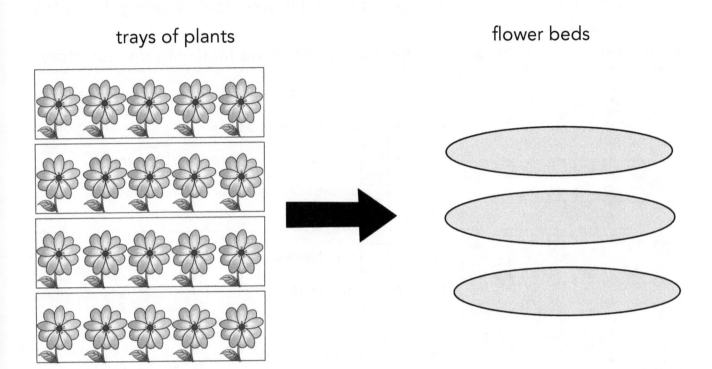

trays of plants

flower beds

Resource 3.2.11b

Strawberry arrays

Cut out the cards. Match the arrays to the equivalent division story. Fill in any missing numbers.

One array does not have a matching story. Write the matching division story.

	There are 31 strawberries. There are ☐ in each group. There are ☐ groups. There are ☐ strawberries left over.	There are 23 strawberries. There are ☐ in each group. There are ☐ groups. There are 2 strawberries left over.
	There are ☐ strawberries. There are 4 in each group. There are ☐ groups. There are ☐ strawberries left over.	
	There are ☐ strawberries. There are ☐ in each group. There are 4 groups. There are ☐ strawberries left over.	
	There are ☐ strawberries. There are ☐ in each group. There are 3 groups. There are ☐ strawberries left over.	
	There are ☐ strawberries. There are ☐ in each group. There are ☐ groups. There are 4 strawberries left over.	There are ☐ strawberries. There are ☐ in each group. There are 5 groups. There are ☐ strawberries left over.

Balances 1

We can see how many cubes are in one pan. How many are in the other?
Tick the correct label.

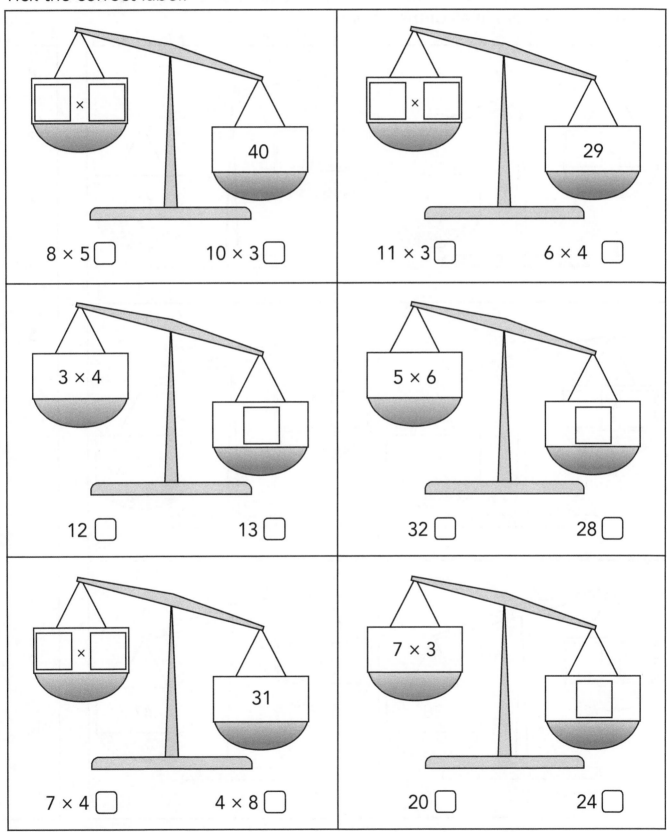

8 × 5 ☐   10 × 3 ☐

11 × 3 ☐   6 × 4 ☐

3 × 4

12 ☐   13 ☐

5 × 6

32 ☐   28 ☐

7 × 4 ☐   4 × 8 ☐

7 × 3

20 ☐   24 ☐

Balances 2

The pan balances have a number of cubes on each pan. We can see how many are on one pan. Use each number from the column on the right once to show how many cubes are in all of the pans.

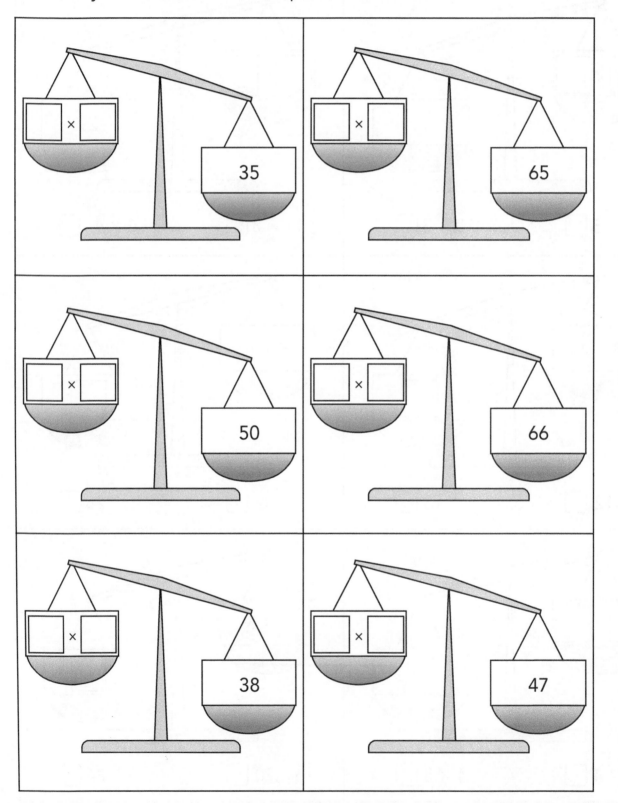

Pencil pots

If 8 pencils are put in each pot, then how many pencils are there in 7 pots?	If 8 pencils are put in each pot, then how many full pots can be made from 50 pencils? How many pencils are left over?	If 7 pencils are put in each pot, then how many full pots can be made from 50 pencils? How many pencils will be left over?
	$50 \div 7 = 8 \text{ r } \square$	$8 \times 7 = \square$

Match up word problems

Write a word problem to match each picture.

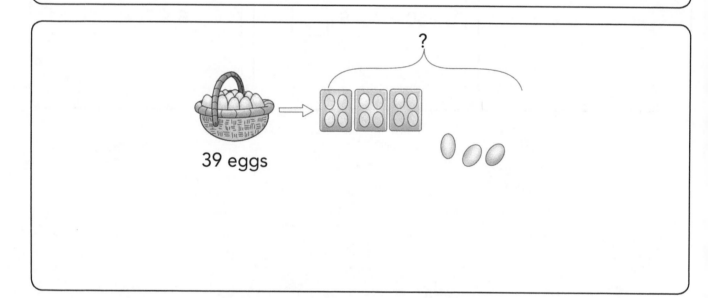

Multiples of 6

Missing information

1

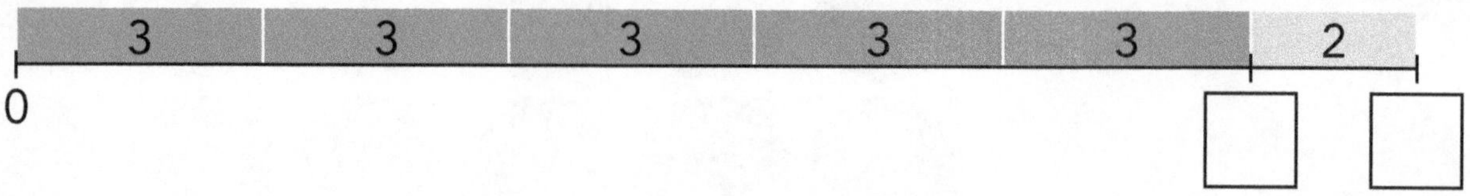

| 3 | 3 | 3 | 3 | 3 | 2 |

0

$17 ÷ 3 = \boxed{}$ r2

2

| 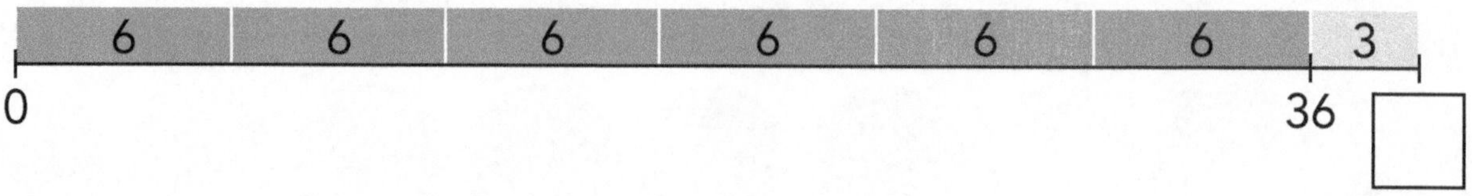 6 | 6 | 6 | 6 | 6 | 6 | 3 |

0   36

$\boxed{} ÷ 6 = 6$ r3

3

| 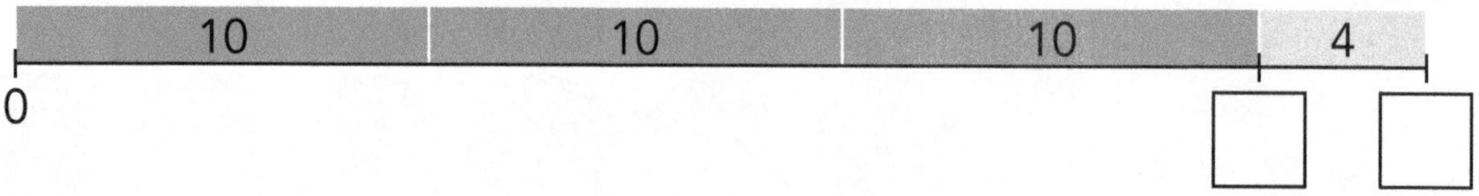 10 | 10 | 10 | 4 |

0

$\boxed{} ÷ 10 = 3$ r4

4. Draw your own rods to show $27 ÷ 4 = 6$ r3

Resource 3.2.12c

Number sentences

Use each number only once to make all of these number sentences and problems correct.

| 3 | 3 | 3 | 4 | 5 | 6 | 7 | 8 | 8 | 9 | 12 | 43 | 44 | 38 | 29 | 59 |

1

$7 \times \boxed{} + 3 = \boxed{}$

2

$36 \div \boxed{} = \boxed{} \ r4$

3

$27 \div \boxed{} = \boxed{} \ r3$

4

5 oranges are put in each bag.

$\boxed{}$ oranges are put in 8 bags with 3 oranges left over.

5

$6 \times \boxed{} + 5 = 41$

6

$\boxed{} \div 5 = \boxed{} \ r4$

7

38 stickers were shared between $\boxed{}$ children equally.

They each get $\boxed{}$ stickers.

There are 2 stickers left over.

8

$\boxed{} \div 8 = \boxed{} \ r3$

9

$\boxed{}$ biscuits are put on a plate.

10 people have two biscuits each.

There are $\boxed{}$ biscuits left over.

Spot arrays

Word problems

Complete the following word problems.
Use the pictures each time to help you.

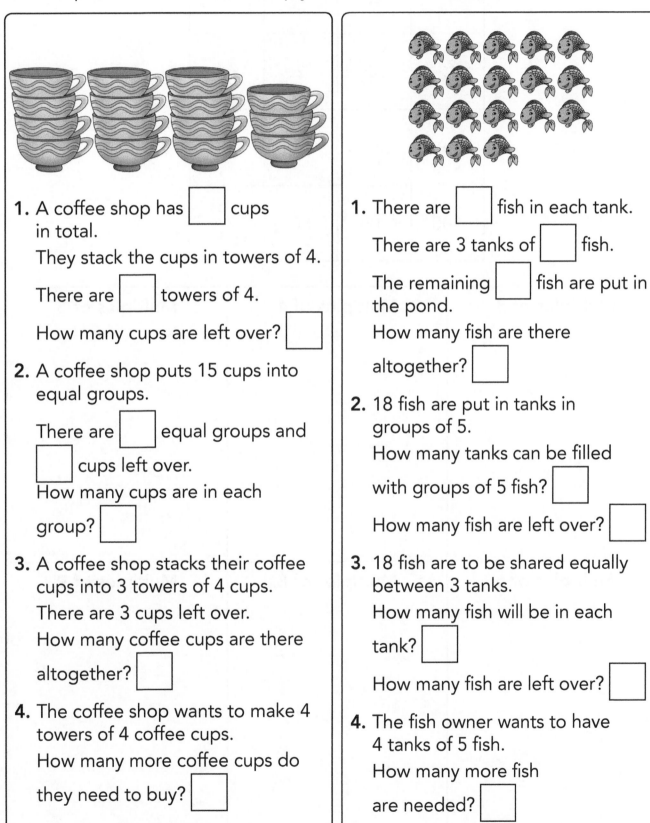

1. A coffee shop has ☐ cups in total.

   They stack the cups in towers of 4.

   There are ☐ towers of 4.

   How many cups are left over? ☐

2. A coffee shop puts 15 cups into equal groups.

   There are ☐ equal groups and ☐ cups left over.

   How many cups are in each group? ☐

3. A coffee shop stacks their coffee cups into 3 towers of 4 cups.
   There are 3 cups left over.
   How many coffee cups are there

   altogether? ☐

4. The coffee shop wants to make 4 towers of 4 coffee cups.
   How many more coffee cups do

   they need to buy? ☐

1. There are ☐ fish in each tank.

   There are 3 tanks of ☐ fish.

   The remaining ☐ fish are put in the pond.
   How many fish are there

   altogether? ☐

2. 18 fish are put in tanks in groups of 5.
   How many tanks can be filled

   with groups of 5 fish? ☐

   How many fish are left over? ☐

3. 18 fish are to be shared equally between 3 tanks.
   How many fish will be in each

   tank? ☐

   How many fish are left over? ☐

4. The fish owner wants to have 4 tanks of 5 fish.
   How many more fish

   are needed? ☐

Number grid and table

41	25	36	30
12	32	?	63
48	40	24	45
100	18	80	27

Multiples of 3	Multiples of 4	Multiples of 5
Multiples of 6	Multiples of 8	Multiples of 9

Remainders

Shade in the white squares to show the smallest and the largest remainder each time. Write the matching division sentences.

1. Smallest remainder is ☐

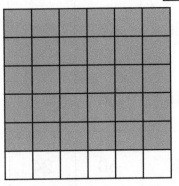

Matching division sentence

_____

Largest remainder is ☐

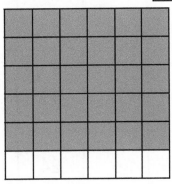

Matching division sentence

_____

2. Smallest remainder is ☐

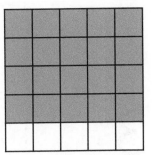

Matching division sentence

_____

Largest remainder is ☐

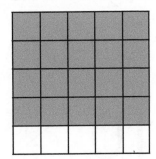

Matching division sentence

_____

3. Smallest remainder is ☐

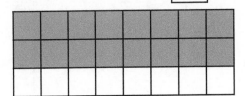

Matching division sentence

_____

Largest remainder is ☐

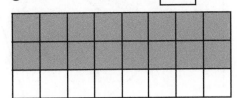

Matching division sentence

_____

Grids

Grid 1

9	12	20	8
18	30	14	10
16	5	28	15
25	40	6	24

Grid 2

39	42	64	28
81	30	48	50
36	45	56	27
25	40	60	24

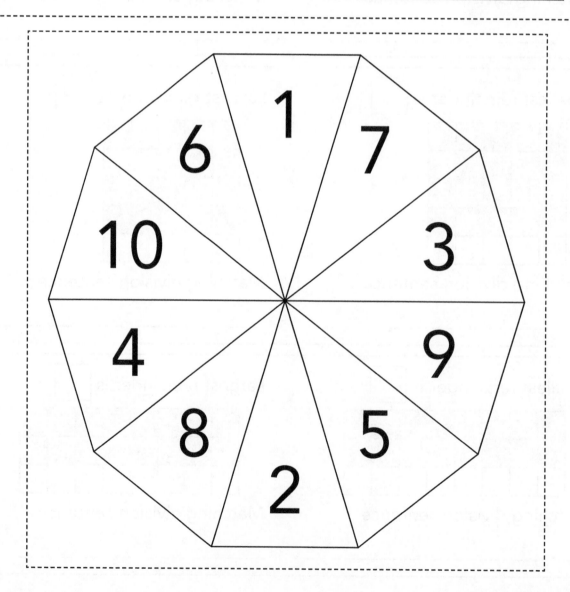

Food masses

Here are some items of food with the mass in grams shown on them.

Find and write the mass of each item in the table.

Biscuits	Beans	Honey	Tuna	Coffee

Don't forget to write the 'g' for grams!
Look at the mass and answer these questions.

1. Biscuits: What is the tens digit? _____

2. Beans: How much more than 400 g does the tin weigh? _____

3. Honey: What does the zero in the ones column mean? _____

4. Tuna: What is the hundreds digit? _____

5. Coffee: How many lots of 10 g are there in the jar of coffee? _____

6. What is the difference in mass between the coffee and the biscuits?

_____

7. How much do two tins of beans weigh?_____

Make up and answer three questions of your own. Try them out on a friend.

8. _____

9. _____

10 _____

Look at some real items of food to find out what they weigh.

Sequences

For each sequence fill in the missing number, add the next two numbers in the sequence and describe the pattern of the numbers. The first one has been done for you.

1. 458, 459, 460 , 461, 462 , 463, 464 , 465

The sequence is going up in ones.

2. 346, [ ], 348, [ ], 350, 351, [ ], [ ]

_____

3. [ ], 999, 998, [ ], 996, 995, [ ], [ ]

_____

4. 90, 95, [ ], [ ], 110, 115, [ ], [ ]

_____

5. [ ], 900, 800, [ ], 600, [ ], [ ]

_____

6. 670, 680, 690, [ ], [ ], [ ], [ ]

_____

Sequencing errors

There is a mistake in each of these sequences. Sometimes a number is missing and sometimes a number is incorrect.

Use an arrow to show where the missing number should be or circle the incorrect number. Then write the sequence out correctly and explain the error. The first one has been done for you.

1. 160, ↑ 180, 190, 200, 210

   <u>160, 170, 180, 190, 200, 210</u>   <u>170 is missing.</u>

2. 232, 234, 235, 236, 237, 238

   _____

3. 380, 390, 400, 401, 420, 430

   _____

4. 820, 800, 780, 760, 740, 730

   _____

5. 150, 250, 350, 450, 650, 750

   _____

6. 615, 610, 605, 600, 590, 585

   _____

Number questions

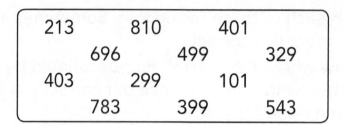

213		810		401	
	696		499		329
403		299		101	
	783		399		543

Fill in the table using the numbers from the box.
Some numbers can be used more than once.

Numbers with 3 in the ones position
Numbers greater than 600
Numbers with an even digit in the tens place
Numbers less than 300
Numbers that come immediately before and after 400
Numbers with an odd digit in the hundreds place
Numbers between 400 and 500

Which number appears four times? _____

Number sequences

Start each number sequence with the number 290 and write the next five numbers in the pattern.

1. Increasing in ones

290, _____

2. Increasing in tens

290, _____

3. Decreasing in fives

290, _____

4. Increasing in twenties

290, _____

5. Increasing in hundreds

290, _____

6. Decreasing in ones

290, _____

Number patterns

Look at the number patterns, write the next three numbers and explain the rule. The first rule has been done for you.

1. 340, 345, 350, 355, ┃360┃, ☐, ☐.

The rule is _add 5._

2. 840, 830, 820, 810, ☐, ☐, ☐.

The rule is _____

3. 508, 506, 504, 502, ☐, ☐, ☐.

The rule is _____

4. 795, 815, 835, 855, ☐, ☐, ☐.

The rule is _____

5. 951, 851, 751, 651, ☐, ☐, ☐.

The rule is _____

6. 682, 684, 686, 688, ☐, ☐, ☐.

The rule is _____

Partitioned numbers

Use place value arrow cards to make each number.
Write the number and the partitioned number in a number sentence.

1. 435 = <u>400 + 30 + 5</u>

2. 609 = _____

3. 216 = _____

4. 398 = _____

5. 723 = _____

6. 453 = _____

7. 192 = _____

8. 540 = _____

9. 389 = _____

10 Order the numbers, from greatest to least.

_____

Making 3-digit numbers

Use place value arrow cards to make each of the following numbers.
Write each number and the partitioned number in a number sentence.

435 = <u>400 + 30 + 5</u>

609 = _____

216 = _____

723 = _____

192 = _____

540 = _____

389 = _____

Order the numbers, from largest to smallest.

_____

Finding 3-digit numbers

Use three of the four cards to answer the following questions.
Circle TRUE or FALSE.

1. The largest possible even 3-digit number is 978.   TRUE   FALSE

2. The smallest possible even 3-digit number is 362.   TRUE   FALSE

3. The largest possible 3-digit number is 987.   TRUE   FALSE

4. The largest possible even 3-digit number is 286.   TRUE   FALSE

5. The smallest possible odd 3-digit number is 263.   TRUE   FALSE

6. The smallest possible 3-digit number is 236.   TRUE   FALSE

7. The largest odd 3-digit number is 879.   TRUE   FALSE

8. The smallest odd 3-digit number is 379.   TRUE   FALSE

9. Only four answers are correct. Write the correct 3-digit numbers for the ones that are incorrect.

_____   _____

_____   _____

Resource 3.3.5

Hundreds, tens and ones

Write where the extra dot needs to be placed to make the new number. The first one has been done for you.

1.

Hundreds	Tens	Ones
●●●●●	●	●

The new number is 521.

The extra dot needs to be placed in the _tens_ column.

2.

Hundreds	Tens	Ones
●●●●●●	●●●●●	●●●

The new number is 654.

The extra dot needs to be placed in the _____ column.

3.

Hundreds	Tens	Ones
●●●	●●●●●●	●●●●

The new number is 374.

The extra dot needs to be placed in the _____ column.

4.

Hundreds	Tens	Ones
●●	●●●●●●	●●●●●●●●

The new number is 367.

The extra dot needs to be placed in the _____ column.

5.

Hundreds	Tens	Ones
●●●●●●●●	●●●●	●●●●●●●

The new number is 857.

The extra dot needs to be placed in the _____ column.

6.

Hundreds	Tens	Ones
●●●●●●●●	●●●●	●●●●●●●

The new number is 848.

The extra dot needs to be placed in the _____ column.

7. Make up and answer a question of your own.

Hundreds	Tens	Ones

The new number is _____.

The extra dot needs to be placed in the _____ column.

3-digit numbers

The digit sum for every 3-digit number in this question is 6.
Find a number and draw a place value chart to fit the following criteria:

1. Has only even digits _____

Hundreds	Tens	Ones

2. Has two digits the same _____

Hundreds	Tens	Ones

3. Has three digits the same _____

Hundreds	Tens	Ones

4. Has 0 in the tens place _____

Hundreds	Tens	Ones

5. Has 0 in the ones place _____

Hundreds	Tens	Ones

6. Has 5 in the hundreds place _____

Hundreds	Tens	Ones

The target is 500!

The challenge is to remove two dots from the place value diagram to make the number 500 or, when this is not possible, the number closest to 500. Draw the dots and write the new number. The first one has been done for you.

1.

Hundreds	Tens	Ones
• • • • • • •		•

Hundreds	Tens	Ones
• • • • •		•

The new number is <u>501</u>

2.

Hundreds	Tens	Ones
• • • • •	•	• •

Hundreds	Tens	Ones

The new number is _____

3.

Hundreds	Tens	Ones
• • • • • • •	•	•

Hundreds	Tens	Ones

The new number is _____

4.

Hundreds	Tens	Ones
• • • • •	• • • • • • • • •	• • • • • • • •

Hundreds	Tens	Ones

The new number is _____

5.

Hundreds	Tens	Ones
• • • • • •	• •	• • •

Hundreds	Tens	Ones

The new number is _____

6.

Hundreds	Tens	Ones
• • • • •	• • • • • • • • •	

Hundreds	Tens	Ones

The new number is _____

Start with 400

Start with 400 and then follow the instructions. Calculate the numbers and partition them.
The first one has been done for you.

1. Add the number of letters in the alphabet. $\underline{400 + 26 = 426}$

$\underline{426 = 400 + 20 + 6}$

2. Add your age.

$\underline{400 +}\underline{=}$

3. Subtract your age.

4. Add the number of pupils in your class.

5. Subtract the number of letters in your first and last names.

6. Make up a question of your own.

Toast!

Pupils collected some data about their favourite spread to put on toast. Fill in the missing information in the table, pictogram and block diagram. Work out the number of children in Year 3.

Table   Year 3's favourite spreads for toast

Type of spread	honey	peanut butter	jam	chocolate spread	marmalade
Number of children		18			3

Pictogram   Year 3's favourite spreads for toast

Key: ▆ represents 2 children (▪ is 1 child)

Block diagram   Year 3's favourite spreads for toast

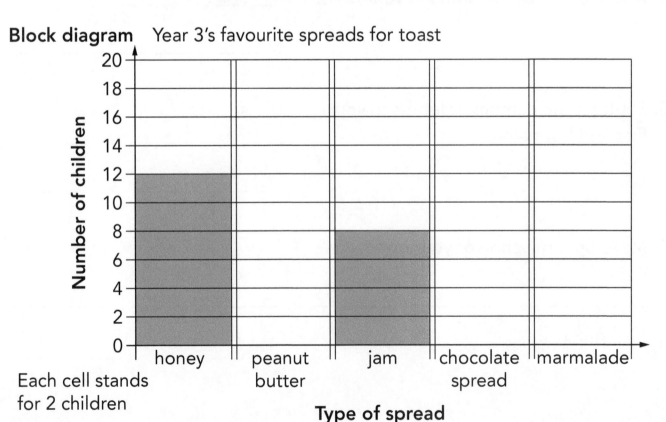

Each cell stands for 2 children

Favourite subjects

Pupils in Class 3 voted on their favourite school subject.

Table

Favourite subject	Art	Literacy	Maths	PE	Science
Number of pupils	8	5	5	4	6

Use the information in the table to complete the bars in the bar chart.

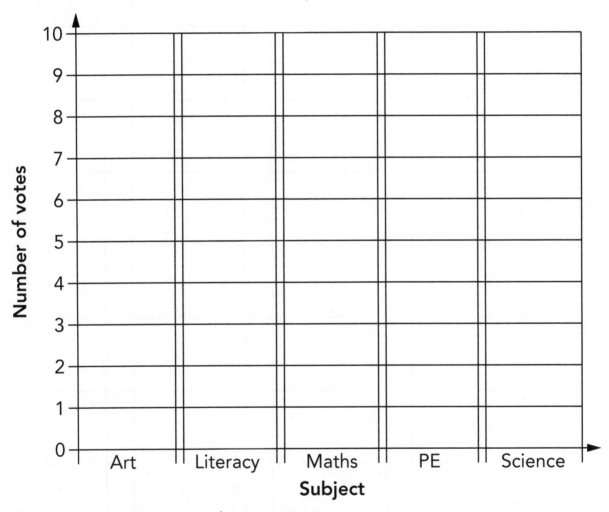

Use the bar chart to answer the questions.
- Which subject has the most votes?
- Which subject has the least number of votes?
- What is the difference between the most popular and least popular subject?
- Which two subjects have the same number of votes?
- How many pupils chose Science as their favourite subject?
- How many pupils are in Class 3?

Blank bar chart

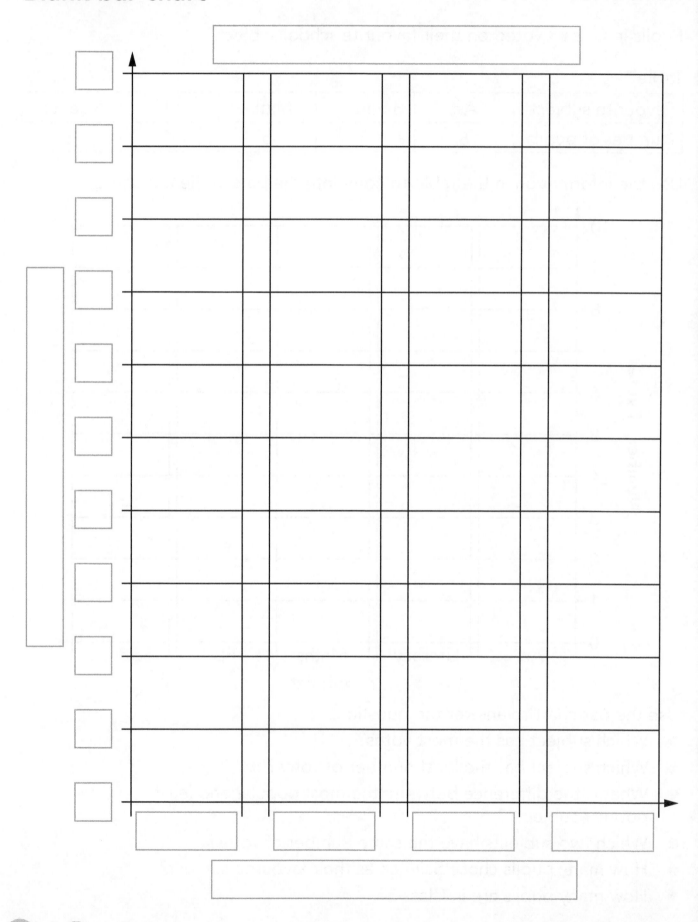

Colour charts

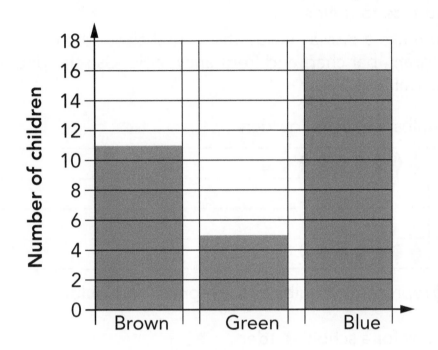

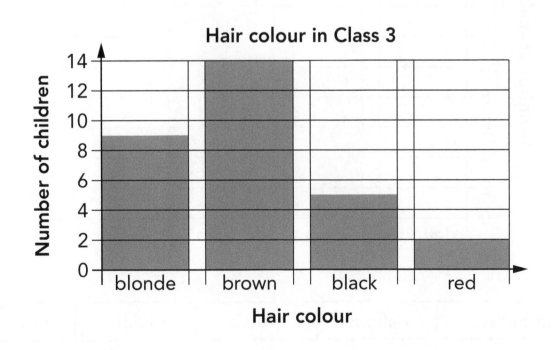

Spring flowerbed

The school council has been given a selection of flower bulbs to make a spring flowerbed at the entrance to their school.

The children want to make a display inside the school about the bulbs and start to draw a pictogram, bar chart and frequency table. Use the data to complete each statistical tool.

Pictogram: Flower bulbs for school garden

daffodil	◆◆◆◆◆◆◆◆▼
crocus	
tulip	
hyacinth	◆◆◆◆◆◆◆▼

Key: ◆ represents 2 bulbs and ▼ represents 1 bulb

Bar chart: Flower bulbs for a school garden

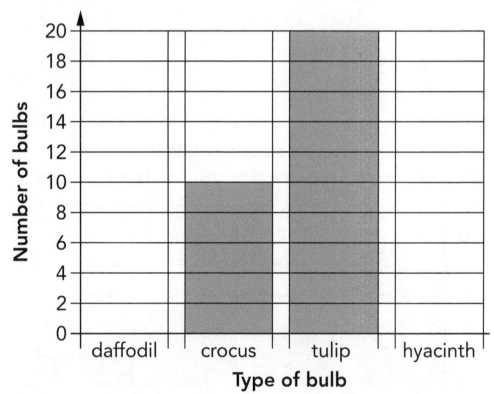

Table

Type of bulb	daffodil	crocus	tulip	hyacinth
Number of bulbs				

Out of the frying pan

Look at the picture and write a number sentence to represent it.

_____

Multiplication tables

$2 \times 1 = 2$	$4 \times 1 = 4$	$6 \times 1 = 6$
$2 \times 2 = 4$	$4 \times 2 = 8$	$6 \times 2 = 12$
$2 \times 3 = 6$	$4 \times 3 = 12$	$6 \times 3 = 18$
$2 \times 4 = 8$	$4 \times 4 = 16$	$6 \times 4 = 24$
$2 \times 5 = 10$	$4 \times 5 = 20$	$6 \times 5 = 30$
$2 \times 6 = 12$	$4 \times 6 = 24$	$6 \times 6 = 36$
$2 \times 7 = 14$	$4 \times 7 = 28$	$6 \times 7 = 42$
$2 \times 8 = 16$	$4 \times 8 = 32$	$6 \times 8 = 48$
$2 \times 9 = 18$	$4 \times 9 = 36$	$6 \times 9 = 54$
$2 \times 10 = 20$	$4 \times 10 = 40$	$6 \times 10 = 60$

What's the same? What's different?

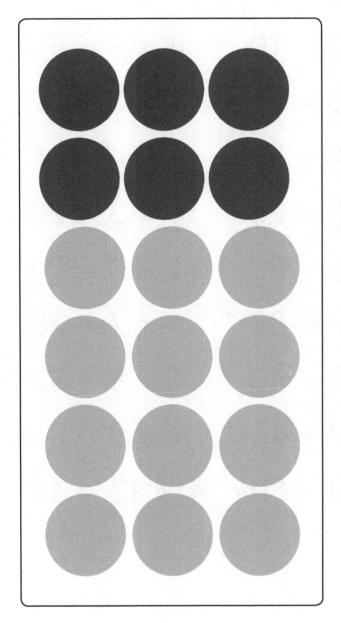

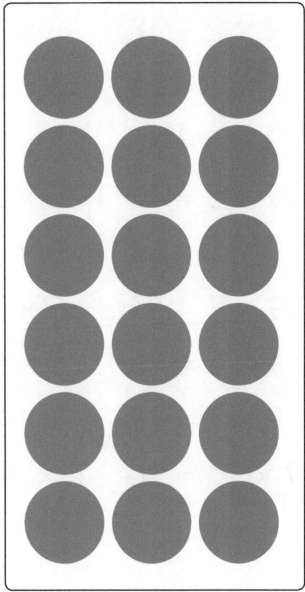

More multiplication tables

9 × 1 = 9	5 × 1 = 5	4 × 1 = 4
9 × 2 = 18	5 × 2 = 10	4 × 2 = 8
9 × 3 = 27	5 × 3 = 15	4 × 3 = 12
9 × 4 = 36	5 × 4 = 20	4 × 4 = 16
9 × 5 = 45	5 × 5 = 25	4 × 5 = 20
9 × 6 = 54	5 × 6 = 30	4 × 6 = 24
9 × 7 = 63	5 × 7 = 35	4 × 7 = 28
9 × 8 = 72	5 × 8 = 40	4 × 8 = 32
9 × 9 = 81	5 × 9 = 45	4 × 9 = 36
9 × 10 = 90	5 × 10 = 50	4 × 10 = 40

Resource 3.6.2b

Multiplication and subtraction puzzles

Cut out the triangles. Can you put the matching pieces back together?

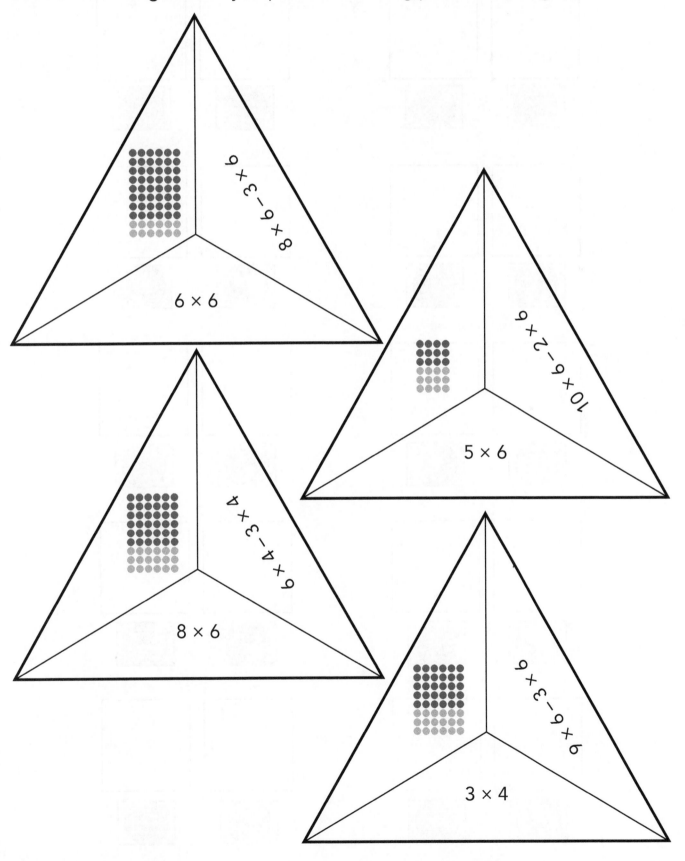

Classroom

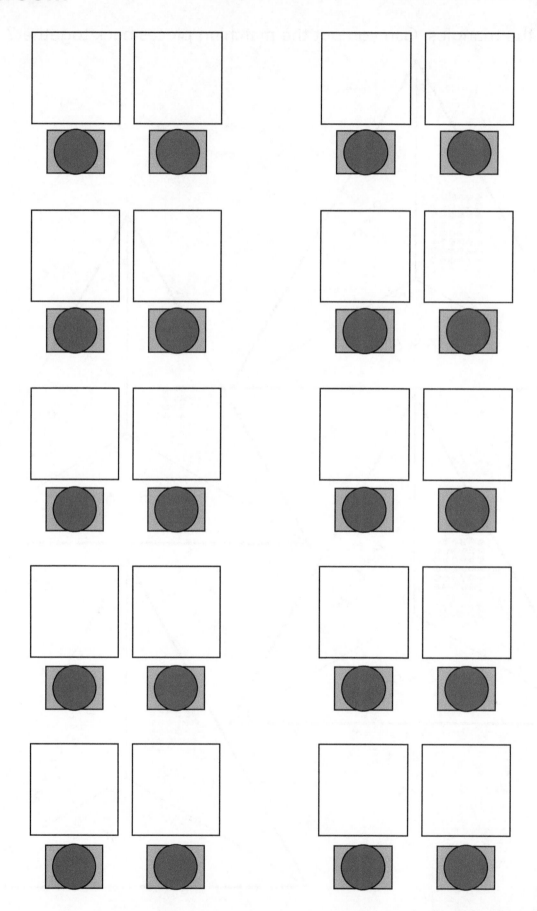

Books and shelves

Cut out the cards. Can you link each fact with the correct question and number sentence?

There are 5 shelves on a bookshelf. Each shelf contains 10 books.	How many books are there on each shelf?	$5 \times 10 = \square$
There are 50 books altogether. There are 5 shelves.	How many books are there altogether?	$50 \div 9 = \square$
There are 50 books altogether. There are 9 shelves.	How many books are on each shelf?	$50 \div 5 = \square$

Posing a question

Look at the image below. Can you think of some linked facts to pose a multiplication and division question?

Multiplication question:

Number sentence: _____

Answer: _____

Division question:

Number sentence: _____

Answer: _____

Resource 3.6.4a

Investigate 12

Here is a diagram of 12 double-sided counters arranged to show 7 and 5.

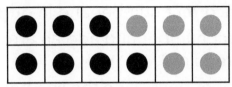

$$7 + 5 = 12 \qquad O + O = E$$

Use the diagrams below to record as many different ways as possible to split 12 into two parts. Write the number sentence below, together with a sentence using O and E, where O represents an odd number and E, an even one.

1.

_____

2.

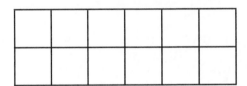

_____

3.

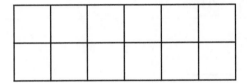

_____

4.

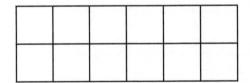

_____

5.

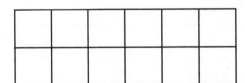

_____

© HarperCollins*Publishers* 2017

Odd and even number patterns

- The following odd and even number patterns have an error in them.
- In some there is an incorrect number and in others a number is missing.
- Cross out the incorrect number and add an arrow to show where a missing number is required.
- Write the correct number pattern on the line below and add the next three numbers. The first one has been done for you.

1. 1, 3, 5, 7, 8, 11, 13, 15, 19, _____, _____, _____

1, 3, 5, 7, 9, 11, 13, 15, 19, 21, 23, 25 _____

2. 50, 48, 46, 42, 40, 38, 36, _____, _____, _____

_____

3. 31, 29, 26, 25, 23, 21, 19, _____, _____, _____

_____

4. 102, 104, 106, 108, 110, 121, 114, 116, _____, _____, _____

_____

5. 467, 465, 463, 461, 459, 457, 455, 451, _____, _____, _____

_____

6. 220, 222, 224, 226, 228, 232, 234, 236, _____, _____, _____

_____

True, false or sometimes true

Decide whether the statements are true, false or sometimes true and underline your answer.

Give three examples as evidence.

1. When you double a number, the answer is always even.

True   False   Sometimes true

_____

_____

_____

2. When you halve an even number, the answer is always even.

True   False   Sometimes true

_____

_____

_____

3. If you count on 2 from an odd number, the answer is even.

True   False   Sometimes true

_____

_____

_____

4. The number that comes after an even number is always odd.

True   False   Sometimes true

_____

_____

_____

5. An odd number added to odd number gives an odd number.

True   False   Sometimes true

_____

_____

_____

Magic square

1. Complete this 4 × 4 magic square, which has a magic number of 34.

7		1	14
2			
			10
	6	15	4

2. What do you notice about the four numbers shaded in the top left-hand corner?

_____

3. Try to find more sets of four numbers that total 34.

_____

4. Mathematicians have known about magic squares for hundreds of years. Ask your teacher if you can research magic squares on the internet.
This is the Lo Shu magic square. Look at it carefully and describe it.

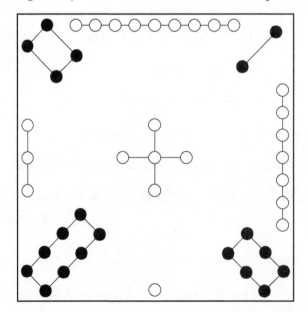

Resource 3.6.6a

Number pairs 1

- Cut out and shuffle the cards.
- Place them face down. Take turns to turn over two cards. If they are a matching pair, keep them and have another turn. If they are not a pair, turn them back over, keeping them in the same position.
- Continue playing until all the cards have been collected. The winner has the greater number of cards.

3000	1426	7089	2954
5600	3098	5309	4444
three thousand	one thousand four hundred and twenty-six	seven thousand and eighty-nine	two thousand nine hundred and fifty-four
five thousand six hundred	three thousand and ninety-eight	five thousand three hundred and nine	four thousand four hundred and forty-four

Number pairs 2

Resource 3.6.6c

Number questions

1. Write any two numbers that come between the two numbers shown.

    a) 3459, _____, _____, 3467

    b) 4993, _____, _____, 5012

    c) 10 000, _____, _____, 10 020

    d) 999, _____, _____, 1111

2. Write the numbers described:

    a) 6 thousands, 7 hundreds, 1 ten and 8 ones   _____

    b) 4 thousands, 2 hundreds, 3 tens and 5 ones   _____

    c) 9 thousands, 6 hundreds and 3 tens   _____

    d) 3 thousands, 4 tens and 5 ones   _____

3. Decide whether the statements are true or false and circle your answer.

    a) 100 tens is 1000. . . . . True   False

    b) Six thousand and nine is written in numerals as 6090. . . . . . True   False

    c) Counting from the right, the fourth digit is
       the thousands digit. . . . . True   False

    d) There are 24 tens in 2400. . . . . True   False

Number values

1. Circle the number that has the same value as the first number.

a) **200**   2   20 tens   200 tens

b) **4500**   450 hundreds   45 hundreds   45 thousands

c) **560**   56 tens   50 tens   56 hundreds

d) **7100**   710 hundreds   71 thousands   710 tens

e) **12 000**   12 hundreds   120 hundreds   120 thousands

2. Make up three similar questions of your own.

a) _____   _____   _____   _____

b _____   _____   _____   _____

c) _____   _____   _____   _____

Try your questions on a friend.

Blank tables square

×										

Resource 3.6.7b

Number grid

one thousand	seven hundred	and seventy	two
and thirty	two thousand	three	nine hundred
six hundred	six	three thousand	and twenty
eight	and fifty	four hundred	five thousand

4-digit numbers

Look at these numbers.

Use them to make ten different 4-digit numbers and write the number sentence showing how the number is made.
For example, 9000 + 300 + 50 + 2 = 9352.

1. _____

2. _____

3. _____

4. _____

5. _____

6. _____

7. _____

8. _____

9. _____

10. _____

5-digit numbers

Look at these numbers.

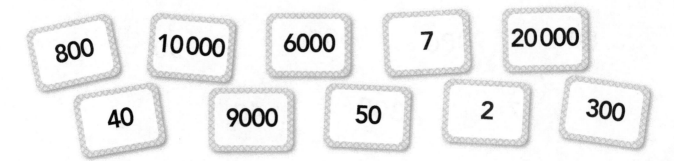

Use them to make ten different 5-digit numbers and write the number sentence showing how the number is made.

For example, 20 000 + 6000 + 800 + 40 + 7 = 26 847.

1. _____

2. _____

3. _____

4. _____

5. _____

6. _____

7. _____

8. _____

9. _____

10. _____

Greater than or less than

Write these numbers in numerals and add > or < to compare the numbers.

1. four thousand six hundred and thirty-four ☐ three thousand nine hundred and twenty

_____

2. six thousand and ninety ☐ six thousand four hundred and twenty-two

_____

3. one thousand nine hundred and forty-eight ☐ one thousand seven hundred and forty-seven

_____

4. three thousand and seventy ☐ four thousand

_____

5. eight thousand eight hundred and eighty-eight ☐ eight thousand nine hundred

_____

Place value

1. Make the number on the left with base 10 blocks or place value counters.
How many tens or hundreds are needed to make each number?
Circle the two correct answers.

a) 4 hundreds are equal to   40 ones   400 tens   40 tens   400 ones

b) 27 tens are equal to   270 tens   270 ones

                                277 ones   2 hundreds and 7 tens

c) 3 thousands are equal to   30 hundreds   30 tens

                                3000 ones   30 thousands

2. Draw the correct number of dots in each box to show the following numbers and write the number in numerals.

a) three thousand two hundred and six

Thousands	Hundreds	Tens	Ones

In numerals _____

b) five thousand and twenty

Thousands	Hundreds	Tens	Ones

In numerals _____

c) six thousand and three

Thousands	Hundreds	Tens	Ones

In numerals _____

Resource 3.6.7g

Making numbers

Use the cards to make the following numbers in numerals and words.

1. Two odd 4-digit numbers

_____   _____

2. Two 4-digit numbers with zeros in the hundreds and ones columns

_____   _____

3. Two 4-digit numbers less than one thousand five hundred

_____   _____

4. Two even 4-digit numbers greater than six thousand

_____   _____

5. Two 4-digit numbers with zeros in the hundreds and ten columns

_____   _____

6. Which two questions have the same numbers as answers?
Can you explain why?

_____

_____

_____

Answers

Chapter 1 Revising and improving

Unit 1.1

1. **(a)** 72 **(b)** 99 **(c)** 68
   (d) 35 **(e)** 50 **(f)** 73
   (g) 27 **(h)** 40 **(i)** 63
   (j) 71 **(k)** 16 **(l)** 18
   (m) 54 **(n)** 40 **(o)** 7
   (p) 100 **(q)** 37 **(r)** 91

2. **(a)** 94 **(b)** 93
   (c) 80 **(d)** 83
   (e) 104 **(f)** 92
   (g) 89 **(h)** 77
   (i) 91 **(j)** 83

3. **(a)** 27 **(b)** 9
   (c) 24 **(d)** 5
   (e) 6 **(f)** 27
   (g) 25 **(h)** 34
   (i) 15 **(j)** 34

4. **(a)** 70 **(b)** 61
   (c) 19 **(d)** 74
   (e) 42 **(f)** 97
   (g) 42 **(h)** 27
   (i) 61 **(j)** 68

5. **(a)** 15 **(b)** 27
   (c) 75 **(d)** 35
   (e) 51 **(f)** 68

6. **(a)** 18 + 25 = £43
   (b) 50 − 46 = £4
   (c) 46 − 18 = £28

7. **(a)** 1 + 2 − 3 + 5 − 4 = 1
   (b) 1 + 2 + 3 − 4 + 5 − 6 = 1

Unit 1.2

1. **(a)** 43 **(d)** 55
   (b) 59 **(e)** 100
   (c) 82 **(f)** 51
   (g) 47 **(j)** 55
   (h) 11 **(k)** 0
   (i) 85 **(l)** 46

2. **(a)** 84 **(b)** 57 **(c)** 87

3. **(a)** 27 + 18 = 45 (fruits) **(b)** 45 − 27 = 18 (pears)
   (c) 25 + 6 = 31 (girls) **(d)** 38 − 29 = 9

4. **(a)** 56 − 19 = 37 (eggs)
   (b) 27 + 14 = 41 (pages)

5. **(a)** 76 − 47 = £29
   (b) 2 (toys) Answers may vary, for example, 47 + 24 = £71, yes 100 − 71 = £29

Unit 1.3

1. **(a)** 22 **(d)** 67
   (b) 94 **(e)** 90
   (c) 81 **(f)** 62
   (g) 66 **(j)** 44
   (h) 18 **(k)** 63
   (i) 46 **(l)** 18

2. **(a)** 81 **(e)** 64 **(i)** 64
   (b) 24 **(f)** 19 **(j)** 74
   (c) 14 **(g)** 39 **(k)** 84
   (d) 17 **(h)** 59 **(l)** 41

3. **(a)** 15 + 10 = 25 (butterflies) **(b)** 38 − 12 = 26 (chicks)
   (c) 45 − 16 = 29 (pears) **(d)** 46 − 17 = 29 (crates)

4. **(a)** 32 + 68 = £100 A scooter and a skateboard
   (b) 24 + 68 = £92
   (c) 54 − 48 = £6

5. M = 7, A = 9, T = 1

Unit 1.4

1. **(a)** 2 **(b)** 10 **(c)** 6
   (d) 9 **(e)** 5 **(f)** 4

2. **(a)** 65 **(b)** 82
   (c) 39 **(d)** 47
   (e) 65 **(f)** 39
   (g) 83 + 2 − 2 40 + 43 = 83
   (h) 24 + 1 + 1 64 − 40 = 24

3. **(a)** 12 52 **(b)** 59 29
   (c) 16 50 66 **(d)** 96 50 46
   (e) 46 30 **(f)** 29 74 44 (answer may vary)
   (g) 23 + 70 = 22 + 71 = 21 + 72 = 20 + 73
   (h) 70 − 32 = 68 − 30 = 78 − 40 = 80 − 42 (answers may vary)

4. lines drawn from 55 + 28 to 53 + 30, 13 + 48 to 10 + 51, 62 − 15 to 60 − 13 and 74 − 36 to 78 − 40

5. 74

Unit 1.5

1. **(a)** 29 **(b)** 38 **(c)** 64
   (d) 29 **(e)** 38 **(f)** 64
   (g) 34 **(h)** 28 **(i)** 68
   (j) 0 **(k)** 100 **(l)** 39
   (m) 30 **(n)** 92 **(o)** 34

2. **(a)** 36 + 18 = 54 18 + 36 = 54 54 − 18 = 36 54 − 36 = 18
   (b) 59 59 59
   (c) 85 85 85
   (d) 29 29 29 29 29

3. **(a)** 72 − 34 = 38 (fish)
   (b) 15 + 36 = 51 (children)
   (c) 21 − 12 = 9 (birds)

4. **(a)** > **(b)** < **(c)** <

Unit 1.6

1. **(a)** Table correctly completed
   (b) Multiplication facts with repeated numbers circled, for example, 2 × 2 = 4, 5 × 5 = 25
   (c) Multiplication facts correctly coloured
   (d) The products are all multiples of 2. (answer may vary)
   (e) Answer may vary

2. 8 × 9 to '8 times 9 is 72' to 72 ÷ 9
   24 ÷ 8 to '3 times 8 is 24' to 3 × 8
   60 ÷ 10 to '6 times 10 is 60' to 60 ÷ 6
   6 × 10 to '6 times 10 is 60' to 60 ÷ 6

3. Answers may vary

4. There are several ways, for example:
   (a) 2 × 5 = 1 × 10
   (b) 4 × 2 = 2 × 4 = 1 × 8
   (c) 6 × 3 = 3 × 6 = 2 × 9, and so on.

Unit 1.7

1. **(a)** There are 3 times as many ◇ as ☆.
   9 ÷ 3 = 3
   (b) 5 2 10 10 ÷ 2 = 5
   (c) 2 × 6 + 5 = 17 5 × 3 + 2 = 17
   3 × 6 − 1 = 17 6 × 3 − 1 = 17

2. **(a)** 24 ÷ 6 = 4 (days)
   (b) 24 ÷ 3 = 8 (days)
   (c) 24 ÷ 7 = 3 (days) r 3

3. **(a)** 2 4 × 2 = 8
   (b) 40 8 × 5 = 40

4. 35 children

Chapter 1 test

1. **(a)** 65 **(b)** 9 **(c)** 10
   (d) 2 **(e)** 25 **(f)** 97
   (g) 48 **(h)** 4 **(i)** 21
   (j) 89 **(k)** 40 **(l)** 7

2 **(a)** 62 **(b)** 6
(c) 55 **(d)** 76
(e) 39 **(f)** 72
(g) 33 **(h)** 31
(i) 68 **(j)** 63

3 **(a)** 17 **(b)** 22
(c) 79 **(d)** 31
(e) 40 **(f)** 0
(g) 10 **(h)** 96
(i) 4 **(j)** 14
(k) 2 **(l)** 18

4 **(a)** 65 − 28 = 37 37
(b) 30 − 16 = 14 14 (pears)
(c) 24 + 38 = 62 62 (sweets)
(d) 45 + 16 = 61 61 (storybooks)
(e) 72 − 37 = 35 35 (lilies)
(f) 40 ÷ 4 = 10 10
(g) 72 ÷ 9 = 8 8 (metres)

5 **(a)**
$$\begin{array}{r} 31 \\ +46 \\ \hline 77 \end{array}$$
(b)
$$\begin{array}{r} 85 \\ -45 \\ \hline 40 \end{array}$$
(c)
$$\begin{array}{r} 24 \\ +39 \\ \hline 63 \end{array}$$
(d)
$$\begin{array}{r} 66 \\ -17 \\ \hline 49 \end{array}$$

Chapter 2 Multiplication and division (II)

Unit 2.1

1 fourteen
twenty-eight
fifty-six
seven
seven
Three
Seven nine (or vice versa)
seven
seventy-seven
Twelve

2 The following linked:
'Three times seven is twenty-one' to
7 × 3 to 21 ÷ 7
'Seven times eight is fifty-six' to 7 × 8
to 56 ÷ 7
'Six times seven is forty-two' to 7 × 6
to 42 ÷ 6
'Five times seven is thirty-five' to 5 × 7
to 35 ÷ 5

3 **(a)** 7 **(b)** 28
(c) 5 **(d)** 7
(e) 0 **(f)** 63
(g) 2 **(h)** 8
(i) 77 **(j)** 84
(k) 12 **(l)** 10
(m) 7 **(n)** 21
(o) 11 **(p)** 70

4 **(a)** 14 ÷ 7 = 2
(b) 3 × 7 = 21
(c) 7 × 5 = 35 (pupils)
(d) 35 + 6 = 41 (pupils)

5 Answers may vary, for example,
6 × 7 = 42 or 14 ÷ 2 = 28 ÷ 4

Unit 2.2

1 **(a)** fifteen 3 × 5 = 15 5 × 3 = 15 15 ÷
3 = 5 15 ÷ 5 = 3
(b) Three eleven 3 × 11 = 33
11 × 3 = 33 33 ÷ 3 = 11 33 ÷ 11 = 3
(c) six 3 × 6 = 18 6 ×3 = 18 18 ÷ 3 = 6
18 ÷ 6 = 3

2 **(a)** 18 **(b)** 24 **(c)** 27
(d) 33 **(e)** 21 **(f)** 36
(g) 10 **(h)** 15 **(i)** 12
(j) 6 **(k)** 1 **(l)** 3
(m) 4 **(n)** 5 **(o)** 33

3 3 21 3 3 36 10 0

4 **(a)** 3 × 3 = 9
(b) 15 ÷ 3 = 5
(c) 3 × 7 = 21

5 **(a)** 3 × 8 = £24 50 − 24 = £26
(b) 6 × 4 = 24 (ducks) 24 + 6 = 30
(ducks)

6 6

7 **(a)** Younger baby monkey: 24 ÷
(2 + 1) = 8 (peaches)
(b) Older baby monkey: 8 × 2 = 16
(peaches)

Unit 2.3

1 **(a)** 6 × 3 = 18 3 × 6 = 18 Three times
six is eighteen.
(b) 6 × 4 = 24 4 × 6 = 24 Four times six
is twenty-four.
(c) 6 × 5 = 30 5 × 6 = 30 Five times six
is thirty.
(d) 6 × 8 = 48 8 × 6 = 48 Six times eight
is forty-eight.

2 **(a)** = **(b)** >
(c) = **(d)** <

3 **(a)** 12 = 2 × 6 = 3 × 4 = 6 × 2
(b) 36 = 4 × 9 = 6 × 6 = 9 × 4
(c) 18 = 2 × 9 = 3 × 6 = 6 × 3 (answers
may vary)

4 **(a)** 30 **(b)** 42
(c) 54 **(d)** 4
(e) 21 **(f)** 72
(g) 0 **(h)** 9
(i) 33 **(j)** 66
(k) 72 **(l)** 6
(m) 60 **(n)** 12
(o) 36 **(p)** 66

5 **(a)** 6 × 8 = 48 (hours) 5 × 8 = 40
(hours),
(b) 48 − 40 = 8 (hours)

6 8

7 25

Unit 2.4

1 **(a)** nine 9 9
(b) twenty-seven 3 × 9 = 27 9 × 3 = 27
(c) thirty-six 4 × 9 = 36 9 × 4 = 36
(d) ninety-nine 11 × 9 = 99 9 × 11 = 99
(e) Six times nine 54 ÷ 9 = 6 54 ÷ 6 = 9
(f) Two times nine 18 ÷ 9 = 2 18 ÷ 2 = 9

2 **(a)** 9 **(b)** 2 **(c)** 2
(d) 4 **(e)** 10 **(f)** 5
(g) 2 **(h)** 6 **(i)** 11

3 3 × 7 = 21 5 × 9 = 45 9 × 10 = 90
7 × 3 = 21 9 × 5 = 45 10 × 9 = 90
21 ÷ 3 = 7 45 ÷ 5 = 9 90 ÷ 9 = 10
21 ÷ 7 = 3 45 ÷ 9 = 5 90 ÷ 10 = 9

4 **(a)** 54 ÷ 6 = 9 (carrots)
(b) 36 ÷ 4 = 9 (chocolate bars)
(c) 9 × 3 = 27 (bananas)
(d) 18 ÷ 9 = 2 (boxes)

5 **(a)** 36 20 **(b)** 63 127

6 **(a)** 18 ÷ (3 − 1) = 9 (seconds)
(b) 9 × (8 − 1) = 63 (seconds)

Unit 2.5

1 **(a)** 6 × 3 = 18 3 × 6 = 18
(b) 3 × 6 = 18 6 × 3 = 18
(c) 2 × 9 = 18 9 × 2 = 18

2 **(a)** 6, 3, 2
(b) 9, 6, 12
(c) 6, 9, 18

3 **(a)** 7 × 6 = 42
(b) 36 ÷ 4 = 9
(c) 24 ÷ 3 = 8 24 ÷ 6 = 4

4 **(a)** 9 × 2 = 18 (tulips)
(b) 18 + 9 = 27 (flowers)
27 ÷ 9 = 3 (bouquets)
(c) 6 × 6 ÷ 3 = 12 (balloons)

Answers

5 6

6 36

Unit 2.6

1 Table correctly completed

2 **(a)** twenty-eight $4 \times 7 = 28$ $7 \times 4 = 28$
   $28 \div 4 = 7$ $28 \div 7 = 4$
   (b) forty $5 \times 8 = 40$ $8 \times 5 = 40$ $40 \div 5 = 8$ $40 \div 8 = 5$
   (c) sixty-six $6 \times 11 = 66$ $11 \times 6 = 66$ $66 \div 6 = 11$ $66 \div 11 = 6$
   (d) four $3 \times 4 = 12$ $4 \times 3 = 12$ $12 \div 3 = 4$ $12 \div 4 = 3$
   (e) six $4 \times 6 = 24$ $6 \times 4 = 24$ $24 \div 4 = 6$ $24 \div 6 = 4$
   (f) Four $4 \times 9 = 36$ $9 \times 4 = 36$ $36 \div 4 = 9$ $36 \div 9 = 4$

3 **(a)** $10 \div 2 = 5$ 10, 2, 5, $10 \div 2 = 5$
   (b) $3 \times 5 = 15$ 5, 3, 15, $15 \div 3 = 5$
   (c) $8 \times 3 = 24$ 24, 8, 3, $24 \div 3 = 8$
   (d) 3 $9 \times 3 = 27$ 9, 3, 27, $27 \div 3 = 9$

4 **(a)** $2 \times 2 - 2 \times 2 = 0$ **(b)** $2 \times 2 \div 2 \div 2 = 1$
   (c) $2 \div 2 + 2 \div 2 = 2$ **(d)** $2 \times 2 - 2 \div 2 = 3$ (answers may vary)

Unit 2.7

1 Answers may vary

2 **(a)** 16 pupils took part in a relay race $16 \div 4 = 4$ (pupils)
   (b) 25 pupils took part in the rope skipping $25 \div 5 = 5$ (classes)
   (c) 42 played football. How many pupils played table tennis? $42 \div 7 = 6$ (pupils)

3 **(a)** Holly was the first runner.
   (b) $100 + 20 = 120$ (metres)

4 $16 \div 2 \div 2 = 4$ (apples)

Unit 2.8

1 Answers may vary

2 **(a)** $10 \times 4 = 40$ (origami cranes)
   (b) $9 \times 5 = 45$ (metres) $45 + 9 = 54$ (metres)
   (c) $10 \times 3 = £30$
   (d) $6 \times 5 = 30$ (birds) $30 - 18 = 12$ (birds)

3 $60 \div (3 - 1) = 30$ (seconds) $30 \times (6 - 1) = 150$ (seconds)

4 $16 \div 2 \div 2 = 4$ (metres)

Unit 2.9

1 **(a)** 18, $3 \times 5 + 3 = 18$, $3 \times 4 + 6 = 18$ (answers may vary)
   (b) 34, $4 \times 8 + 2 = 34$, $8 \times 4 + 2 = 34$ (answers may vary)

2 **(a)** 7   **(b)** 4
   (c) 5   **(d)** 3
   (e) 5   **(f)** 5
   (g) 4, 2   **(h)** 7, 7
   (i) 10, 3   **(j)** 5, 1
   (k) 9, 6   **(l)** 4, 1
   (m) 6, 5   **(n)** 9, 4
   (o) 11, 3   **(p)** 9, 7

3 **(a)** $19 = 2 \times 9 + 1$ **(b)** $32 = 3 \times 10 + 2$
   (c) $19 = 3 \times 6 + 1$ **(d)** $32 = 4 \times 8 + 0$
   (e) $19 = 4 \times 4 + 3$ **(f)** $32 = 5 \times 6 + 2$
   (g) $19 = 5 \times 3 + 4$ **(h)** $32 = 6 \times 5 + 2$
   (i) $19 = 6 \times 3 + 1$ **(j)** $32 = 7 \times 4 + 4$
   (k) $19 = 7 \times 2 + 5$ **(l)** $32 = 8 \times 4 + 0$
   (m) $19 = 8 \times 2 + 3$ **(n)** $32 = 9 \times 3 + 5$
   (o) $19 = 9 \times 2 + 1$ **(p)** $32 = 10 \times 3 + 2$
   (q) $19 = 10 \times 1 + 9$ **(r)** $32 = 11 \times 2 + 10$
   (s) $19 = 11 \times 1 + 8$ **(t)** $32 = 12 \times 2 + 8$
   (answers may vary)

4 Answers may vary

5 36

Unit 2.10

1 **(a)** 7, 1, $7 \times 3 + 1 = 22$
   (b) 4, 2, $4 \times 5 + 2 = 22$

2 6, 3, $27 \div 4 = 6$ r 3

3 **(a)** 2 r 1   **(b)** 18, 3 r 3
   (c) 11, 3, 3 r 2

4 **(a)** 5 r 2   **(b)** 3 r 2
   (c) 6 r 2   **(d)** 7 r 3
   (e) 7 r 1   **(f)** 5 r 7
   (g) 5 r 6   **(h)** 6 r 7
   (i) 9 r 2   **(j)** 6 r 4

5 **(a)** $9 \div 4 = 2$ (bottles) r 1 (bottle)
   (b) $7 \times 6 + 5 = 47$ (peaches)
   (c) $50 \div 6 = 8$ r 2

6 43

Unit 2.11

1 **(a)** 20
   (b) 3, 6
   (c) 2
   (d) $20 \div 3 = 6$ r 2

2 **(a)** 7
   (b) 3
   (c) $45 \div 6 = 7$ r 3

3 **(a)** 5   **(b)** 5   **(c)** 8
   (d) 4   **(e)** 8   **(f)** 9
   (g) 4   **(h)** 10   **(i)** 8

4 **(a)** ✓   **(b)** ✗
   (c) ✓   **(d)** ✓
   (e) ✗   **(f)** ✓

5 **(a)** $31 \div 7 = 4$ (weeks) r 3 (days)
   (b) $50 \div 6 = 8$ (children) r 2 (sweets)
   (c) $49 \div 5 = 9$ (coats) r 4 (buttons)

6 **(a)** $26 \div 4 = 6$ (boats) r 2 (children)
   $6 + 1 = 7$ (boats)
   (b) $26 \div 6 = 4$ (boats) r 2 (children)
   $4 + 1 = 5$ (boats)
   (c) Answers may vary

Unit 2.12

1 **(a)** 4 r 1   **(b)** 9 r 2   **(c)** 5 r 3
   (d) 2 r 5   **(e)** 8 r 5   **(f)** 8 r 6
   (g) 8 r 2   **(h)** 4 r 3   **(i)** 4 r 1
   (j) 3 r 2   **(k)** 9 r 5   **(l)** 10 r 3

2 **(a)** 28   **(b)** 44   **(c)** 46
   (d) 35   **(e)** 23   **(f)** 53

3 **(a)** 2, 3, $15 \div 6 = 2$ r 3
   (b) $8 \times 6 + 4 = 52$
   (c) 4, 4, $24 \div 5 = 4$ r 4
   (d) 1 - 5 5 47
   (e) 7, 5 r 5

4 **(a)** 3, 1   **(b)** $26 \div 4 = 6$ r 2
   4, 1 $26 \div 6 = 4$ r 2

5 **(a)** $25 \div 6 = 4$ (bouquets) r 1 (flower)
   (b) $58 \div 7 = 8$ (oranges) r 2 (oranges)
   $7 \times 9 - 58 = 5$ (oranges)

6 3, 76

Unit 2.13

1 **(a)** 20   **(b)** 8
   (c) 7   **(d)** 54
   (e) 4   **(f)** 7 r 2
   (g) 32   **(h)** 48
   (i) 2   **(j)** 38
   (k) 50   **(l)** 79
   (m) 33   **(n)** 71

2 **(a)** 41 5   **(b)** 19 4
   (c) 31 3   **(d)** 21 1

3 **(a)** 19 17
   (b) 35 28

4 **(a)** $3 \times 4 = 12$ (chicks)
   (b) $23 \div 4 = 5$ (hutches) r 3 (rabbits)
   $5 + 1 = 6$ (hutches)
   (c) $50 \div 9 = 5$ (hamsters) r £5
   (d) 3, 2

5 **(a)** 6, 5   **(b)** 6, 9 (or 9, 6)
   (c) 8, 4 (or 4, 8) **(d)** 7, 8 (or 8, 7)

6 25

Chapter 2 test

1 **(a)** 49   **(b)** 23
   (c) 66   **(d)** 12
   (e) 40   **(f)** 8
   (g) 4 r 2   **(h)** 11
   (i) 4   **(j)** 0
   (k) 7   **(l)** 100
   (m) 77   **(n)** 9 1 (answer may vary)
   (o) 2 4 (answer may vary)
   (p) 4 5 (answer may vary)

2 **(a)** 71   **(b)** 27
   (c) 83   **(d)** 57

3 **(a)** 45 – 29 = 16
(b) 12 × 2 = 24
(c) 3 × 3 × 3 = 27
(d) 8 × 8 + 3 = 67

4 **(a)** **(i)** 5 × 4 = 20 (balloons)
(ii) 4 × 3 = 12 (pens)
(iii) 19 + 23 = 42 (baskets)
(b) **(i)** 50 ÷ 8 = 6 (boxes)
(ii) 2 (cartons)
(c) 3 × 7 = 21 (days)
(d) 8 × 4 + 4 = 36 (flowers)
(e) 5 × 9 = 45 (cupcakes) 72 – 45 = 27 (cupcakes)

5 **(a)** **(i)** 4, 40 ÷ 10 = 4
(ii) 12 ÷ 3 = 4, 12, 4, 3
12 ÷ 4 = 3, 12, 4, 3
(iii) 6 6 × 2 = 12 answers may vary
(b) **(i)** 5   9
10 11
(ii) 6, 2, 1 (answer may vary)
2 × 5 (answer may vary)
(iii) 44 – 40 = 4
(iv) less 3
(v) 39
(vi) 25
(c) **(i)** ✗
(ii) ✓
(iii) ✗
(d) **(i)** B
(ii) C
(iii) D

Chapter 3 Knowing numbers up to 1000

Unit 3.1

1 **(a)** 49   **(b)** 8   **(c)** 27
(d) 5 r 5   **(e)** 80   **(f)** 40
(g) 57   **(h)** 10 r 4   **(i)** 0
(j) 100   **(k)** 63   **(l)** 7

2 **(a)** **(ii)** 306 three hundred and six
(iii) 420 four hundred and twenty
(iv) 404 four hundred and four
(v) 1000 one thousand
(b) **(i)** 605 six hundred and five
(ii) 824 eight hundred and twenty-four

3 **(a)** 8, 5, 6
(b) ones second fourth
(c) seven hundred and seven 7, 7, 693
(d) 403 four hundred and three

4 **(a)** 400 + 60 + 2
(b) 1000 + 0 + 50 + 0
(c) 700 + 80 + 8
(d) 390
(e) 808

5 60

6 8, 8

Unit 3.2

1 **(a)** 56   **(b)** 26
(c) 165   **(d)** 39
(e) 12   **(f)** 62
(g) 29   **(h)** 1
(i) 48   **(j)** 50

2 **(a)** 334
(b) 505

3 **(a)** six hundred and thirty-five
(b) three hundred and two
(c) 936
(d) one thousand
(e) 400

4 **(a)** 408 three 4 8
(b) 460
(c) 1000 5
(d) 299 301
(e) 47

5 **(a)** 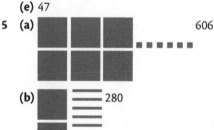 606
(b) 280

6 4 420 > 402 > 240 > 204

Unit 3.3

1 **(a)** Numbers correctly marked on the number line
(b) A = 457 B = 479 C = 462 D = 500
E = 491 F = 505

2 **(a)** **(i)** 277 279
(ii) 998 1000
(iii) 405 407
(b) **(i)** 380 400
(ii) 450 460
(iii) 780 790
(c) **(i)** 600 700
(ii) 400 500
(iii) 700 800

3 **(a)** 570 571
(b) 410 430
(c) 740 739
(d) 350 450

4 **(a)** 439 499
(b) 888
(c) 499 501
(d) 1000 > 888 > 654 > 501 > 499 > 439 > 328 > 92

5 **(a)** 499 500 501 502 503 504
(b) 100 200 300 400 500 600 700 800
(c) 100 111 122 133 144 155 166 177 188 199

6 492 663 834

7 20 120 22

Unit 3.4

1 **(a)** Numbers correctly marked on the number line
(b) 3, 5, 9, 1, 8, 8
(c) 13, 65, 1, 71, 98, 18
(d) B < D < A < C < E < F

2 **(a)** 300 295 280 275
(b) 490 492
(c) 423 523 623

3 **(a)** 751 > 715 > 517 > 175 > 157 > 117
(b) 668 < 689 < 869 < 886 < 898 < 969
(c) 967 > 867 > 767 > 667 > 567 > 467 > 367 > 267 > 167

4 **(a)** <   **(b)** >   **(c)** <
(d) <   **(e)** >   **(f)** >
(g) <   **(h)** <
(i) =

5 **(a)** 5   **(b)** 9   **(c)** 9
(d) 2   **(e)** 5   **(f)** 9

6 198 – 99 = 99

7 70 ÷ 8 = 8 r 6

Answers

Unit 3.5

1. (a) 734 Seven hundred and thirty-four
   (b) 430 Four hundred and thirty
   (c) 301 Three hundred and one
   (d) 400 Four hundred

2. (a)

Hundreds	Tens	Ones
••• ••		•••• •••

   (b)

Hundreds	Tens	Ones
•••• •••		

3. 364 Three hundred and sixty-four

Hundreds	Tens	Ones
•••	••••••	••••

274 Two hundred and seventy-four

Hundreds	Tens	Ones
••	•••••••	••••

25 Two hundred and sixty-five

Hundreds	Tens	Ones
••	••••••	•••••

4. 8
5. 1, 1

Unit 3.6

1. (a) 300 Three hundred

Hundreds	Tens	Ones
•••		

   (b) 210 Two hundred and ten

Hundreds	Tens	Ones
••	•	

   (c) 201 Two hundred and one

Hundreds	Tens	Ones
••		•

   (d) 120 One hundred and twenty

Hundreds	Tens	Ones
•	••	

   (e) 111 One hundred and eleven

Hundreds	Tens	Ones
•	•	•

   (f) 102 One hundred and two

Hundreds	Tens	Ones
•		••

2. 153 One hundred and fifty-three

Hundreds	Tens	Ones
•	•••••	•••

(a) 63 Sixty-three

Hundreds	Tens	Ones
	••••••	•••

(b) 54 Fifty-four

Hundreds	Tens	Ones
	•••••	••••

(c) 243 Two hundred and forty-three

Hundreds	Tens	Ones
••	••••	•••

(d) 144 One hundred and forty-four

Hundreds	Tens	Ones
•	••••	••••

(e) 252 Two hundred and fifty-two

Hundreds	Tens	Ones
••	•••••	••

(f) 162 One hundred and sixty-two

Hundreds	Tens	Ones
•	••••••	••

3. (a) 600 + 40 + 3
   (b) 338
   (c) 300 + 0 + 2
   (d) 909
4. 24
5. 13

Chapter 3 test

1. (a) 35   (b) 40
   (c) 27   (d) 9
   (e) 5   (f) 32
   (g) 680   (h) 31
   (i) 810   (j) 79
   (k) 75   (l) 70
2. (a) One hundred and sixty-five

   (b) Six hundred and eight

3. (a) 1000 One thousand
   (b) 166 One hundred and sixty-six
4. (a) 8 × 2 − 10 = 6
   (b) 31 − 15 + 27 = 43
   (c) 7 + 7 = 14

5. (a) 698 702
   (b) (i) Numbers correctly marked on
           the number line
       (ii) 483 501 539 578 550
   (c) three hundreds 2 4 7
   (d) 99 199
   (e) (i) 13
       (ii) 50
       (iii) 742
   (f) 380
   (g) 1000 > 968 > 806 > 405 > 380 > 45
6. (a) ✗
   (b) ✓
   (c) ✓
   (d) ✓
7. (a) C
   (b) D
   (c) B
   (d) D
8. 6 × 9 = 54 54 (books)
9. 8 × 2 + 3 × 4 = 28 28 (legs)
10. 30 ÷ 4 = 7 (cars) r 2 (children) 7 + 1 = 8
    (cars)
11. 34 + 18 = 52 (volleyballs) 52 + 28 = 80
    (basketballs)

Chapter 4 Statistics (II)

Unit 4.1

1 (a) Values correctly inserted in the table: 20 under Group A, 30 under Group B and 10 under Group C
   (b) Pictogram with 2 circles in Group A, 3 circles in Group B and 1 circle in group C
   (c) A block diagram with shaded cells, 4 for Group A, 6 for Group B and 2 for Group C
2 Values correctly inserted in the table: 6 under science, 8 under story book, 12 under comic and 10 under puzzle
3 (a) 1
   (b) Football swimming 5
   (c) Basketball tennis
   (d) 35
4 Answers may vary

Unit 4.2

1 (a) 2
   (b) 16, 12
   (c) 4
   (d) 16
   (e) 2

2 (a) 16, 6, 10, 8, 20
   (b) toy aeroplane puppy 14
   (c) 60
3 Bar chart with correct height bars as given in the table
4 (a) Maya Ella
   (b) 7
   (c) 8

Unit 4.3

1 (a) (i) 2
      (ii) 11
      (iii) 10
   (b) 23
   (c) Correctly constructed and labelled bar chart
2 (a) Joe Sarah Asif Lila Tom
   (b) Correctly constructed and labelled bar chart
3 (a) 2
   (b) football badminton
   (c) 33
   (d) 1

Chapter 4 test

1 (a) 1 child 5 children
   (b) 2 metres 12 metres
   (c) 1 unit 5 units
2 (a) 2
   (b) bus 14 bicycle 6
   (c) 26   (d) 42
3 (a) Values correctly inserted in the table: 7 under Brand A, 14 under Brand B, 15 under Brand C and 14 under Brand D
   (b) Brand C Brand A
   (c) Brand B Brand D
   (d) 50
4 (a) 4 28 4
   (b) bar 3 and a half units high drawn for strawberry
   (c) Answer may vary

Chapter 5 Introduction to time (III)

Unit 5.1

1 (a) 60
   (b) 60
   (c) 30
   (d) 600
2 Times matched to correct clock faces
3 (a) seconds
   (b) hour
   (c) hours
   (d) minute
4 Answers may vary
5 (a) 30
   (b) 300
   (c) 45
   (d) 2
   (e) 100
   (f) 150
   (g) 1 30
   (h) 1 40
6 (a) 3:30
   (b) 7:35
   (c) 11:33
   (d) 7:47
7 11:32 reasons may vary

Unit 5.2

1 (a) 24
   (b) 12
   (c) 12
   (d) morning
   (e) afternoon
2 1, 10, 2, 6, 11
   12, 7, 3, 5, 4, 8
3 8:05 10:31 3:07 12:00
4 08:00 1:36 p.m. 11:58 p.m. 24:00
5 (a) 01:20 13:20
   (b) 07:17 19:17
   (c) 11:00 23:00
6 7:30 10:30, 16:30 19:30 or 4:30 p.m. 7:30 p.m. 6
7 Items linked: 19:15 to 'a quarter past seven'
   22:30 to 'half past ten'
   16:25 to 4:25
   9:25 to 21:25

Unit 5.3

1 (a) ✓
   (b) ✓
   (c) ✗
   (d) ✓
   (e) ✗
2 Table correctly completed: 31 28 31 30 31 30 31 31 30 31 30 31
3 (a) 30 June September November
   (b) 25 December January March May July August October
   (c) February 28
   (d) 365
4 (Answer may vary) Years with 365 days have 28 days in February and years with 366 days have 29 days in February.
5 (a) 28
   (b) 29
   (c) 365 366
   (d) 2020 2024
6 (Answer may vary) for example, twenty-ninth of February 2004

Unit 5.4

1. (a) >   (b) >
   (c) >   (d) <
   (e) >   (f) <
2. (a) 4
   (b) 45
   (c) 2 10
   (d) 22:05
3. (a) 8:30 a.m. 9:00 p.m.
   (b) 12 30
   (c) yes 40 minutes
4. 17 days
5. (a) Second promotion 10 days
   (b) First promotion 8 days
   (c) 27 days
6. 42 25

Chapter 5 test

1. (a) 2:23   (b) 4:41
   (c) 11:52   (d) 10:09
2. Hands correctly drawn on clock faces
3. (a) 120
   (b) 48
   (c) 70
   (d) 1 40
   (e) 90
   (f) 1 25
4. (a) =
   (b) =
   (c) >
   (d) >
   (e) <
   (f) <
5. (a) ✗
   (b) ✓
   (c) ✗
   (d) ✗
6. (a) 3:38 p.m.
   (b) 11:40 a.m.
   (c) 18:30
   (d) 84
7. $2 \times 7 - 2 = 12$ (days)
8. (a) 4:00   (b) 4:25   (c) 4:40
   (d) 5:05
   (e) 25 minutes
   (f) 15 minutes
   (g) 25 minutes

Chapter 6 Consolidation and enhancement

Unit 6.1

1. (a) 4 2 6 18
   (b) 3 2 3 2 6 2 12
2. Correctly completed table:
   2 times: 2 4 6 8 10 12 14 16 18 20
   4 times: 4 8 12 16 20 24 28 32 36 40
   6 times; 6 12 18 24 30 36 42 48 54
   60 6 answer may vary, for example,
   $2 \times 5 + 4 \times 5 = 6 \times 5 = 30$
3. (a) 8 48
   (b) 9 36
   (c) 2 20
   (d) 9 63
4. (a) 48   (b) 63
   (c) 90   (d) 18
5. (a) 1   (b) 2
   (c) 3   (d) 7, 7
6. $2 \times 9 + 3 \times 9 = £45$
7. (a) 5   (b) 5
8. (a) $9 \times 5 = 8 \times 5 + 1 \times 5$
   (b) $9 \times 5 = 7 \times 5 + 2 \times 5$
   (c) $9 \times 5 = 6 \times 5 + 3 \times 5$
   (d) $9 \times 5 = 5 \times 5 + 4 \times 5$
   (e) $9 \times 5 = 3 \times 5 + 3 \times 5 + 3 \times 5$
   (answers may vary)

Unit 6.2

1. (a) 6 3 3 6
   (b) 6 3 2 3 4 3 12
2. Correctly completed table:
   9 times: 9 18 27 36 45 54 63 72 81 90
   5 times: 5 10 15 20 25 30 35 40 45 50
   4 times: 4 8 12 16 20 24 28 32 36 40
   4 answer may vary, for example, $9 \times 6 - 5 \times 6 = 4 \times 6 = 24$

3. (a) 3, 18
   (b) 1, 4
   (c) 3, 15
   (d) 3, 7, 21
   (e) 6, 5, 30
4. (a) 16
   (b) 42
   (c) 81
   (d) 18
5. (a) 2
   (b) 3
   (c) 10
   (d) 7 7 4
6. $8 \times 9 - 3 \times 9 = 5 \times 9 = £45$
7. (a) 7, 8
   (b) 4, 5
   (c) 2
   (d) 2, 3, 2
8. (a) $3 \times 6 = 4 \times 6 - 1 \times 6$
   (b) $3 \times 6 = 5 \times 6 - 2 \times 6$
   (c) $3 \times 6 = 6 \times 6 - 3 \times 6$
   (d) $3 \times 6 = 10 \times 6 - 5 \times 6 - 2 \times 6$
   (answers may vary)

Unit 6.3

1. (a) 24   (b) 56
   (c) 45   (d) 100
   (e) 6   (f) 8
   (g) 5   (h) 10
   (i) 4   (j) 7
   (k) 9   (l) 0
2. (a) 8   (b) 6
   (c) 9   (d) 7
   (e) 5   (f) 9
3. (a) $42 \div 7 = 6$ 6 (rows)
   (b) $42 \div 6 = 7$ 7 (sets)
   (c) $6 \times 7 = 42$ 42 (sets)

4. (a) $6 \times 6 = 36$ (deer)
   (b) $10 \times 4 = 40$ (models) 40 > 32 the boys have made more
   (c) $27 \div 4 = 6$ (taxis) r 3 (people)
   $6 + 1 = 7$ (taxis)
   (d) 4 3 31 ÷ 7 = 4 r 3
5. 21 cupcakes
6. 32

Unit 6.4

1. (a) even   (b) even
   (c) odd   (d) odd
2. (a) $8 + 2 = 10$
   (b) $6 + 4 = 10$
   (c) $6 + 5 = 11$
3. (a) 11 13 15
   (b) 12 14 16
   (c) 20 18 16
   (d) 10 13 12 15
4. (a) 5 1
   (b) 3 1
5. (a) Answer may vary, for example, 1, 3, 5, 7, 9
   (b) Answer may vary, for example, 2, 4, 6, 8, 10
   (c) 31 33 35 37 39
   (d) 60 62 64 66 68
   (e) Answers may vary, for example, 20 22 24 26
6. (a) 4
   (b) 9
   (c) 4, 16
   (d) 5, 5, 25
   (e) 9, 11, 6, 6, 36

7 **(a)** 9 **(b)** 10 **(c)** 10
 (d) 11 **(e)** 12 **(f)** 12
 (g) 13 **(h)** 14 **(i)** 14
 (j) 15 **(k)** 16 **(l)** 16

8 **(a)** odd number
 (b) even number
 (c) even number

Unit 6.5

1 All the sums are 15.

2 ✗ ✓

3 3 9 1 2 7   8 5 4 2

4 15:4 9 7 8 1 6   18:7 2 4 3 10 5
 21:6 11 5 7 9 3

5 5 7   3 5

6 (Answers may vary)

Unit 6.6

1 **(a)** 69 **(b)** 36
 (c) 8 **(d)** 197
 (e) 0 **(f)** 82

2 ten thousands thousands hundreds

3 3000 5000 7000 9000 11 000

4 **(b)** 0 9 9 9 0 **(c)** 1 0 0 0 8 **(d)** 0 2 0 0 6

5 **(a)** Two thousand two hundred and
 two 2202
 (b) Ten thousand 10 000
 (c) Five thousand two hundred and
 thirty 5230
 (d) Four thousand and fifty-three 4053

6 **(a)** 10, 100, 1000, 10 000
 (b) ones hundreds ten thousands
 (c) 7523
 (d) 6, 60, 17
 (e) 9999, 10 000, 10 001

7 **(a)** 5080, 5081
 (b) 5656, 6767
 (c) 9970, 9960, 9950

Unit 6.7

1 **(a)** 80 **(b)** 12
 (c) 51 **(d)** 15
 (e) 85 **(f)** 5
 (g) 50 **(h)** 143
 (i) 38 **(j)** 42

2 6348, 5050, 13 004
 Nine thousand and eight
 Four thousand four hundred
 and fifteen
 Nineteen thousand and six

3 **(a)** Four thousand six hundred and
 thirty-two 4000 + 600 + 30 + 2
 (b) Two thousand five hundred and
 forty-seven 2000 + 500 + 40 + 7
 (c) Six thousand and three
 6000 + 0 + 0 + 3
 (d) Two thousand and thirty
 2000 + 0 + 30 + 0

4 **(a)** 1812
 (b) 4050
 (c) 6500
 (d) 5006

5 **(a)** > **(b)** > **(c)** >
 (d) > **(e)** > **(f)** <
 (g) < **(h)** < **(i)** <

6 **(a)** B
 (b) C
 (c) C

7 **(a)** 209 < 367 < 627 < 736
 (b) 7800 < 8007 < 8070 < 8700

8 **(a)** 95 400
 (b) Answers may vary, for example,
 95 400, 94 500, 45 900
 (c) Answers may vary, for example,
 90 045, 90 054, 95 004
 (d) Answers may vary, for example,
 90 540, 95 400, 94 500

Chapter 6 test

1 **(a)** 24 **(b)** 45
 (c) 4 r 4 **(d)** 24
 (e) 63 **(f)** 8 r 2
 (g) 100 **(h)** 56
 (i) 6 **(j)** 0
 (k) 25 **(l)** 27
 (m) 36 **(n)** 6

2 **(a)** 6 **(b)** 5
 (c) 4 **(d)** 2

3 **(a)** + **(b)** × **(c)** ÷
 (d) × **(e)** − **(f)** ÷

4 **(a)** = **(b)** > **(c)** <

5 **(a)** 9, 72
 (b) 6, 36
 (c) 6 (answers may vary)
 (d) 56
 (e) 2 (answers may vary)

6 **(a)** 8 × 8, 6 × 7, 7 × 8, 6 × 8
 (b) 1, 8, 7, 2, 9, 4

7 **(a)** 5, 25
 (b) 6, 35
 (c) 72

8 **(a)** 3
 (b) 32
 (c) 5

9 **(a)** 4005
 (b) 6800
 (c) 1052
 (d) 10 039

10 **(a)** 1, 5, One thousand and five
 (b) 105 < 501 < 1005 < 1050 < 5001

11 **(a)** 4200
 (b) 2004
 (c) Answers may vary, for example,
 4020, 2004
 (d) 4200, 4020, 2040, 2400

12 **(a)** 6 × 5 = 30 (cups)
 (b) 32 ÷ 8 = 4
 (c) 27 ÷ 6 = 4 (boats) r 3 (people)
 4 + 1 = 5 (boats)
 (d) 2 × 9 = £18 18 > 17 no, not enough
 (e) 6 × 5 = 30 (apples) 30 < 32 the girls
 have more apples

13 **(a)** Correctly drawn and labelled bar
 chart (answer may vary)
 (b) 6
 (c) softball and gymnastics
 (d) swimming 3 badminton
 (e) 48